Work-Integrated Learning in Engineering, Built Environment and Technology:

Diversity of Practice in Practice

Patrick Keleher
Central Queensland University, Australia

Arun Patil
Central Queensland University, Australia

R. E. Harreveld
Central Queensland University, Australia

Information Science
REFERENCE

Senior Editorial Director:	Kristin Klinger
Director of Book Publications:	Julia Mosemann
Editorial Director:	Lindsay Johnston
Acquisitions Editor:	Erika Carter
Development Editor:	Joel Gamon
Production Coordinator:	Jamie Snavely
Typesetters:	Natalie Pronio & Milan Vracarich, Jr.
Cover Design:	Nick Newcomer

Published in the United States of America by
Information Science Reference (an imprint of IGI Global)
701 E. Chocolate Avenue
Hershey PA 17033
Tel: 717-533-8845
Fax: 717-533-8661
E-mail: cust@igi-global.com
Web site: http://www.igi-global.com/reference

Library of Congress Cataloging-in-Publication Data

Work-Integrated Learning in Engineering, Built Environment and Technology: Diversity of Practice in Practice / Patrick Keleher, Arun Patil, and R.E. Harreveld, editors.
p. cm.
Includes bibliographical references and index.
Summary: "This book provides a snap shot of a global perspective of the diversity of practices adopted to successfully develop, implement, deliver and re-invigorate undergraduate and postgraduate Work Integrated Learning (WIL) focused programs in engineering, the built environment and information technology"--Provided by publisher.
ISBN 978-1-60960-547-6 (hardcover) -- ISBN 978-1-60960-548-3 (ebook) 1. Engineers--In-service training. 2. Construction industry--Employees--Training of. 3. Career education. I. Keleher, Patrick, 1952- II. Patil, Arun S. III. Harreveld, R. E., 1952-
T65.W896 2011
620.0071'1--dc22

2010054440

British Cataloguing in Publication Data
A Cataloguing in Publication record for this book is available from the British Library.

Table of Contents

Detailed Table of Contents

This chapter presents a domain-independent computational environment which supports work-integrated learning at the professional workplace. The Advanced Process-Oriented Self-Directed Learning Environment (APOSDLE) provides learning support during the execution of work tasks (instead of beforehand), within the work environment of the user (instead of within a separate learning system), and repurposes content which was not originally intended for learning (instead of relying on the expensive manual creation of learning material). Since this definition of work-integrated learning might differ from other definitions employed within this book, a short summary of the theoretical background is provided. Along the example of the company Innovation Service Network (ISN), a network of SME's, a rich and practical description of the deployment and usage of APOSDLE is given. The chapter provides the reader with firsthand experiences and discusses efforts and lessons learned, backed up with experiences gained in two other application settings, namely EADS in France and a Chamber of Commerce and industry in Germany.

Practical training in companies has been recognized for many years as an important component of the education of new engineering graduates all over the world. The format used to provide this education to students varies widely not only across national boundaries, but also within each country. This chapter deals first with the state of industrial training in engineering education in Spain, both in the old engineering degrees and in the new ones, following the European higher education area requirements, which are now in the process of introduction. An analysis is also carried out on the evaluation and assessment of industrial training when this activity is a part of first cycle engineering curricula, and the role played by the tutor is discussed. Finally, some weak points of industrial training in Spanish curricula are discussed, as well as some proposals to overcome that situation and to move towards a global approach of industrial training in engineering education.

This chapter presents the curriculum and training process applied in the IESP department, standing for "Ingénieur d'Exploitation des Systèmes de Production" of the Ecole Polytechnique Universitaire de Lille (Polytech'Lille, F59 Lille, France), with an emphasis upon the WIL training process for exploitation engineers of production systems. The IESP department is dedicated to lifelong learning and apprenticeship leading to a master degree (French Engineer level) in the production field. It is an accredited program. This practice relies upon a background experience of 18 years with close partnerships with industry from many sectors, such as: energy, metallurgy, food industry, automotive, chemical engineering, and aeronautics. The graduates from Continuing Vocational Education and Training (CVT) that are already employed in the company improve their position. Younger graduates from Initial Vocational education and Training (IVT) that also validated their abilities are much more employable. They can work in any industrial sector dealing with engineering production having a strong technical and managerial skill base. In this chapter, the historical setting up and evolution is elaborated in the legal French education framework. The training model based on the IESP professional profile is presented. This model presents an academic curriculum with WBL that integrates a real formative work situation in the industry. The tools and methods developed all along the training process are also focused within a sustainable development policy. Finally, success and difficulties or challenges with mobility due to the globalization of the economy and innovation with respect to the economical crisis are also discussed.

This chapter describes Georgia Institute of Technology's (Georgia Tech) model for producing globally competent engineers. It details two aspects that Georgia Tech thinks are vital to its success: (1) the need for institutional support and resources and (2) making international experiential education a part of an institution's culture.

The Chemical Engineering Practice School (ChEPS) at King Mongkut's University of Technology Thonburi (KMUTT) in Bangkok is a two-year international curriculum modeled after Massachusetts Institute of Technology's School of Chemical Engineering Practice. The aim of this Master's program is to produce professional chemical engineers with strong fundamentals, practical experience, and a good command of English. The program's uniqueness lies in its strong linkage with the industrial sector. This chapter contains a history of ChEPS and details how KMUTT operates the program. The key factors contributing to the success of the program are identified. Moreover, critical analyses gleaned from the faculty, the alumni, and the industrial sponsors are carried out to examine the current strengths of ChEPS and to identify areas for improvement. Key challenges still facing the program are also outlined. Finally, potential solutions to these challenges are recommended.

Instead of only staying in the university to carry out research, postgraduate research students nowadays are expected to gain knowledge and experience through work-integrated learning. The advantages of this kind of learning include better support and facilities for research and more comprehensive and in-depth experience in the research area. The learning also provides an opportunity for students to gain other research experience and explore other research interests. However, sometimes such kind of learning opportunity is not available or not the best available locally, therefore work-integrated learning is necessary or better to be carried out in remote regions. Taking regional collaboration of work-integrated learning for postgraduate design research students between the Chinese mainland and Hong Kong as a case study, this chapter discusses the advantages, merits, issues, and problems of regional collaboration. The chapter then identifies possibilities for improvement and directions for further investigation.

Central Queensland University conducts a suite of postgraduate programs in maintenance management. There is an emphasis upon Work Integrated Learning, and the programs are delivered in a flexible mode by academics and lecturer-practitioners so that learners are provided with an authentic learning experience. The learners are mature aged, experienced practitioners who are either graduate engineers or trades qualified, working in the maintenance management area of their organisation. Study materials have been designed and developed through the collaboration and consultation with industry, university, and practitioner stakeholders to ensure the student's learning and the assessment of that learning is incorporated into the tasks and responsibilities of learners in their workplace. A blended learning delivery model includes the opportunity for students to attend a two day residential school. Ongoing improvement of the academic program, consultancy, and research opportunities arises from engagement with stakeholders through a number of mechanisms including a conference to showcase innovative practices of physical asset and maintenance management.

Three of the professions directly related to the construction industries by which Central Queensland University's undergraduate Built Environment programs are accredited, form the foci of this chapter. The students enrolled in those programs are working in the relevant industries during their part-time external studies. Although learning while working 'on the building site' has been known since human beings first began constructing shelter, relevant background theories of philosophy and psychology have been introduced here and utilised to provide substantive support for a debate regarding the mixture of formal and informal opportunities for work-integrated learning to which these students are introduced. The strengths and weaknesses of flexible external studies, as well as the nature of their workplace-based learning in these programs, are discussed at length. Greater emphasis is placed on the need to capitalise on the many opportunities for reinforcement of, and reflection about principles and practices introduced in either or both their employment and/or undergraduate studies, than merely on the advantages or disadvantages of flexible external studies. These students' truly work-integrated learning experience may be considered to be a 'self-paced flexible learning while earning' process.

This chapter concerns the design, delivery and management of a unique part-time postgraduate MSc in Construction Project Management (MScCPM) programme through an industry-academia partnership in Ireland during the period 2007-2010. The partners are BAM Contractors, part of the wider Royal BAM Group based in The Netherlands, and Waterford Institute of Technology (WIT). There are many innovative Work Integrated Learning (WIL) aspects to this bespoken programme, including the blending of teaching and assessment by both WIT lecturers and senior BAM staff. There is also a blend of traditional classroom activities and e-learning technologies to suit the geographically dispersed participants. All stakeholders in this programme have benefited from their participation. These benefits and the associated lessons learned are described in the hope that they may be of use to those developing WIL postgraduate programmes in the future.

The chapter provides an overview of Work Integrated Learning (WIL) into the Information Technology (IT) programme offered at the Baden-Württemberg Cooperative State University in Ravensburg, Germany. The opinions and debates of leading role-players in WIL are featured. The university's pose and the operations adopted in managing this programme will be presented. These include the structure of the IT programme, its accreditation process, strengths, and weaknesses. The chapter concludes with the programmes direction for the future.

Foreword

The Order of New Work Integrated Learning in Universities

The integration of work and learning is not new, as many of the writers in this book will attest and illustrate. Industries and businesses function in a technically and socially changing knowledge environment, simultaneously generating and requiring new knowledge and skills. Universities are an integral element of societies' knowledge generation and transfer systems. How they meet this mission is under challenge and pressure to change due to significant reforms in government agendas, social expectations and technological innovation. These forces have produced an invigorated and more formal focus in universities on the relationship between learning and work and among industries, professional regulatory bodies, and universities. Universities have always had both formal and informal programmes in which students engaged in work outside of the classroom in order to learn. What is new is the public inclusion of work integration into institutional missions, targets, and strategic plans making it visible and contestable in higher education and subject to quality assurance and systematic research.

Work Integrated Learning has a broad agenda driven by contrasting but often tacit philosophies which cause researchers to question the interplay and balance of work and learning and to ask just where is the 'integration' and its value. Explications of philosophies, however, are largely overlooked in the need to attend to governments' drive for graduate competencies to meet workforce demands and to comply with professional standards requirements.

Despite these overriding pressures, universities, curriculum developers, and researchers have generated an exciting new field of endeavour for higher education to foster innovation in work integrated learning. In this book, the researchers and curriculum and institutional innovators rightly focus attention on the adequacy in the preparation of the workplace and the students for optimal learning through, and from, experience, noting that experience alone is not enough. It is important to be reminded that theory and practice are not unrelated. At best, they are essential and symbiotic elements of the whole. Practice will not change unless informed by theory, and theory will not develop unless challenged by real world experience. How this transfer between knowledge development and practice development occurs is an important site of inquiry.

The new order of Work Integrated Learning as a higher education enterprise involves the full scope of curriculum concerns. It has a role in developing graduates' global and cultural awareness, as well as their capacities to be ethical and responsible citizens and as knowledge workers. These following chapters explore these important concerns, reflecting their national diversity yet similarity in their endeavour for innovative programs that use research and evaluation to assure high quality education.

Janice Orrell
Education Consultant, Australia

Janice Orrell *(Education Consultant, Australia) Janice has been an educator for over 40 years. In the 1960s and 1970s she taught in rural, remote and urban Schools in Western Australia and South Australia and was a Director in early childhood education. She also worked in graduate careers guidance, establishing links between industry and higher education programs for academics, students, graduates and employers. In 1984-87 she taught in an International School in Southern India, at the same time established a non-formal education program for young tribal and dalit women in remote Palani Hills villages that now, 23 years later, is a Community Education College. Returning to Australia, she taught in Aboriginal Teacher Education and Nursing Education before establishing and coordinating the Academic Development Unit at Flinders University. In 1986 she became a foundation Australian Learning and Teaching Council (ALTC) Director of Discipline Networks. Janice is now a private higher education consultant and holds a number of adjunct associate professorships (Flinders University, University of New South Wales and Sydney University). Janice has recently co-published a book: Cooper, L. Orrell, J. and Bowden, M. (2010). Work Integrated Learning: A Guide to Effective Practice, Routledge: London.*

Preface

This text provides a snap shot of a global perspective of the diversity of practices adopted to successfully develop, implement, deliver and re-invigorate undergraduate and postgraduate Work Integrated Learning (WIL) focussed programs in Engineering and the Built Environment and Information Technology. The initial chapters serve to explore definitive features of the work and learning interface by categorising WIL as a series of nine models, highlighting the importance of continuous professional development and by demonstrating an innovative mechanism by which to monitor the learning being undertaken, in situ, during the time that new knowledge is being acquired.

Work Integrated Learning manifests in many guises including as Work Based Learning, Workplace Learning or Practice Based Learning. The learning occurs within a real-life, work or 'work-like' context, simulated work conditions and location or an actual workplace or a combination of these. The mechanisms (training, fieldwork, practicum, immersion, placement, or sandwich programs) used to implement it are married to the delivery mode (face-to-face, blended learning) used to impart the knowledge and skills being developed with an emphasis upon the practices (authentic learning, reflective practice, mentoring) to gauge or assess the acquisition of the skills, knowledge and experiences, and the level of self-awareness or self-development overseen by internal (university committees and processes) and external (professional bodies, stakeholder advisory committees) quality assurance processes.

The range of theoretical frameworks cited as underpinning WIL includes experiential learning (Argyris & Shon, 1976), (Kolb, 1984), including problem based learning (Wilkerson & Wim, 1996), lifelong learning (Ballou, Bowers, Boyatzis & Kolb, 1999) and experiential education, including service learning or service-based learning (Dewey 1916, 1933, 1938), (James, 1916), constructivist theory (Piaget, 1926), Vygotsky (1962), Papert (1980), and co-operative education (Smollins, 1999), Grubb and Villeneuve (1995), Groenewald (2004), communities of practice (Lave & Wenger, 1990), informal learning in the workplace (Colley, et al. 2002), Eraut (2000) and activity theory (Rubinshtein, 1973), Leont'ev (1978), Engeström (1987).

While proponents of WIL, and its many guises, highlight and nominate focus upon specific theoretical frameworks or approaches, there is no uniformly or centrally focussed or recognised theoretical body that encapsulates the field of Work Integrated Learning. This is not to suggest a 'unified field theory' of Work Integrated Learning is emerging or indeed, necessary. Instead, it emphasises the diversity of theoretical practice, the purpose or reason for its specific adoption over any other and that the implementation of opportunities for Work Integrated Learning are more likely to be driven by pragmatic or practical considerations. Consequently, the enrichment associated by the diversity is more purposeful than any uniformly 'beige' theoretical or implementation approach and so one is wise to embrace the notion of *viva la difference!*

In its broadest sense, WIL is most often described as an attempt to provide support for the employability of employees by initial skill development and knowledge acquisition (paradigm: 'school-to-work', 'work ready') or by upskilling or retraining them (paradigm: continuing professional development, 'lifelong learning'). What seems to be evident is that a pragmatism or practicality arises from a complex dynamic between the needs of the niche market or specific client (industry, commercial, government), the adopted educational practices (theoretical approach and delivery mode), the quality assurer (university, stakeholder, professional body), policy initiatives (government, professional bodies) and the learners' needs. The interactions are not inclusive of all aspects or players. As Calway and Murphy, from Australia, in Chapter 1, highlight in their chapter, 'in some cases the employer is a key stakeholder of the processes and in others the employer may be incidental or irrelevant.' At other times, some programs are not accredited by a professional body, while for others it is imperative for graduates to achieve professional recognition.

All authors highlight WIL provides a successful framework for contextual learning and the effective transfer of learning to undertake practice and operate as a competent practitioner. Although, in Chapter 2, Murphy and Calway, from Australia, contend that 'practice alone is generally insufficient to ensure knowledge is effective and up to date' and is a mechanism by which practitioners are able to be accredited, meet compliance requirements, support deeper learning, and enhance transference of learning. They emphasise the importance of continuing professional development in acquiring revised, new, or deeper knowledge and in ensuring currency and relevancy or in establishing a specialisation. Interestingly, most of the programs described by the authors fit into Models 7 and 8 as categorised by Calway and Murphy. The transferability of skills is highlighted by most authors and in particular there appears to be an emphasis, whether stated explicitly or inferred implicitly, upon communication, problem solving and team work.

Operating within a WIL context is all very well but how does one monitor what is being learnt? In Chapter 3, Lindstaedt and Christl, from Austria, demonstrate an innovative domain-independent computational environment which supports WIL at the professional workplace. They outline how their Advanced Process-Orientated self-Directed Learning Environment (APOSDLE) is designed as a learning support mechanism so people can learn during the execution of the work task as opposed to learning first and then applying the newly acquired knowledge. Their mechanism shows that they have 'enabled relatively inexperienced knowledge workers to efficiently improve their knowledge in various ways.'

In Chapter 4, Dominguez and Magdaleno, from Spain, deal with the state of industrial training in undergraduate engineering in Spain and highlight the impacts of introducing European Higher Education Area requirements. They go on to review the structure of the new degrees and how industrial training is considered in them and some proposals are discussed on the methods to improve the evaluation of students and the outcomes of WIL experiences, with special focus upon mechanical engineering degrees in Spain.

Chapter 5 sees Nuninger and Chatelet, from France, discuss the curriculum and training processes adopted to upskilling graduates from vocational and continuing vocational education and training programs. They nominate the opportunities their programs provide in offering degrees, to Master's level, for students studying in the field of production with the benefit of the integration of a real informative work situation in industry.

Akins and Gulick, from the USA, in Chapter 6, highlight the extension of their institutions long running co-operative program to include a work abroad program. This program serves to focus upon students undertaking international experiential education, developing graduates with a global competence. Students elect to study, research, or work abroad.

In Chapter 7, Ku and Thonglek, from Thailand, highlight the successes that are evident from having conducted, for the past 14 years, a premier Master's program aimed at graduate chemical engineers. Their practice school approach has been adapted from the Massachusetts Institute of Technology (MIT) model to meet Thai industry requirements. It has been through the industry-government-university nexus engagement and commitment that they have been able to ensure students are provided with a meaningful learning experience. Their success has inspired many spin-off programs, both internal and external to King Mongkut's University of Technology, and have aided in changing the face of work contextualised learning in Thailand.

A contribution from China by Siu, in Chapter 8, outlines the challenges and successes encountered in providing postgraduate research students in Industrial and Product Design and Product Engineering access to industry to undertake research projects. Siu highlights how regional collaboration between mainland China and Hong Kong, supported by a university-industry partnership, provides WIL experiences to ensure students conduct research that has a stronger connection to industry. The mechanism has helped to build research networks between universities and also has been used as a barometer for establishing a larger scale and more in-depth collaboration between participating organisations.

In Chapter 9, Keleher, Patil, and Chattopadhyay, from Australia, outline the successful industry-university nexus that has developed a suite of postgraduate programs in Maintenance Management. A blended learning model is adopted with the majority of lecturers delivering the courses being experienced practitioners currently working in the field of Asset or Maintenance Management. Learning experiences and their assessment follows an authentic learning philosophy.

Chapter 8, 9 and 11 highlight the adoption of WIL aspects in terms of a research experience. With the creation of new knowledge, this demonstrates the emergence of what could be legitimately be termed Work Integrated Research (WIR) grounded in or arising from a WIL philosophy or context. The authors of these chapters highlight how their programs provide the basis for knowledge creation and the development of new business ideas, initiatives innovations and efficiencies (financial, temporal, human resources, physical resources). The Waterford Institute of Technology and BAM Construction have approved their first cohort of research projects in December 2009. CQUniversity has been conducting the Master's in Maintenance Management since 1998 and has over 30 graduates since its inception. The Hong Kong Polytechnic University's program had its inception in 2003 and has provided the opportunity for research design students to engage in industry based research in a Work Integrated Learning context. Stewart and Chen (2009) discuss developing a framework for Work Integrated Research Higher Degrees exploring and reporting on this approach in the context of three case studies that are of a Doctor of Philosophy context and not in a WIL context as is the philosophy adopted by CQUniversity, The Hong Kong Polytechnic University, and Waterford Institute of Technology.

Clark, from Australia, in Chapter 10, describes a suite of undergraduate Built Environment Programs (Construction Management, Building Design, and Building Surveying and Certification) delivered by flexible mode. The model provides a learning experience where more than half of the courses are taught by external professions working full time in the construction industry with the course co-ordination overseen by university academics. The active presence of professions working with academics to delivery these courses is seen as an integral part of the strength of providing students with an authentic learning experience.

Thomas, Wall, Graham, Troy, Crowe and O'Connell, from Ireland, in Chapter 11, describe the valuable experiences gained from the industry-academia partnership which collaborated in designing and establishing the delivery of a two year, part-time postgraduate Master's in Construction Project Management. The major challenges facing the Irish construction industry has been the ramping up of the National Qualifications Framework focus upon 'learning outcomes,' the impact of European Union's greater emphasis upon lifelong learning and the establishment of the European Qualifications Framework. They demonstrate how aligning learning outcomes with the business objectives of their partner organisation, BAM contractors, a subsidiary of the Royal BAM Group of the Netherlands, that they have been able to successfully establish a blended learning model for educating innovative construction executives.

With a German perspective, Reinhard and Singh, in Chapter 12, discuss their undergraduate Information Technology program. Their program serves to address the issues faced by the Europeanisation of Germany and neighbouring countries in providing students with learning experiences that empower them as graduates as proactive contributors to their internationally orientated employer organisations.

The authors in this text provide a diverse perspective on the manner in which the WIL philosophy can be adopted and adapted to meet the requirements of undergraduate and postgraduate modes of study. They emphasise a learning environment which creates or supports meaningful learning through a contextual lens. Mirvis and Hall (1996) succinctly and pertinently observe that workers need to *learn a living rather than earn a living* and this is borne out by the diversity of practice practiced by these contributors.

Patrick Keleher
Lead Editor

REFERENCES

Argyris, C., & Shon, D. A. (1976). *Theory in practice: Increasing professional effectiveness*. San Francisco, CA: Jossey-Bass Publishers.

Ballou, R., Bowers, D., Boyatzis, R. E., & Kolb, D. A. (1999). Fellowship in lifelong learning: An executive development program for advanced professionals. *Journal of Management Education*, *23*(4), 338–354. doi:10.1177/105256299902300403

Colley, H., Hodkinson, P., & Malcolm, J. (2002). *Non-formal learning: Mapping the conceptual terrain. A consultation report*. University of Leeds Lifelong Learning Institute. Leeds. Retrieved August 08, 2010, from http://www.infed.org/ archives/e-texts/ colley_informal_learning.htm

Dewey, J. (1916). *Democracy and education*. New York, NY: The Free Press.

Dewey, J. (1933). How we think: A restatement of the relation of reflective thinking to the educative process . In *The later works of John Dewey* (*Vol. 8*, pp. 105–352). Carbondale, IL: Southern Illinois University Press.

Dewey, J. (1938). *Experience and education*. New York, NY: Collier Books.

Engeström, Y. (1987). *Learning by expanding: An activity-theoretical approach to developmental research*. Helsinki, Finland: Orienta-Konsultit Oy.

Eraut, M. (2000). Non-formal learning and tacit knowledge in professional work. *The British Journal of Educational Psychology*, *70*(1), 113–136. doi:10.1348/000709900158001

Groenewald, T. (2004). Towards a definition for cooperative education. In R. K. Coll & L. Eames (Eds.), *International handbook for cooperative education: An international perspective of the theory, research, and practice of work-integrated learning.* (pp. 17-25). Boston, MA: World Association of Co-operative Education.

Grubb, W. N., & Villeneuve, J. C. (1995). *Co-operative education in Cincinnati.* Berkeley, CA: National Center for Research in Vocational Education.

James, W. (1916). *Talks to teachers on psychology, and to students on some of life's ideals.* New York, NY: Holt.

Kolb, D. A. (1984). *Experiential learning: Experience as the source of learning and development.* Englewood Cliffs, NJ: Prentice Hall.

Lave, J., & Wenger, E. (1990). *Situated learning: Legitimate peripheral participation.* Cambridge, UK: Cambridge University Press.

Leont'ev, A. (1978). *Activity, consciousness, and personality.* Englewood Cliffs, NJ: Prentice-Hall.

Mirvis, P. H., & Hall, D. T. (1996). New organisational forms and the new career . In Hall, D. T. (Ed.), *The career is dead: Long live the career* (pp. 72–100). San Francisco, CA: Jossey-Bass.

Papert, S. (1980). *Mindstorms. Children, computers and powerful ideas.* New York, NY: Basic Books.

Piaget, J. (1926). *The language of thought of the child.* London, UK: Routledge & Kegan Paul Ltd.

Rubinshtein, S. L. (1973). *Problems of general psychology.* Moscow, Russia: Academic Science.

Smollins, J. P. (1999). The making of the history: Ninety years of Northeastern Co-op. *Northeastern University Magazine, 24*(5).

Vygotsky, L. S. (1962). *Thought and language.* New York, NY: Wiley. doi:10.1037/11193-000

Wilkerson, L., & Wim, H. G. (1996). *Bringing problem-based learning to higher education: New directions for teaching and learning.* San Francisco, CA: Jossey-Bass.

Chapter 1
A Work-Integrated Learning Philosophy and the Educational Imperatives

Bruce A. Calway
Swinburne University of Technology, Australia

Gerald A. Murphy
Swinburne University of Technology, Australia

ABSTRACT

The Work-Integrated Learning (WIL) philosophy, as theorised, can be expressed as a number of generic WIL models, imperatives and learning frameworks. The models are the result of using the lens of a content study first completed in 2006 (reported in Calway and Murphy, (2006) and repeated in 2009/10. The studies investigated WIL models using resources such as: cooperative education literature, continuing professional development, and published public and higher education policy. These models reflect a concentration mainly upon education policy of a school-to-work paradigm. From associated studies, a set of learning imperatives was also proposed (Calway & Murphy, 2007) together with a pre and post graduation WIL approach for both higher education and continuing professional practice (Murphy & Calway, 2008).

The chapter explores these WIL findings as: embedded life-long learning, continuing professional development, and the various learning modes that employ the work and learning interface. Further, any understanding of a WIL philosophy must incorporate a learner environment for individual 'action learning' and institutional (i.e. educational as well as workplace) 'active learning' as an initiative that addresses the specific educational imperatives and models outlined.

DOI: 10.4018/978-1-60960-547-6.ch001

INTRODUCTION

Background

Work-Integrated Learning found its genesis in engineering in the early 20th century, and during the past century the initial model has evolved into a multiplicity of variations and been transferred and implemented across a vast array of disciplines, institutions and workplaces. Work-integrated learning is a broad church of approaches, across many levels of education worldwide, that incorporates knowledge and skills acquisition with 'real-world' experience. The World Association for Cooperative Education (WACE) is the peak organisation worldwide and the promoter of the work-integrated learning paradigm, as a deep learning model at school level and the workplace (WACE, 2010). Broadly speaking WIL can be expressed as "… educational activities that integrate theoretical learning with its application in the workplace." And "should provide a meaningful experience of the workplace application that is intentional, organised and recognised by the institution, in order to secure learning outcomes for the student that are both transferable and applied". (Griffith University, 2006)

Much of the literature examined expresses the perceived benefits of WIL as a form of learning. We instead concentrated on the educational and public policy understanding relative to the usage of WIL as expressed in education institutions (i.e. pre and post graduation) and also expressed by professional associations and in Standards Legislation. As an example relative to WIL pre graduation: 'cooperative education' represents one educational form which has broad recognition worldwide among employers, students and tertiary institutions. This example of cooperative education forms a significant part of literature reporting WIL and incorporates hands-on work experience in a real-world setting. Cooperative education assumes a level of explicit knowledge/skills on the part of the student and the exchange of tacit knowledge/skills from the employer to the student. Alternatively an example of post graduation WIL would be continuing professional development, aimed at competency assessment, satisfying compliance requirements and/or knowledge/career development.

In its broadest sense WIL is most often described as an attempt by educators to provide a 'schooling-to-work' pathway to support the employability of graduates. The cooperative arrangements, between stakeholder groups, have for decades shaped programs of study at many tertiary education institutions, and certainly for the best part of the last century in the Western economies. Such approaches endorse a workforce and competence imperative for education and learning which, while consistent with the broader expressions in public policy, offer less encouragement or engagement in career-based professional development education and learning.

Government Perspectives

At all times, the major stakeholders in WIL have been seen as industry, students and the education institutions. However, a wider relationship exists that includes: Professional Associations; the Government; and more widely the community, which may extend over time as a lifelong career development paradigm. Successive Western governments have expressed a social-economic imperative for 'work-ready' and 'life-long learners' as part of the labour policy. An examination of public policy recorded in Australia over three decades, undertaken by Calway and Murphy (2007), found that education policy was being driven by workforce-ready imperatives and risk minimisation (i.e. competence and compliance) strategies. This was also consistent with worldwide policy directions detailed in:

- OECD reports (OECD, 2002, 2003, 2004, 2005-06);

- National submissions to the International Symposium for Career Development and Public Policy (CICA, 2006),
- The emphasis on employability issues in the submissions to, and reports of, various national educational reviews (Bologna Process, 1999 - 2007; Kohler, 2004) (ACCI/BCA, 2005; BIHECC, 2007; CHSRF - Canadian Health Services Research Foundation, 2007; CICA, 2005; Dearing, 1997; Department of Education South Africa, 1995; DEST, 2002, 2007; Haines, et al., 2006; Jones, et al., 2002; Scottish Executive Health Department Management Executive, 1999; West, 1997); and
- Quality Assurance (ENQA, 2007).

Further, for an analysis of learning and work transition policies in Canada and Germany see Heinz and Taylor (2005). The overwhelming emphasis of these policies is the need for, 'Skilling for the workforce'. 'Skilling for the workforce' is the common thread and emphasis on training which is based on 'learning from doing' to support academic programs. The concept of individual professional development, particularly learning which is initiated and driven by the learner, professional associations and workplaces does not feature prominently. Rather analysis suggests learning is often prescribed narrowly through an emphasis on defined competencies and their compliance assessment.

The third International Symposium on Career Development & Public Policy (CICA, 2006), also illustrates the significance of labour policy on learning. The policy themes being: Human capital; Labour supply; Employability skills; Career development services for workforce development; older workers; and the information base for public policy making. The Symposium is another illustration of the significance of a need for broader policy on the pre and post graduation learning using work integration models.

Analysis of successive Australian Government's policies, up to the present i.e. 2010, provide evidence of other than economic policies as the basis for higher education; i.e. "It is critical that undergraduate education is fully effective as a foundation for life-long learning." (Kemp, 1998) Also, Higher education is to fulfil significant functions in our society, valuing learning throughout life, and promoting the pursuit, preservation and transmission of knowledge. Such policies extol the value of research, both 'curiosity-driven' and 'use-inspired', enabling personal intellectual autonomy and development, and providing skills formation and educational qualifications to prepare individuals for the workforce and helps position Australia internationally (Nelson, 2002). The Australian policy of the past several decades is also representative of what can be seen in other nations.

Following a change in government in 2007, Gillard (2008) and O'Connor (2008) stipulated that in addition to technical skills, the generic skills specified by West (1997) and others such as Dearing (1997), generic skills were critical in building "critical thinking and reflection and the capability to better understand our place in the world." (Gillard, 2008 n.p.) Government policies were needed to produce "advanced technical skills, high quality research and the best analytical thinking must go together to improve the way our businesses operate. And that together this skill and knowledge builds prosperity, more jobs and wealthier households." (Gillard, 2008 n.p.) O'Connor (2008) stated that "effective Work Integrated Learning partnerships not only deliver personal and career benefits to the individual student, but also contribute to learning and teaching excellence by developing skills that make students more attractive to employers". These views are consistent with a wider social capital paradigm; however, the WIL literature substantially retails a labour force focus.

WIL Philosophy Framework

As part of our review of cooperative education literature, and WIL as a learning philosophy, we questioned:

- Whether WIL, as reported, was meeting the needs of public policy broadly; and
- The relationship that active and action learning has with WIL as both learning theory and learning methodology.

Our analysis provides an understanding of a WIL philosophy that should:

- Use active and action learning methodologies and focus upon a broader individual and corporate professional development approach:
  - *Action learning* as an overt contrivance by the learner and strongly embedded in the learner's culture and learning desires - a continuous process of purposed learning and reflection, centred about the need to find the solution to a real problem. Learning is initiated and driven by the learner (Knowles, 1975; Revans, 1991). Our analysis shows that this is not how WIL is reported in literature; and
  - *Active learning* that is concerned with learning from doing, and/or taking action. It can involve reflection and/or include student course materials, constructed activities, case studies, group projects, etc. This sees WIL as an integration of work and learning, contrived not by the learner but by other parties (educators, employers, etc.) (Meyers & Jones, 1993);
- Express WIL as *six educational imperatives* as a social construction (enculturation) (Calway & Murphy, 2007), i.e.:
  - *Work* (labour force) *ready* graduates – a vocational and skills/competency focus in the proliferation of degrees with specialized major studies;
  - *A Continuing Professional Development Culture*;
  - *Life-long learning*;
  - *Knowledge transfer and exchange* that occurs through "linkage and exchange" - the interaction, collaboration, and exchange of ideas;
  - *Human and Social potential*; and
  - *Internationalisation* - international relevance and collaboration; and
- Be the general term given to learning that occurs through undertaking a component of industry/professional practical experience while studying whether studying for an accredited program of tertiary level studies or not and expressed in one of the nine models defined below.

With these points in mind the chapter now explores WIL as embedded life-long learning, continuing professional development and theorises the various WIL models that employ the work and learning interface.

WORK-INTEGRATED LEARNING MODELS

Perhaps part of the answer lay in the various WIL models. In 2006 we studied approximately 900 articles (Calway & Murphy, 2007) reporting work-integrative and cooperative learning projects, from prominent journals and publications that were collected as abstracts (note: abstracts report a holistic synopsis of the content of an article by definition) and content analysed along multiple dimensions. The groupings of experiential education were consolidated into the first eight models (refer Table 1). We conducted a further review in 2009 that repeated the original study on the wider

Table 1.

Stakeholders Models	**Student/***Learner* **Expectations**	**Educational Institution** *and/ or Professional* **Expectations**	**Employer** *and/ or Industry Workplace* **Expectations**	*Enrolment Conditions*	*Benefits to Students*	*Duration*	*Related fields – examples*
M1 – Pre-course Experience	• Skill establishment • Work experience • Enhanced Employment opportunities • Advanced Qualification	• Students entering the course have a certain level of practical knowledge • Prior knowledge/ experience	• Student is employee • Higher level of expertise through doing the course	*Related experience may be necessary*	*• Extension of experience*	*Unspecified*	
M2 – Project-Based Experience	• Independent knowledge application • Field experience • Enhanced employment opportunities	• Industry relationships • Students with practical skills	• Project completion • Business Improvements through 'consultation'	*Undertake a course specific project*	*• Apply theory into practice.* *• Apply real-life setting*	*Short to medium term*	*Natural sciences, Leadership, Business/ICT*
M3 – Vocational Education	• Practical/occupational skill experience • Enhanced employment opportunities	• Students who are skilled and work force ready	• Students who have to trade skills to complete specific tasks • Good, cheap, dependable workers	*Classroom studies*	*• Preparing students for work in specific fields* *• Trade focused*	*Medium term*	*Plumbing, Carpentry, etc*
M4 – Contextual Learning	• Enhanced learning style • Able to apply theory to practice • Application of information / knowledge	• Students who are independent learners • Contextualised knowledge • Students who are more interested in learning	• N/A	*Classroom setting*	*• Contextualised learning*	*Unspecified*	*Any discipline*
M5 – Work Experience	• A taste of the work place • Generic skills	• High schools develop relationships with industry • Students who have had some 'work force' experience • Community service (Depending on type of experience)	• Students get a taste of company • Students are to comply with company guidelines	*Job shadowing, upper secondary students*	*• Workplace mentoring and job shadowing* *• Real world taster* *• Pre-tertiary and no experience required*	*1-2 weeks duration*	*Any discipline*
M6 – Supervised Experience	• Enhanced employability • Marketable job skills • Exposure to current industry practices • Clarification of career goals	• Industry partnerships • Enhanced employment opportunities • Development of reputation • Student's application of knowledge	• High level competence • Minimal or no cost	*Supervision in workplace, classroom integration is minimal, compulsory for course completion*	*• Career focused* *• Study for credit*	*Short to long term*	*Any discipline*

literature and that included professional associations, not previously studied, as major stakeholders and continuing professional development as the learning paradigms. This has highlighted the need to include continuing professional development as a post graduation learning opportunity or mode/ model of WIL.

WIL Models Literature Study Analysis

The models derived are now explained and then followed by an extended representative stakeholder table that outlines the stakeholder expectations follow by the benefits (Table 1). The data collection that was content analysed provided the basis of the model descriptions and Table 1 details matrices:

Model One was named the 'Pre-course experience' model. Pre-course experience was not included in a group with the other terminology used in the area as it may be a prerequisite for course entry rather than a work based experience during the term of a course. It assumes a certain level of competence upon acceptance into a course rather than work and learning integration during the course;

Model Two was named 'Project based' and contained the Practicum, independent studies and work-based project models. This model is project based with students completing a research project related to their chosen career, generally at a work site, which offers an opportunity to apply theory learned to the project undertaken:

Model Three 'Vocational' contains the vocational education, technical preparation and apprenticeship models. The models are vocational in focus. Most of the skills are developed on the job with a small amount of theory being learned through course material. These models are 'trade focused' for example: plumber, carpenter:

Model Four identified was named 'Contextual Learning' and contained the models of experiential education, contextual learning, praxis and service learning. This group brings real life experiences into the classroom setting. Learning from doing in a very structured way, this ensures that the student is playing an active learning role in their own education. These models ensure that the curricula is not studied in isolation but that ideas, skills and insights learned in a classroom are tested and experienced in real life. In this instance 'context' is interpreted to mean 'real life';

Model Five, named 'Work Experience' contained the models: Work experience and job shadowing. These models give students a sample of what it feels like in the workplace. They do not necessarily have to have any relevance to course material as the students are generally 'observers' simply attempting to gaining an understanding of what a job entails. These are generally undertaken during the middle years of high school as students attempt to plan for their future careers, the placement is generally short-term;

Model Six was named 'Supervised Experience' and contained the models externship, field studies, internship, cognitive apprenticeship, professional practice and 'preceptorships'. Supervised experience in a focused field of study. These models are generally 'built' into the course rather than a student making an active choice to participate in industry learning, for example a medical internship. These are skill based, in a professional field;

Model Seven was named 'Work-Based Learning' and contained the models: Work based learning; cooperative education; organisational learning; industry based learning; sandwich courses; and practice oriented education. These are a form of education that integrates periods of academic study with periods of work experience related to the student's studies. Academic credit is offered for generally 6 to 12 month placements. These models are generally optional and are therefore the result of student's initiative and commitment to study. Students learn first and then enter the work place to apply knowledge;

Model Eight was named 'Joint Industry / University Courses' and contains cooperative programs and joint industry/university courses. These models are a partnership between industry and university where industry can move into the classroom to ensure that students have the necessary skills to be employable by the industry; and

Model Nine was named 'Continuing Professional Development' (CPD) and represents learning designed for enhancing knowledge (traditionally university based although knowledge may be expanded through other methods including both Non-formal and Informal learning); or Vocational Practice (Praxis); or Competency. CPD is necessary for professionals to keep up to date within their profession in areas of technical, legal, conceptual and/or social change. Professionals use CPD to acquire new knowledge which may be incremental, transformative, or supplantive (Atherton, 1999) and can be achieved through Formal, Non-formal and/or Informal Learning (Eraut, 2004). Many professional associations specify a minimum level of CPD for their members to ensure fitness to practice and to minimise risk to the consumers of professional services.

Learning Benefits

All WIL models analysed reported benefits and WIL should be designed and implemented so that the learning is maximised (Andresen, et al., 2000; Billett, 2001). In some cases the employer is a key stakeholder of the process and in others the employer may be incidental or irrelevant. In all models it is possible to construct a learning environment which challenges learners to reflect on the practices and processes of the workplace and to critically review their own and the operational performance of the business.

Dewey (1938); Kolb (1983); Brown, Collins and Duguid (1989); Lave and Wenger (1991) and others have written on the learning benefits of linking work and learning. For example, learners involved in WIL are able to move from observation to participation. Throughout a work placement, this transition may take place on a number of occasions as learners are exposed to new challenges. Each transition enables the learner to become involved in informal learning. Further, this informal education is a way of helping people to learn which involves elements such as conversation and experience in any setting (Smith, 1997). Informal education provides an opportunity for broader learning than can be obtained just in a class room. Learners in WIL are forced to move from a strategic/pragmatic approach to positive and active problem solving learning (Calway, 2005).

In WIL, learners become part of a community of practice within the workplace. WIL provides learners with both the opportunity and the incentive to reflect on and examine the way knowledge is enlarged, reorganized or transformed. For this to happen, they need to understand and locate sources of knowledge and see how this knowledge may be used. They should also understand the critical issues of access to knowledge, including internal politics associated with knowledge, power and prestige within an organisation. Learners should be encouraged to actively pursue gathering knowledge within the organisation and this is best achieved when educators and professional associations are seen to be concerned with the learner's ability to apply the knowledge gained to real world situations and actively promote this learning (Moore, 1999).

NEED FOR WIL EMBEDDED LIFE-LONG LEARNING

In many countries students contribute financially toward their education and are employed part or full-time while studying (Anderson, 2006; DEST, 2002; James & Beckett, 2001; McInnes, et al., 2000). They come to the learning environment with a variety of skills and understandings. There is little evidence that this employment is widely drawn upon in the learning process by learners or teachers to enrich either individual learning or class room experiences. There is a need for additional study to advance the understanding of the relationship between work and learning and the relationship to adult and professional learning processes.

We would argue that any WIL philosophy should encompass a learner as freely moving between an 'active learning' and an 'action learning' education paradigm (described earlier), particularly when a student may have prior work experience. Anderson (2006) reported a growing number of students/learners with independent work and we argue that WIL promoters must seek to join the contrived workforce focused environment of institutionalised 'real-world' active learning, which contains all the structural elements necessary to promote student participation and assessment, to that of learner contrived action learning paradigms in order to achieve human and social potential. WIL can provide an environment in which learning is centred on the student/learner. WIL could also enact more emphasis upon action learning, which then would focus upon the professional/personal career of the learner. The moment that action learning moves to an institutional framework it is likely to result in a work-ready paradigm at the expense of individual professional development.

Explicit institutionalised knowledge is not sufficient for a professional practitioner. "Very little of real working life is run on agreed common definitions" (Lambe, 2002 n.p.). The work of professionals involves approximations and most of practice is "highly interpreted, time and place contingent, and constantly shifting" (Lambe, 2002 n.p.). Knowledge is dynamic in connectivity and not made up only of discrete pre-formed units, whereas education systems have been designed mainly involving assessment of discrete pre-formed units of manageable problems which lend themselves to pre-specified solutions – a competency or sufficiency approach.

The professional who gains entry into a professional association is acknowledged to have acquired a body of knowledge, a skill, or ability at the sufficiency level defined by the professional body (a minimal level of knowledge sufficient to practice within a particular profession). This is generally recognised by professions and the public that the professional is competent (Painter, 2006; Professions Australia, 2006). The Committee for Economic Development Australia's study on life-long learning found that "the skills needed for work throughout a lifetime are diversifying. Jobs which once changed little in the course of a worker's career are now changing every few years. Qualifications achieved at age 20 are more and more likely to be out of date by the time you turn 50". (Dimopoulos & Walker, 2005)

Simply maintaining the sufficiency level of knowledge held at entry into the profession is inadequate. Equally, practice alone is generally insufficient to ensure knowledge is effective and up to date. New knowledge required to continue to practice within a profession evolves over time as a consequence of legal changes, technological advances or conceptual development. Professionals therefore have a requirement for professional development to obtain knowledge which reflects current practice within the profession and workplace as well as society norms.

Universities, as higher education providers; work places; and professional associations, as keepers of a body-of-knowledge; need to provide the foundation for deep and life-long learning combined with a capacity for directed and self-directed WIL activity. This extension of learning can and should be taken up by professional associations and by the individual learners/professionals and their employers. Deep learning by students/learners is dependent on the learning opportunities provided by the employer, the ability and commitment of individual students to achieve learning from the work experience and the commitment of the education authority to support the learning process (Coll & Eames, 2004; Van Gyn, 1996; Young, 1997).

In cooperative education deep learning is more likely to occur when workplace supervisors understand the learning objectives and provide meaningful work experiences which offer the students appropriate challenges and when students know how to reflect on their experiences (Crebert, 1995; Van Gyn, 1996). Moore (Moore, 1999) advocates additional research to advance the understanding of the relationship between work and learning and the relationship to adult and professional learning processes, a point with which we concur based upon our study of WIL.

Therefore, WIL purports and supports greater depth of learning as learners are exposed to the multiple cultures of learning and work. Through WIL, learning can be placed in context providing learners with:

- Greater understanding of the technical aspects of their profession;
- Broader understanding of their profession and their workplace; and
- The incentive to become active learners rather than strategic/pragmatic participants.

Unfortunately for many students/learners the first exposure to problems without predetermined solutions comes after graduation. This may be the cause of the many requests over the years for work-ready graduates (CIO Executive Council, 2006; IBM, 2004; Jancauskas, et al., 1997). Education providers need to strike a balance between:

- Purely vocational education, focusing on standards and which is competency based; and
- A pure education philosophy which does not extend the student beyond the class room.

Many professional associations provide entry level programs, once learners have graduated, to assess the graduate's skills, knowledge, understanding and competency in order to practice within the profession, with many of these programs based on technical competency. In order to fully equip the new professional, professional associations should build WIL models into their professional development programs. Through WIL learners are given a context for learning providing insights into relevance of the material covered. Learners are exposed to practical, real-world (ill-structured and complex) problems which provide an understanding of problem solving processes needed by future leaders.

When professionals are challenged to apply knowledge to meaningful situations there is a need for self-directed learning and reflection to extend their formal education. Education, as a consequence, therefore needs to structure and enrich learning so that students/learners are equipped for careers which require more than just technical competence. We argue that unless learning is put into context, professionals will not obtain the knowledge, skills and understanding specified as being required by Dearing (1997); West (1997); Jones, et al. (2002); and the Bologna Process (2004) (Table 2).

LEARNING OUTCOMES INVESTIGATION

The study of public policy statements has highlighted an emphasis which differed from the graduate attributes set out in Dearing (1997); West (1997); DEST (2002); Jones, et al. (2002) and in the Bologna Process (1999 - 2010). Higher education institutions are being directed to respond to these policy directions with a change of emphasis. The learning outcomes specified for graduates include generic skills; core knowledge which is domain specific for the profession; and the ability to engage in self-actualised and life-long learning.

With this in mind, there is an expectation that professionals possess both discipline specific and generic skills. To be admitted into a profession, professionals may have only been tested on one of these and even that may have only been at a superficial or sufficiency level. The bodies of knowledge specified by professional associations,

Table 2.

DEST (2002) (Australia)	Dearing (1997) (United Kingdom)	Bologna Process (Europe) (Kohler, 2004)
Skills including:	**Concept of '*Graduateness*'**	**Cognitive, emotive and value-oriented Soft Skills competences including:**
• initiative and enterprise; • information literacy and management; • capacity for life-long learning; • ability to: o be adaptable o 'learn-to-learn' in jobs and roles yet to be envisaged; o work efficiently in multidisciplinary contexts.	• Knowledge; • Understanding; • Dispositions; • Attitudes; • Values.	• Knowledge-related; o languages, o basics of law o economics o ICT • Methodological; o problem solving o integrative thinking, o decision making, o time management • Personal; o individual values o social interaction, o reliability, o initiative, o willingness to work • Social o empathy, o ability to: - cooperate, - lead, - bear conflict.

for entry into professions, mostly concentrate on discipline specific issues. Our wider research project examined this in more detail, but the preliminary examination of a sample of professional associations' websites supports this contention (ACS, 2010). Table 3 shows the entry and other credentialing requirements for a range of professional associations in Australia specifically examining the need for practice-based learning or work-integrated learning. The table also examines the types of specialisations available within these professions. All specialisations required further study and assessment and sometimes include a WIL component.

Further, professional association's policy statements, including Professions Australia (2006) Blueprint for National Registration of the Professions, also revealed contrasting expectations for graduates in the statements of government policy and those of professions (Murphy & Calway, 2007). On one hand there is a call for specialist technical knowledge based on demonstrating competency for the needs of presumed job requirements, on the other there is a significant body of evidence calling for broader work related knowledge and understanding through enriched professional development. The first concentrates on a sufficiency level of learning and encourages learners to adopt a strategic pragmatic approach to learning, while the second demands a broader and/or deep level of knowledge. Engineers Australia (2008),for example, requires professional development to keep up to date technically. The Australian Library Association (2008 n.p.) state that professional development enhances personal and professional knowledge and skills outside regular job responsibilities. Table 4 sets out the essential elements of professional and career development.

Professional development commences during post-secondary education where participants learn the technical and generic skills relevant to their

Table 3.

Professional Membership				Specialist Membership		
Discipline	**Entry to Occupational Assoc.**	**WIL Required or experience requirement**	**Registration/ Licensing/ Accreditation**	**Specialisation (Examples)**	**WIL required or experience requirement**	**Registration/ Licensing/ Accreditation**
Medical	Degree (Doctor)	Internship	Yes	Ophthalmology, Anaesthetist, Cardio/Thoracis et al.	Internship	Yes
Nursing	Degree	Internship	Yes	Midwifery, Intensive Care et al.	Internship	Yes
Engineering – Civil/ Construction	Degree	No WIL. 3 years experience for full membership	Yes	Project Management	No	Certification available
Engineering – Manufact'g/ Production/ Chemical etc.	Degree	No WIL. 3 years experience for full membership	No			
Accounting	Degree	No WIL. 1 year experience for full membership	No	Tax, Financial Planning Auditing,	No	Yes
				Company Secretary Corporate	No	Limited
				Management Finance	No	Limited
ICT	Degree (Various types)	No WIL. 4 years experience for full membership	No	Project management,	No	Certification available[1]
				Network administration,	No	No
				e-Business	No	No
				Knowledge Management, Security	No	Under discussion
Teaching	Degree	Practicum	Yes			
Insurance Broker	Certificate	No WIL	Yes			

[1]There is Project Management certification available from a number of bodies. Certification is not a requirement to practice as a Project Manager. (Source – our survey of professional association websites)

chosen profession and where they can develop the cultural, motivational and attitudinal skills necessary for their careers (see Table 5).

The skills specified by DEST (2002) underlie the reality that a new graduate (novice professional) must recognise a professional commitment to life-long learning, competence and productivity which incorporates both career development and professional development. However, cultural and motivational commitment impact learning and may limit the ability of post-secondary education to influence individual's professional level career development and professional development.

Further, Shulman (2002) suggests that the learning necessary to be a professional is more than just intellectual endeavour. Professionals must learn not only to think in certain ways but also to perform particular skills, and to practice or act in ways consistent with the norms, values, and conventions of the profession. This cannot all be learned within universities. Shulman's table of learning categories (i.e. Engagement and Mo-

Table 4.

Life-long learning	Involves: professional associations and professional organisations as learning organisations individuals government
Conducted for Compliance	Required for credentialing, risk minimisation by: professional associations professional organisations individuals Government
Conducted for Productivity	Required for work readiness and just-in-time by: professional organisations – topic specific individuals – topic specific Government
Process	• self-directed, life-long learning and just-in-time learning • formal award courses through education providers e.g. universities and TAFE's • PD providers – topic specific, sometimes opportunistic • WIL to increase the relevance of the learning

Table 5.

Skills required by graduates:	**Supports**
initiative and enterprise;	life-long learning, career development and professional development
information literacy and management;	life-long learning, career development and professional development
capacity for life-long learning;	career development and professional development
ability to be adaptable;	career development
ability to 'learn-to-learn' in jobs and roles yet to be envisaged;	life-long learning, career development and professional development
ability to work effectively in multidisciplinary contexts.	career development

tivation; Knowledge and Understanding; Performance and Action; Reflection and Critique; Judgment and Design; and Commitment and Identity), identify the characteristics which professionals are expected to possess. His views are consistent with the concept of graduateness proposed by Dearing (1997). WIL can provide an environment in which learning is centred on the student. WIL could also enact individual action learning which would focus upon the professional/personal career of the learner. The moment that action learning moves to an institutional framework it is likely to result in a rationalist work-ready paradigm at the expense of individual professional development.

Shulman does not propose that his table of learning follows strict sequential patterns or has a beginning and an end. He considers that 'Commitment and Identity' is most likely to be followed by new 'Engagement and Motivation' as professionals engage in life-long learning and continually need to re-commit to learning in order to maintain levels of practical knowledge or to extend their knowledge base for further career development. An examination of Shulman's Table of Learning (2002), in relation to WIL, provides the foundation for a learning taxonomy relevant to post graduation paradigms and professional development (see Table 6).

Table 6.

Shulman's Elements	Relevance of Work-Integrated Learning
Engagement and Motivation	Is enhanced as learners are involved in their chosen careers prior to completing their studies
Knowledge and Understanding	Learners see practical application of the theory that they have studied, plus they will see possibly for the first time the critical issues of knowledge, power and prestige
Performance and Action	A key element of WIL is that the learners do real work and solve real business problems. The more challenging the problems, the broader level of learning likely to be achieved (even if the learner is not always successful). The learner is exposed to the culture of work
Reflection and Critique	WIL should always require the learner to engage in critical reflection. This is an important element of both Action and Active learning. It is essential to achieve deep learning.
Judgment and Design	Whether learners achieve this level may depend on individual employers and possibly the way the university structures learning objectives for WIL. Significantly WIL provides a vehicle by which learners can become involved in projects which do not have pre-set solutions and which contain the complexities of business operations.
Commitment and Identity	Learners start to be treated as a peer or colleague. This is different to normal teacher/student relationships. Learners start to identify with their profession and see what it means to practice within that profession.

LEARNING THROUGH CONTEXT & TRANSFER OF KNOWLEDGE

Learning through context provides a framework to equip participants for professional practice. The knowledge which professionals need to solve real world problems requires that the professional is capable of constructing meaning in a given situation using both domain knowledge and experience (Brown & Duguid, 1991). The concept of standard solutions to standard problems does not fit the needs of professionals in practice (Lambe, 2002). Shulman (2002) argues that we need to go beyond teaching and assessing for understanding in order to foster judgment and design. It is our contention that when education focuses on competency (particularly technical competence) then learning is likely to be at a superficial level. WIL, as represented in literature broadly, is likely to provide the self-directed life-long learning necessary for both career and professional development.

The DEST report on employability skills states – "Students' employability skills will also be strengthened where students have access to relevant work experience through quality work-integrated learning programs" (DEST, 2007, p. 37). Greater depth of learning can be obtained through learning in context in which the learner can relate the theory in the learning to practical situations (e.g. Brown, et al., 1989; Dewey, 1938; Kolb, 1983; Lave & Wenger, 1991). WIL involves both formal and informal learning and is a tested learner setting which can provide a higher education response to the need for undergraduate and postgraduate professional development. An examination of the knowledge requirements of graduates in the workplace shows that they are required to work on problems which may be unique. The ability to remember details learned from a text is only a building block in a graduate's career. Professionals are therefore valued for their ability to apply their knowledge to novel situations.

In relation to learning, the concept of 'Transfer of Learning' is most relevant. Learning is transferred when the knowledge and skills learned in the classroom will transfer to a range of situations in the outside world (Knapper, 2004). This concept specifically opposes the perception, held by many educators and institutions that because something is taught that it will be absorbed and can be applied by the learner. Employers using cooperative education models identified three categories of skills and knowledge developed by students throughout their placements (Young, 1997), viz: cognitive; psycho-motor; or behavioural. The top six skills and knowledge identified in a study of employers were all either cognitive or behavioural, viz: teamwork; capacity to learn; initiative; communication skills oral; adaptability; and interpersonal skills. Similarly studies by Shaw (1992) on employers' perceptions of students in WIL programs and the longitudinal study of WIL graduates (Murphy, et al., 1997) reported a high level of generic skills developed in WIL programs.

SOFT SKILLS RELEVANCE

Sharma, et al. (1994) defined, measured and tested generic and discipline skill for a specific case study of a WIL program. The project surveyed students, recent graduates (three years or less), academic staff and employers attitudes to 28 generic skills. Significantly all stakeholders rated the importance and the accomplishment associated with these attributes as above average. On a five point Likert scale, all attributes were above three with the one exception of Preparation for future study which was rated 2.96 students, 2.98 graduates, 3.13 staff and 2.86 employers. Significantly:

- The mean and the median scores of all the generic skills outcomes for graduates were rated higher by all stakeholders for graduates than for students;
- Proficiency in solving workplace problems increased from 4.38 students (seventh in importance) to the top position of the graduates ratings 4.62; and
- The next elements in importance in the graduates results: capacity to think logically, capacity to function effectively as part of a team, self-confidence, initiative, ability to listen carefully to people, ability to plan, ability to elicit accurate information from others, capacity to work with minimum supervision, ability to manage time were all rated more important by graduates than students.

The results in this study suggest that while students may see these generic skills as important, involvement in a WIL program can increase this awareness.

In contrast the Graduate Careers Council of Australia (2007) in their survey of students across disciplines and programs reported that students considered problem-solving, initiative and enterprise as among their weakest abilities. Similarly only 19% of employers surveyed by the Association of Graduate Employers (Australian Association of Graduate Employers (2008)) were satisfied with the problem solving skills of graduates. A study of non-WIL students by Calway (2005) using Knowles (1975) taxonomy of self-directed learning skills reported that the lowest ranked skills that students believed they had were the ability to: collect data on performance through self observation and feedback from others; assess your present performance using that data; translate learning needs into learning goals, plans, and activities; and maintain continuous self-motivation. These are results of analysing pre graduation and therefore preparation for self-directed learning appears to be limited.

Competence Focused Learning

Professional associations are the guardians of professional standards. As such they require members to engage in continuing professional development in order to retain membership and any associated credentials. Professional development is necessary for professionals to keep themselves up to date within their profession in areas of: technical; legal; conceptual; and/or social change. Continuing professional development programs should foster development of a learning culture which encourages continual growth of knowledge and the professionals' ability to apply that knowledge.

Professional Standards legislation, enacted in legislation in Australia broadly from 2003, seems to reinforce a focus of professional development on competency and standards. The legislation emphasises a need to ensure that professionals comply with the requirements of an 'approved scheme' as a means of protecting the public through risk minimisation policies (PSC, 2009; Standards Australia, 2007). This emphasis upon standards therefore focuses less on the development of new or deeper career knowledge or on the acquisition of specialisation or integration knowledge (Smith, 2005; Standards Australia, 2007).

In the Department of Education, Science and Training (DEST) report (Striving for quality: learning, teaching and scholarship) (DEST, 2002), there is a recognition that technical competencies should not be the sole determinant of policy in Australia. However the report noted a confusion of terminology in the use of terms such as outcomes, attributes and skills and that the development of graduate attributes has shadowed the adoption of the concept of key competencies within the vocational education sector. The report identified 'emerging skills and knowledge that have not been previously a focus of higher education curricula' (DEST, 2002:1).

Competency is defined as a combination of skills, abilities, and knowledge needed to perform a specific task (Jones, et al., 2002, p. 8). However, Smith (2005 n.p.) argues that "in much current usage the notion of competence has been whittled down to the ability to undertake specific tasks; it has been largely stripped of its social, moral and intellectual qualities". The current emphasis of professional development, specified by professional associations and legislation, appears to concentrate on developing and assessing competency (or a sufficiency standard of knowledge for professionals). While Professional Standards Legislation specifies a requirement for professional development to improve standards, it does not suggest what this involves or how it may be achieved. Rather, it relegates professional development to individual professional associations under schemes of self-regulation. Applications by professional associations, applying for 'registration of a scheme' under the legislation, have concentrated on limiting the civil liability of professionals rather than improving standards (PSC, 2007). Either way, to develop professionals beyond a sufficiency level requires an appropriate learning paradigm.

Role for Professional Associations and WIL

The professional is required to solve real world problems and to take account of the peculiarities of each situation or case. Individual problems are often conceptually complex and require more than the ability to retrieve from memory intact pre-existing knowledge. Conceptual knowledge needs to be able to be applied to novel situations using reasoning and inference through an ability to flexibly reassemble pre-existing knowledge to adaptively fit the needs of each new situation (Spiro, et al., 1996). Problems which are complex and ill-structured require the professional to draw on a knowledge base, derived from both prior studies and prior experiences, to solve each new situation. Often cases will vary from previous cases with individual solutions being non-repeatable.

The ill-structured nature of the knowledge domain of a professional is characterised by multiple, wide-application, conceptual structures (multiple schemas, perspectives, organizational principles, and so on), for which the applicability of concepts and solutions are inconsistent from case to case (Spiro, et al., 1996). The professional has to construct meaning in each situation. Constructivist theory (Bruner, 1996; Spiro, et al., 1996) views learning as the establishment of knowledge rather than committing information to mind. It involves going beyond information presented in a course to encourage comprehension or an understanding of the issues involved. In a learning situation mere knowledge of the text is insufficient; it must be combined with knowledge built through experience. The schemas held in the head of a professional can be changed, enlarged and made more sophisticated though the assimilation and accommodation of new concepts to enable the construction of knowledge which is relevant for a variety of situations (Clark, 1999).

This growing need for people who are deeply knowledgeable about their domain areas and who have well developed generic skills is not in question. Commonly recognised generic skills identified in studies of experiential education include the ability to be flexible, adaptable, creative, think critically and problem-solve. Important additions to these generic skills identified are the ability to have the skills required for learning, and to transfer learning to new and different contexts (Doyle, 2002; Murphy, et al., 1997; Shaw, 1992; Young, 1997). These studies also demonstrated that WIL programs are effective in developing these skills.

Many professional associations specify an experience requirement for admission to membership. The integration of that experience with learning is seldom or fleetingly addressed. Ongoing integration of work with learning is unlikely to be considered in the design of professional development programs which have compliance as the main driving force. Evidence shows that educational systems which integrate work with learning provide a more meaningful context not only at a skills level, where relevance adds to the motivation and understanding of learners, but also at the conceptual level which is the depth necessary for professional development for professionals.

As the keeper of standards for a profession, professional associations have a role to play to ensure that:

- The professional development requirements they specify are relevant for the profession; and
- The educational systems are designed so that depth of learning is encouraged.

Learning needs to take place in context. Workplace problems are readily available to the professional to enhance learning. The current emphasis of professional development and professional standards legislation is on risk minimisation; the specifying of skill requirements; and the assessing of the specified competency. Professional associations have breadth and depth insight and awareness of their professions which can enable them to provide both the practical understanding of the required domain knowledge and to develop a contextual framework to ensure a deeper level of knowledge within their professions.

The environment in which professionals operate is conducive to the integration of work and learning and we recommend that professional associations act as a conduit to enable this to happen. They have a responsibility of care to ensure the development of their profession. Davenport and Prusak (1998) emphasise the significance of the development of knowledge which originates and is applied in the minds of the knower through a mix of experiences for evaluating and incorporating new understanding and information.

WIL and Transfer of Learning

Perkins and Salomon (1992 n.p.) state that: "Transfer of learning occurs when learning in one context enhances (positive transfer) or undermines (negative transfer) a related performance in another context. Transfer includes near transfer (to closely related contexts and performances) and far transfer (to rather different contexts and performances). Transfer is crucial to education, which generally aspires to impact on contexts quite different from the context of learning." For professional development at a professional level we need to look for positive transfer and far transfer.

Effective transfer of learning is the key to ensuring that education has a positive impact.

Educators assume that transfer of learning always occurs as a result of education and training. They expect that whatever is learned will be retained or remembered over time and used in appropriate situations (Doyle, 2002). Unfortunately conventional educational practices often fail to stimulate students/learners by either using conditions similar to those in the learning context including using well-practiced routines or to search for connections using deliberate abstraction (Perkins & Salomon, 1992).

Learning in its broadest sense takes place when a learner can demonstrate or display that learning later. Passing an exam or skills test can demonstrate that ordinary learning has occurred – transfer learning is always, at least implicitly, contrastive. It assumes learning within a certain context and asks about the impact beyond that context. Human Resources and Skilled Development Canada (HRSDC) refers to the "so what" or "now what" phase of the learning process. However HRSDC states that "abundant evidence shows that the very 'often hoped for' transfer from learning experiences does not occur." HRSDC therefore defines transfer of learning in the context of the workplace as "the effective application by trainees to their jobs of the knowledge and skills gained as a result of attending an educational program." (HRSDC, 2005)

Transfer of Learning is positive when learning in one context impacts on performance in another context (HRSDC, 2005). Further, HRSDC (2005, p 2) cites Baldwin and Ford's (1988) claim that not more than ten percent of expenditures on training and development in North America actually result in transfer to the job. There is no evidence regarding the Australian position, but it is reasonable to assume that Australia is likely to be similar to North America. It is conceivable that with the emphasis on assessment of competency the potential impact of positive and far transfer of learning has been ignored.

Doyle (2002 n.p.) highlighted "... the value of linking learning to real workplace problems and situations". In her study, students were able to apply theory to real life situations with the result that they were engaged in deep learning as they grew in confidence and were able not only to obtain clarity about the actual learning topics, but they were able to identify the future applications of that learning. HRSDC (2005:2) supports this view when referring to Mosel (1957) who identified three conditions for transfer: content must be applicable to the job; the trainee must learn the content; and the trainee must be motivated to change job behaviour by applying what was learned.

Dewey, who was the inspiration for much research on experiential learning, argued that "... all genuine education comes through experience" (Dewey, 1938). He argued that learning from experience should not happen in isolation but that the learner should be encouraged to connect their learning with past and current experiences and to see possible future implications of their new knowledge. According to Kolb (1984p. 41 in (Smith, 2001)): "learning is the process whereby knowledge is created through the transformation of experience. Knowledge results from the combination of grasping experience and transforming it". Here we emphasise the importance of his proposed six characteristic features of experiential learning:

- Learning is best conceived as a process, not in terms of outcomes *(i.e. a direction, a way of life – not a destination);*
- Learning is a continuous process grounded in experience;
- Learning requires the resolution of conflicts between dialectically opposed modes of adaptation to the world *(and world events)*;
- Learning is a holistic process of adaptation to the world;
- Learning involves transactions between the person and the environment; and
- Learning is the process of creating knowledge: '[which] is the result of the trans-

action between social knowledge and personal knowledge' (Kolb 1984, 36 in (Smith, 2001)) *(italics are our comments/ emphasis).*

This is in contrast to a competency/sufficiency approach to education which is:

- Outcome focussed;
- Does not particularly draw on practical and complex real world examples;
- Supports standard solutions to problems; and
- Does not seek to extend knowledge.

Most cooperative education programs are at an undergraduate level. Pre graduation WIL programs should therefore be intentional; organised; real-world and accredited, having an educational structure that accounts for the nature of the student; the roles of the teacher/supervisor; the curricula emphasis; teaching methodologies; and social function of education and institutions (Calway & Murphy, 2006). WIL programs in non-business disciplines are more likely to adopt an apprenticeship model, with students working on a one-to-one basis with a supervisor in largely research-based programs. Whereas in business disciplines students work in an organisation using knowledge obtained from their prior studies. They are expected to extend that knowledge and obtain new knowledge within the constraints of the available opportunities and expertise of the host organisation (Shaw, 1992; Young, 1997). Rowe (2005) considers that post graduate Cooperative Education programs in non-business disciplines are not appropriate, for example, in programs for scientific disciplines as students often complete only the Master's degree and rarely have careers in research.

CONCLUSION

We have theorised a philosophy of Work-Integrated Learning (WIL) that includes the educational imperatives of workforce readiness; a professional development culture (both individual and workforce related); international relevance; life-long learning; knowledge transference; human and social potential. The views of work-based and experiential education are consistent with these six educational imperatives derived for WIL. The connection of structured learning, with past and current experiences, will further enable students/ learners to achieve a transfer of their learning. This philosophy is consistent with the objectives relating to school-to-work and also advanced level professional development for professionals. WIL can be used to provide a focus for the development of relevant professional knowledge which will provide society with greater expertise.

An emphasis on WIL in the pre and post graduate learning pedagogy creates a richer learning environment enhancing the professional's performance and career. It is both feasible and desirable to design higher education and professional development programs which incorporate WIL. WIL is to be a construction by learners, educational practitioners, professional association and associating employers collectively and is not the sole purview of any particular stakeholder or group. The cooperative arrangements between these groups can be developed for programs of study provided under the auspices of professional associations and bring together tertiary education institutions and employers.

Importantly, given the diversity of professions, WIL is not a prescriptive form of education – it supports various models of operation and it can exist within different learning theories. Much of the literature reviewed is focused through real-world experiences of learning implementations and through a diversity of learning theories including; active, experiential, services, vocational, situated, etc. Rather than being specific, WIL can and does make use of and explain a number of the learning theories. As we asserted, WIL should be intentional, organised and real-world, having an active and action educational structure that accounts for:

- The nature of the student;
- Roles of the teacher/supervisor;
- Curricula emphasis;
- Teaching methodologies; and
- Social function of education and institutions.

Most current implementations of WIL are at undergraduate level and are accredited, however, many professionals do not need professional development offerings which lead to a full academic award. Whether studying for an accredited post graduate level program or not, WIL can provide professionals with a deeper level of learning. Knowledge transfer is more likely to occur when learning content is delivered in context and professionals are able to apply their learning. Higher education and professional development programs should be designed to incorporate and engage active and/or action learning methodologies, and also focus upon a broader holistic approach to individual and corporate professional development.

REFERENCES

ACCI/BCA. (2005). *Employability skills for the future*. Retrieved from http://www.acci.asn.au/ text_files/ issues_papers/Employ_Educ /ee21.pdf

ACS. (2010). *About us*. Retrieved April 9, 2010, from http://www.acs.org.au /index. cfm?action=show &conID=aboutacs

Anderson, M. J. (2006). Degree of fit: University students in paid employment, service delivery and technology. *Australasian Journal of Educational Technology*, *22*(1), 88–103.

Andresen, L., Boud, D., & Cohen, R. (2000). Experience-based learning. In Foley, G. (Ed.), *Understanding adult education and training* (2nd ed., pp. 225–239). Allen & Unwin.

Australian Association of Graduate Employers. (2008). *Employer survey*. Melbourne, Australia: AAGE.

Australian Library Association. (2008). *What is PD?* Retrieved February 18, 2010, from http:// www.alia.org.au/education /pd/description.html

BIHECC. (2007). *Graduate employability skills*. Canberra, Australia: DEST.

Billett, S. (2001). *Participation and continuity at work: A critique of current workplace learning discourses. Context, Power and perspective: Confronting the challenges to improving attainment in learning at work.* Paper presented at the Joint Network/SKOPE/TLRP International workshop, Northhampton.

Bologna Process. (1999-2010). *The Bologna process and Declaration: Trends in learning structures in higher education.* Retrieved from http://www.ond. vlaanderen.be/ hogeronderwijs/bologna/ search/ index.htm?cx= 013531861688426506363%3 Afnpyfltxhbu&cof=
FORID%3A11&q=Trends+in+
Learning+Structures+in+Higher+Education&sa=
Zoeken&siteurl=www.ond.vlaan-
deren. be%2Fhogeronderwijs%2Fbologna %2Fsearch%2F#1664

Brown, J. S., Collins, A., & Duguid, P. (1989). Situated learning and the culture of learning. *Education Researcher*, *18*(1), 32–42.

Brown, J. S., & Duguid, P. (1991). Organizational learning and communities of practice: Toward a unified view of working, learning, and innovation. *Organization Science*, *2*(1), 40–57. doi:10.1287/ orsc.2.1.40

Bruner, J. (1996). *The culture of education* (pp. 224 + xvi). Cambridge, MA: Harvard University Press.

Calway, B. A. (2005). *Rethinking a learning environment strategy.* Deakin, Australia: Geelong.

Calway, B. A., & Murphy, G. A. (2006). *WIL theory or philosophy? - A grounded study.* Swinburne University of Technology, Working Paper.

Calway, B. A., & Murphy, G. A. (2007). The educational imperatives for a work-integrated learning philosophy. *Journal of Cooperative Education and Internships*, *41*(2), 95–109.

CHSRF - Canadian Health Services Research Foundation. (2007). *Knowledge transfer and exchange at work.* Retrieved February 28, 2010, from http://www.chsrf.ca/ knowledge_transfer/ work_e.php-

CICA. (2005). *The CA competency map.* Retrieved March 23, 2006, from http://www.cica.ca/index.cfm/ ci_id/17150/la_id/1.htm

CIO Executive Council. (2006). *Situation analysis. Council initiatives-Australia.* Retrieved March 15, 2006, from http://www.cioexecutivecouncil.com/ public/au/ mentoring.html

Clark, D. (1999). *Constructivism.* Retrieved April 4, 2006, from http://www.nwlink.com/~donclark /hrd/history/ constructivism.html

Coll, R. K., & Eames, C. (2004). Current issues in cooperative education. In Coll, R. K., & Eames, C. (Eds.), *International handbook for cooperative education.* Boston, MA: The World Association for Cooperative Education.

Crebert, G. (1995). *Links between higher education and industry: Workplace-based learning programs in Australia: Practices and issues.* Paper presented at the Keynote address, Workplace-based learning conference.

Dearing, R. (1997). *Report of the National Committee into Higher Education (Dearing report).* London, UK: NCIHE.

Department of Education South Africa. (1995). *White Paper on education and training.* Cape Town.

DEST. (2002). *Striving for quality: Learning, teaching and scholarship.* Canberra, Australia: Department of Education, Science and Training.

DEST. (2007). *Graduate employability skills.* Canberra, Australia: BIHECC.

Dewey, J. (1938). *Experience and education.* New York, NY: Macmillan.

Dimopoulos, N., & Walker, D. (2005). Lifelong learning challenges for educators, governments, firms and individuals. *CEDA Australian Chief Executive* (August).

Doyle, D. (2002). *Putting learning to work: The distance learner and transfer of learning.* Retrieved March 23, 2010, from http://www.aare.edu.au/ 03pap/doy03366.pdf

Engineers Australia. (2008). *Continuing professional development.* Retrieved February 18, 2008, from http://www.engineersaustralia.org.au/education/continuing-professional- development/continuing-professional -development.cfm

Graduate Careers Australia. (2007). *Snapshot: University and beyond.* Carlton.

Griffith University. (2006). *Introduction to Work Integrated Learning (WIL).* Retrieved June 11, 2010 from http://www62.gu.edu.au/ policylibrary.nsf/xmainsearch/ a0d60f04626b7f144a257180 0063dd56?opendocument

Haines, C., Scott, K., & Lincoln, R. (2006). *Australian blueprint for career development.* Retrieved June 11, 2010, from http://www.dest.gov.au/NR/ rdonlyres/B2B0AACE-D719-48AB-8A42-D78203FD4F0D/ 10560/BlueprintTrial Version.pdf

Heinz, W., & Taylor, A. (2005). Learning and work transition policies in a comparative perspective: Canada and Germany. In Leithwood, D. L. K., Cumming, A., Bascia, N., & Datnow, A. (Eds.), *International handbook of educational policy* (*Vol. 2*). New York, NY: Kluwer. doi:10.1007/1-4020-3201-3_45

HRSDC. (2005). *Transfer of learning planning workplace education programs: HRSDC*. Government of Canada.

IBM. (2004). *Services science: A new academic discipline.* Retrieved February 22, 2010, from http://www.almaden.ibm.com/asr/ resources/facsummit.pdf

James, R., & Beckett, D. (2001). *The changing expectations of university students and the implications for learning.* 1-6. Retrieved from http://www.cshe.unimelb.edu.au/ people/staff_pages/James/James&Beckett= Singapore.pdf

Jancauskas, E., Atchison, M., Murphy, G. A., & Rose, P. (1997). *Unleashing the potential of work-integrated-learning through professionally trained academic and industry supervisors.* Paper presented at the WACE Conference, Cape Town 1997.

Jones, E. A., Voorhees, R. A., & Paulson, K. (2002). *Defining and assessing learning: Exploring competency-based initiatives*. Washington, DC: Council of the National Postsecondary Education Cooperative (NPEC), U.S. Department of Education National Center for Education Statistics.

Kemp, D. (1998). *Strategic developments in higher education, address to the OECD thematic review seminar on the first years of higher education.* Sydney. Retrieved June 15, 2010, from http://www.dest.gov.au/ archive/ministers/kemp/ks210498.htm

Knapper, C. (2004). *Research on college teaching and learning: Applying what we know.* Paper presented at the Teaching Professor Conference, Philadelphia.

Knowles, M. (1975). *Self-directed learning-a guide for learners and teachers*. New York, NY: Association Press.

Kohler, J. (2004). *The Bologna process and employabilty: The impact of employability on curricular development.* Paper presented at the A Key Objective of Academic Studies and for Academic Institutions, Bled.

Kolb, D. (1983). *Experiential learning: Experience as the source of learning and development.* Englewood Cliffs: Prentice Hall.

Lambe, P. (2002). *The autism of knowledge management.* Retrieved June 28, 2010, from http://greenchameleon.com/ thoughtpieces/autism.pdf

Lave, J., & Wenger, E. (1991). *Situated learning: Legitimate peripheral interaction*. Cambridge, UK: Cambridge University Press. (Original publication IRL report 90-0013, Palo Alto, CA.: Institute for Research on Learning) (1990).

McInnes, C., James, R., & Hartley, R. (2000). *Trends in the first year experience in Australian universities*. Canberra, Australia: Department of Education, Training and Youth Affairs.

Meyers, C., & Jones, T. B. (1993). *Developing and assessing instructional expertise: Promoting active learning*. San Francisco, CA: Jossey-Bass.

Moore, D. (1999). Toward a theory of work-based learning (IEE brief no.23). *IEE brief.* Retrieved March 21, 2010, from http://www.tc.columbia.edu/ centers/iee/BRIEFS/ Brief23.htm

Murphy, G. A., & Calway, B. A. (2007). *Professional development-the case for deep learning.* Paper presented at the BERA 2007 Conference, London.

Murphy, G. A., Murphy, R. E., & Calway, B. A. (1997). *Career progression of cooperative education graduates.* Paper presented at the 10th World Conference on Cooperative Education, Cape Town.

Nelson, B. (2002). *Higher education at the crossroads: An overview paper.* Retrieved June 15, 2010, from http://www.dest.gov.au/ ministers/ nelson/apr02/ n55_260402.htm

OECD. (2002). *Education policy analysis.* Retrieved May 6, 2010 from http:// www.oecd.org/document/ 42/0,3343,en_2649_ 39263231_1841194 _1_1_1_1,00.html

OECD. (2003). *Education policy analysis.* Retrieved May 6, 2010, from https://www. oecd.org/ document/14 /0,2340,en_2649_ 34511_17735886_ 1_1_1_1,00.html

OECD. (2004). *Education policy analysis.* Retrieved May 6, 2010 from www.oecd.org/.../ 0,3343,en_2649_ 39263231_ 34989090_1_1_1 _1,00.html

OECD. (2005-06). *Education policy analysis.* Retrieved May 6, 2010 from www.oecd. org/.../ 0,3343,en_2649_39263238_ 37693017_ 1_1_1_1,00.html

Painter, J. (2006). *Limited returns: Why management development isn't fully developed.* Retrieved June 1, 2010, from http://www.ceda.com.au/ research/previous-topics/ research/2005/09/28/ dnetto_management.aspx

Perkins, D. N., & Salomon, G. (1992). Transfer of learning. In Tuijnman, A. (Ed.), *International encyclopedia of education* (2nd ed.). Oxford, UK: Pergamon Press.

Professions Australia. (2006). Blueprint for national registration of the professions. Retrieved June 12, 2010, from http://www.professions.com. au/ natreg.html

PSC. (2007). *Application guidelines-guidelines for occupational associations to prepare an application to the Professional Standards Council for the approval or renewal of a scheme in New South Wales.* Retrieved May 21, 2007, from http://www.lawlink.nsw.gov.au/lawlink/ professional_standards_council /psc_ll.nsf/vwFiles/ Application%20Guidelines% 20(NSW).doc/$file/ Application%20Guidelines%20(NSW).doc

PSC. (2009). *Application guidelines for approval of a scheme.* Retrieved June 29, 2010, from http:// www.lawlink.nsw.gov.au/ lawlink/psc/ll_psc. nsf/vwFiles/ Application_Guidelines.pdf/$file / Application_Guidelines.pdf

Revans, R. W. (1991). Action learning: Its origins and practice. In Pedlar, M. (Ed.), *Action learning in practice* (2nd ed.). New York, NY: Gower Press.

Rowe, P. (2005). *Survey of graduate programs with cooperative and internship.* Waterloo.

Scottish Executive Health Department Management Executive. (1999). *Learning together* Retrieved May 2, 2010, from http://www.scotland. gov.uk/ health/mel/mellt-00.asp

Sharma, R. (Ed.). (1994). *Development and implementation of strategies for assuring teaching quality and performance.* Melbourne, Australia: Swinburne University of Technology.

Shaw, J. E. (1992). *An evaluation of the cooperative education program in Information Technology.* Canberra, Australia: DEET.

Shulman, L. (2002). *Making differences: A table of learning.* Retrieved June 14, 2010, from http:// www.carnegiefoundation.org/ elibrary /making-differences- table-learning

Smith, M. K. (2001). David A. Kolb on experiential learning. In *The encyclopedia of informal education.* Retrieved June 16, 2010, from http:// www.infed.org/ biblio/b-explrn.htm

Smith, M. K. (2005). Competence and competency. In *The encyclopedia of informal education.* Retrieved June 14, 2010, from http://www.infed.org/ biblio/b-comp.htm

Spiro, R. J., Feltovich, P. J., & Jacobson, M. J. (1996). *Cognitive flexibility, constructivism, and hypertext*, 2nd ed. Retrieved April 3, 2006, from http://phoenix.sce.fct.unl.pt/ simposio/ Rand_Spiro.htm

Standards Australia. (2007). *What is a standard?* Retrieved March 12, 2010, from http://standards.org.au/ DevelopingStandards/ WhatisaStandard.aspx

Van Gyn, G. (1996). Reflective practice: The needs of professions and the promise of co-operative education. *The Journal of Cooperative Education, 29*(2), 103–131.

WACE. (2010). *Advancing work integrated learning globally.* The World Association for Cooperative Education (WACE). Retrieved June 11, 2010, from http://www.waceinc.org/

West, R. (1997). *Learning for life: Review of higher education financing and policy.* Canberra, Australia: Review Committee on Higher Education Financing and Policy, Department of Employment, Education, Training and Youth Affairs.

Young, J. (1997). *Comparative international study of co-operative education employers: The value of employing co-op students.* Cape Town, Australia: World Association of Cooperative Education.

ADDITIONAL READING

Australian Association of Graduate Employers. (2008). *Employer Survey.* Melbourne: AAGE.

BIHECC. (2007). *Graduate Employability Skills.* Canberra: DEST.

Boud, D., & Solomon, N. E. (Eds.). (2001). *Work-Based Learning: A New Higher Education?* Buckingham: SRHE and Open University Press.

Brown, J. S., & Duguid, P. (1994). *Universities in the Digital Age*, from http://www2.parc.com/ops/ members/brown/papers/ university.html

Coll, R. K., & Eames, C. (Eds.). (2006). *The International Handbook for Cooperative Education.* Boston: WACE.

Davenport, T. H., & Prusak, L. (1998). *Working Knowledge.* Boston: Harvard Business School.

Graduate Careers council of Australia. (2007). *Snapshot: University and Beyond.* Carlton.

James, R. (2001). *Students' Changing Expectations of Higher Education and the Consequences of Mismatches with the Reality.* OECD-IMHE Conference – Management Responses to Changing Student Expectations, Queensland University of Technology, Brisbane.

Marden, B. (2003). High Aims for Professional Standards Legislation. *Law Institute Journal, 77*(11), 28.

Patrick, C.-J., Peach, D., Pocknee, C., Webb, F., Fletcher, M., & Pretto, G. (2008). *The WIL [Work Integrated Learning] report: A national scoping study [Australian Learning and Teaching Council (ALTC) Final report].* Brisbane: Queensland University of Technology.

Wilson, J. W., & Lyons, E. H. (1961). *Work-Study College Programs - Appraisal and Report of the Study of Cooperative Education.* New York: Harper & Brothers.

KEY TERMS AND DEFINITIONS

Active and Action Learning Methodologies: Focus upon a broader individual and corporate professional development approach, i.e.: (1) *Action learning* as an overt contrivance by the learner and strongly embedded in the learner's culture and learning desires - a continuous process of purposed learning and reflection, centred about the need to find the solution to a real problem. Learning is initiated and driven by the learner(Knowles, 1975; Revans, 1991). Our analysis shows that this is not how WIL is reported in literature; and (2) *Active learning* that is concerned with learning from doing, and/or taking action. It can involve reflection and/or include student course materials, constructed activities, case studies, group projects, etc. This sees WIL as an integration of work and learning, contrived not by the learner but by other parties (educators, employers, etc.) (Meyers & Jones, 1993).

Continuing Professional Development (CPD): "The systematic maintenance, improvement and broadening of knowledge and skills, and the development of personal qualities necessary for the execution of professional duties throughout working life". (Professional Development Partnership (The), 2008).

Deep Learning: "Learning that promotes the development of conditionalized knowledge and meta-cognition through communities of inquiry." (Weigel, 2002, p. 5) It implies that learning is related to existing knowledge and experience enabling participants to use patterns and underlying principles and encourages processes which check evidence in order to reach conclusions and recommendations.

Educational Imperatives: Are taken as a social construction (enculturation) (Calway & Murphy, 2007), i.e. (1) *Work* (labour force) *ready* graduates – a vocational and skills/competency focus in the proliferation of degrees with specialized major studies; (2) *A Continuing Professional Development Culture*; (3) *Life-long learning*; (4) *Knowledge transfer and exchange* that occurs through "linkage and exchange" - the interaction, collaboration, and exchange of ideas; (5) *Human and Social potential*; and (6) *Internationalisation* - international relevance and collaboration.

Informal Learning: "Refers to learning that results from daily work-related, social, family, hobby or leisure activities (e.g. the acquisition of interpersonal skills developed through the experience of working as a sales representative)."(Misko, et al., 2007).

Transfer of Learning: In the context of the workplace is defined "as the effective application by trainees to their jobs of the knowledge and skills gained as a result of attending an educational program" (HRSDC (2005) p.1). Transfer of Learning is positive when learning in one context impacts on performance in another context.

Work-Integrated Learning (WIL): "... educational activities that integrate theoretical learning with its application in the workplace.... and should provide a meaningful experience of the workplace application that is intentional, organised and recognised by the institution, in order to secure learning outcomes for the student that are both transferable and applied". (Griffith University, 2006 n.p.) Work-integrated-learning provides a context for learning and in workplace situations the outcomes may not result in formal recognition but in an increased ability to perform in the workplace.

Chapter 2
Continuing Professional Development:
Work and Learning Integration for Professionals

Gerald A. Murphy
Swinburne University of Technology, Australia

Bruce A. Calway
Swinburne University of Technology, Australia

ABSTRACT

Organisations promote their services by stressing that their professional staff are members of the relevant professional association and by listing the certifications/credentials they hold. Practice alone is generally insufficient to ensure knowledge is effective and up to date. Continuing professional development (CPD) is a requirement for professionals to develop knowledge to enable them to competently and adequately provide services to clients or employers. Professionals work in environments where technical, legal, conceptual and/or social change mandates that processes, practices, knowledge and understanding need to evolve. Individual professionals, professional associations, and the employers of professionals may have differing objectives for CPD and have vested interests to ensure that CPD is designed to, and meets actual objectives. Integrating work with learning is fundamental, i.e. if learning is not seen as having practical application, it is not valued. The knowledge development of professionals can be enhanced through Work-Integrated Learning which takes into account that they: hold a body of knowledge, are adults, and operate in positions in which learning and work can be related. Informal learning, in particular communities of practice, is significant in the transfer of learning to professionals.

DOI: 10.4018/978-1-60960-547-6.ch002

INTRODUCTION

The learning necessary for a professional is more than just intellectual endeavour. Professionals "... must learn not only to think in certain ways but also to perform particular skills, and to practice or act in ways consistent with the norms, values, and conventions of the profession". (Shulman, 2002 p.2) Learning through context provides a framework for professional practice. Professionals solve real world problems through constructing meaning in a given situation using both domain knowledge and experience (Brown & Duguid, 1991). The concept of standard solutions to standard problems does not fit the needs of professionals in practice (Lambe, 2002). Shulman (2002) argues that "...we need to go beyond teaching and assessing for understanding in order to foster judgment and design". (p.6) Continuing professional development (CPD) needs a focus beyond competency, otherwise professionals' knowledge and understanding is likely to be at a surface level (cf. Beach, 1999; HRSDC, 2005; Perkins & Salomon, 1992; Shulman, 2002)

It is a theme of our research that learning needs to be in context and that the philosophy of Work-Integrated Learning (WIL) provides a means for effective transfer of learning. HRSDC (2005) defines transfer of learning in the context of the workplace "...as the effective application by trainees to their jobs of the knowledge and skills gained as a result of attending an educational program" (p.1). Transfer of Learning is positive when learning in one context impacts on performance in another context.

Much of WIL literature has focussed on either undergraduate programs or on non-professional work-based learning. CPD for professionals can and should focus on integrating work and learning.

Professional associations are often guardians of professional standards; as such they require members to engage in continuing professional development (CPD) in order to retain membership; and for some professions, by government regulations, to maintain practicing certificates. CPD is necessary for professionals to keep up to date within their profession in areas of technical, legal, conceptual and/or social change. Professional recognition (accreditation) is usually based on a body of knowledge (BoK) specified by a professional association (e.g. Professions Australia, 2006). Individual professions have differences in the quantity and quality they specify and monitor for CPD.

CPD programs should also develop learning cultures which encourage continual growth of knowledge and professionals' ability to apply that knowledge. Professional associations and the academics in these associations can play a role to ensure that professional standards are both maintained and improved.

Learning may be designed for Knowledge (traditionally university based); or for Vocational Practice (Praxis); or for competency. Praxis is more than application of applying knowledge to given situations. It involves interpretation, understanding and application in a process involving a continual interplay between end and means rather than following a predetermined process (competency) (M. K. Smith, 1996a). We argue that CPD should not be driven just by risk minimisation strategies and compliance measurements.

CPD programs should incorporate learning theories and methodologies which are appropriate for adult learning and for the needs of the participants. We recognise that the term pedagogy is used in a generic sense in many educational papers but we would like to draw attention to the need for CPD programs to be developed using appropriate adult learning frameworks i.e. andragogy. Andragogy concentrates on specific issues related to adult education and it is concerned with facilitating the acquisition of the content (Clark, 1999).

In this chapter we will use the terms: *professional* to refer to professionally qualified practitioners; *organisations* to professional organisations/practices who employ professionals and offer their services to business/government/the public; and *associations* to professional association, bodies whose members are professionally qualified and who commit to a code of ethics/practice.

We have examined the:

- policy positions, in relation to professional standards and Professional Standards legislation, published by the Professional Standards Association, Professions Australia and individual associations; from which we have derived the knowledge requirements needed by practicing professionals in general;
- underlying learning theories and frameworks which are available to, and used by, associations and organisations for CPD to satisfy government, consumer and professionals' needs; and
- value that organisations place on CPD and the influence that they have on the CPD of their employees.

The chapter documents part of an on-going study which questions whether a policy of minimum compliance to agreed standards is suitable for CPD of professionals and how human capital is optimised, standards improved and consumers protected. Our study is relevant to organisations operating with profit motives e.g. accounting, engineering and technology, rather than helping professions such as social workers and nurses.

The focus of our research of CPD is on determining from organisations the impact that CPD has on the organisation and how the organisation may influence the effectiveness of CPD. Is CPD undertaken for compliance or for development? What value does the organisation place on, and get from professional development?

Questions addressed are:

1. *Do the aspirations, culture and practices of a professional organisation impact the way CPD is conducted and consequently the learning resulting from CPD?*
2. *What are the goals for CPD which motivate professional organisations? Is there a relationship between these goals and the level of learning achieved? What value does the professional organisation place on professional development?*
3. *What Informal Learning is conducted within professional organisations and how does the effectiveness of this compare with more formal learning?*
4. *What role does and can WIL play in developing professionals' competence and expertise?*

BACKGROUND

Within small/medium enterprises (SME) professional practices, maintaining and developing professional's BoK within the organisation is critical to the organisation. Professionals for example in engineering, technology and business working in different environments may approach CPD differently:

- single practitioners may have biases and processes which influence their CPD;
- large professional organisations may employ professionals from multiple disciplines and because of the size and scope of the business, professional's within these organisations may engage in CPD activities which are not directly related to the profession's BoK; and
- professionals working in business and government will tend to require less technical CPD (related to their BoK) and CPD more related to their role in the organisation.

The qualifications of new applicants to a profession are assessed by associations to ensure that they have the necessary skills and knowledge both technical and generic. The general knowledge requirements of graduates have been specified in Australia by West (1997) and confirmed by DEST (2002); in the UK by Dearing (1997) and in Europe under the Bologna Protocol (Kohler, 2004). There are remarkable similarities in all these specifications.

Australian universities in consultation with Australian Council for Educational Research identified a list of generic skills which should result from university study (ACER, 2001 p.27). These generic skills relate more closely to the skills specified in the Bologna process rather than some of the higher level skills implicit in West which are not present in the ACER's list. Learning in context can help achieve these higher levels of knowledge, skills and understanding. (e.g. Brown, et al., 1989; Brown & Duguid, 1991; Lambe, 2002; Lave & Wenger, 1991).

These higher level generic skills are also inherent in Shulman's table of learning which emphasises that learning is the primary outcome of an education process. Shulman's (2002) phases of learning are:

- *Engagement and Motivation;*
- *Knowledge and Understanding;*
- *Performance and Action;*
- *Reflection and Critique;*
- *Judgment and Design; and*
- *Commitment and Identity. (p.2)*

Learners should be encouraged to connect their learning with past and current experiences and to see possible future implications of their new knowledge (Dewey, 1938). According to Kolb (1983, p.41) in (M. K. Smith, 2001):"learning is the process whereby knowledge is created through the transformation of experience. Knowledge results from the combination of grasping experience and transforming it". (p. 41) Kolb proposed that experiential learning has six characteristic features:

- Learning is best conceived as a process, not in terms of outcomes *(i.e. a direction, a way of life – not a destination);*
- Learning is a continuous process grounded in experience;
- Learning requires the resolution of conflicts between dialectically opposed modes of adaptation to the world *(and world events)*;
- Learning is a holistic process of adaptation to the world;
- Learning involves transactions between the person and the environment; and
- Learning is the process of creating knowledge: '[which] is the result of the transaction between social knowledge and personal knowledge'. (Kolb 1984, p.36 in (M. K. Smith, 2001)) *(italics are our comments/ emphasis)*

This is in contrast to a competency/sufficiency approach to learning which is:

- Outcome focussed;
- Does not particularly draw on practical and complex real world examples;
- Supports standard solutions to problems; and
- Does not seek to extend knowledge

Professionals may wish to develop their careers beyond sufficiency knowledge and to develop deeper knowledge in order to become either a specialist; or to integrate their domain knowledge with other domains to develop knowledge over a broader context. This self-actualisation CPD represents the area where professionals grow beyond a competence level and become leaders of their profession and organisations. Real growth in knowledge combined with the ability to apply that knowledge is vital to the future of a nation. We argue that CPD should not be driven just by risk minimisation strategies and compliance measurements.

The specifications by West, Dearing and Bologna make explicit an expectation that professionals possess both technical and non technical skills.

Technical skills have been and always will be an important component of the knowledge, skill and understanding required by professionals. To be admitted into a profession, professionals may have been tested mainly on technical matters. Much assessment may only be at a sufficiency level (Thalheimer, 2007).

The bodies of knowledge specified by associations for entry into professions are most likely to concentrate on technical issues. A sample of associations' websites supports this contention (e.g. ACS, 2007). Emphasis on mere technical competency does not meet the expectations of government and associations for practicing professionals. While technical competence is not the sole component of a BoK, the technical competence prescribed by associations for graduates represents only a portion of the expertise and proficiency expected by the public and specified in government reports. Dadds (2000) discusses the way professionals develop their knowledge, skill and understanding:

The journey of professional growth into new and better practices is often unpredictable; often non-linear; often emotional as well as cerebral. It demands the capacity and strength to ask questions; to analyse and interpret feedback; to describe the emotions generated by self-study; to change established practices in the light of new understanding; to remain interested and emotionally curious. (p.55)

The learning necessary to be a professional is more than just intellectual endeavour. Professionals must learn not only "to think in certain ways but also to perform particular skills, and to practice or act in ways consistent with the norms, values, and conventions of the profession". (Shulman, 2002 p.2) This cannot all be learned within universities even though some universities specify the characteristics that their graduate body should hold.

Learning through context provides a framework for professional practice. Many university courses do not put learning in context. The knowledge which professionals need to solve real world problems requires that the professional is capable of constructing meaning in a given situation using both domain knowledge and experience (Brown & Duguid, 1991; Scottish Executive Health Department Management Executive, 1999). The concept of standard solutions to standard problems does not fit the needs of professionals in practice (Lambe, 2002). Professionals in practice deal with problems which are unlikely to appear in textbooks.

Shulman (2002) argues that "we need to go beyond teaching and assessing for understanding in order to foster judgment and design". (p.6) It is our contention that when education, associations and organisations focus on competency (particularly technical competence) then the:

- student is unlikely to achieve deep learning; and
- level of knowledge and understanding achieved is likely to be at a surface level.

CPD CONCEPTS AND ISSUES

Australia has introduced common Professional Standards legislation across all states and territories (Marden, 2003). Professional Standards legislation is seen by participating associations and organisations as a means of risk minimisation, particularly in reducing professional indemnity insurance fees and claims. There is a risk that this legislation is seen as emphasising that CPD is a way of meeting compliance requirements rather than to expand and develop knowledge. Compliance requirements may take the bulk of an organisation's budget and commitment to CPD.

The Centre for Applied Philosophy and Public Ethics (CAPPE, 2006) found that professionals perceived that the most significant benefit of registration under the scheme was that registration provided evidence that professionals have an (or have improved their) ethico-professional reputation. This benefit exceeded insurance capping as the primary incentive for registration. Nevertheless the Professional Standards Council noted a tendency for some associations to focus on limiting the civil liability of professionals when proposing schemes to the Council. Limiting civil liability on its own is not desirable as it acts to shift risk to consumers. The Council saw that its role was to ensure that schemes **improved standards and protected consumers***[our emphasis]* (PSC, 2006). Standards are egalitarian (self regulation by associations); based on minimising risk for the community and focusing on regulation and compliance (PSC, 2007). Learning methodologies are left to the discretion of individual associations.

In parallel with the introduction of Professional Standards legislation, there has been an on-going emphasis in Australia on competency based training (DEST, 2002). The weight given to compliance may further reinforce an emphasis on competency based training. The danger is that the act will not result in improved levels of knowledge and performance. Standards imply a one-size fits all approach (M. K. Smith, 2005).

CPD issues have been written about and researched, for:

1. businesses in general: (e.g. Billett, 2001; Boud & Middleton, 2003 ; Boud & Solomon, 2001; Boud & Walker, 1998; Brown, et al., 1989; Collins, et al., 1991; Dadds, 2000; DiBella & Nevis, 1998; Eraut, 2004; e.g. Knowles, 1975; Lave & Wenger, 1991; Marton & Säljö, 1997; D. Smith, 2006). These studies look at either single businesses or within an industry base; many of the organisations studied are large and the workforce involved covers a wide spectrum from unskilled workers, trades people, management and professionals.
2. individual professions and/or professionals (e.g. Becher, 1996; Chivers, 2006; Dadds, 2000; Eraut, 1994; Phelan, 2002; P. Ramsden, 2003; Rothwell & Herbert, 2007; Shields, 2004; Webster-Wright, 2006; Wilson & Lyons, 1961). These studies predominantly look at the need that professionals have for CPD and the type of CPD offered.
3. professional associations (e.g. Daley, 2001; Friedman, et al., 2000; Friedman & Mason, 2007; Friedman & Phillips, 2001, 2002, 2004; Gardner & Shulman, 2005). Associations are responsible for the body of knowledge for the profession and should contribute to developing and supporting a learning culture within the profession (Murphy & Calway, 2007b).

Organisations employing professionals may directly influence CPD choices made by individual professionals. CPD is frequently self-directed and opportunistic rather than planned but still may depend on the direction and support, both practical and psychological, given by the organisation. Access varies between positive support, through to active discouragement by the organisation.

The outcomes of CPD within organisations are affected by the policies and practices in operation within the organisation, which include: corporate goals; entrepreneurial intensity; the learning profile; and the factors which facilitate learning within the organisation see Figure 1.

CPD activities for professionals may involve formal, non-formal and informal learning activities. While CPD is fundamentally an individual activity, organisations and associations have vested interests in professionals engaging in CPD.

CPD has four purposes:

1. **Credentialing:** validation of knowledge is essential to provide the public with reassurance that a professional has a sufficiency level of knowledge to practice. Associations recognise the knowledge base of graduate entrants to a given profession through course accreditation processes and often require further development through structured programs and/or the gaining of practical

Figure 1. Components impacting CPD and the Learning Process within Professional Organisations

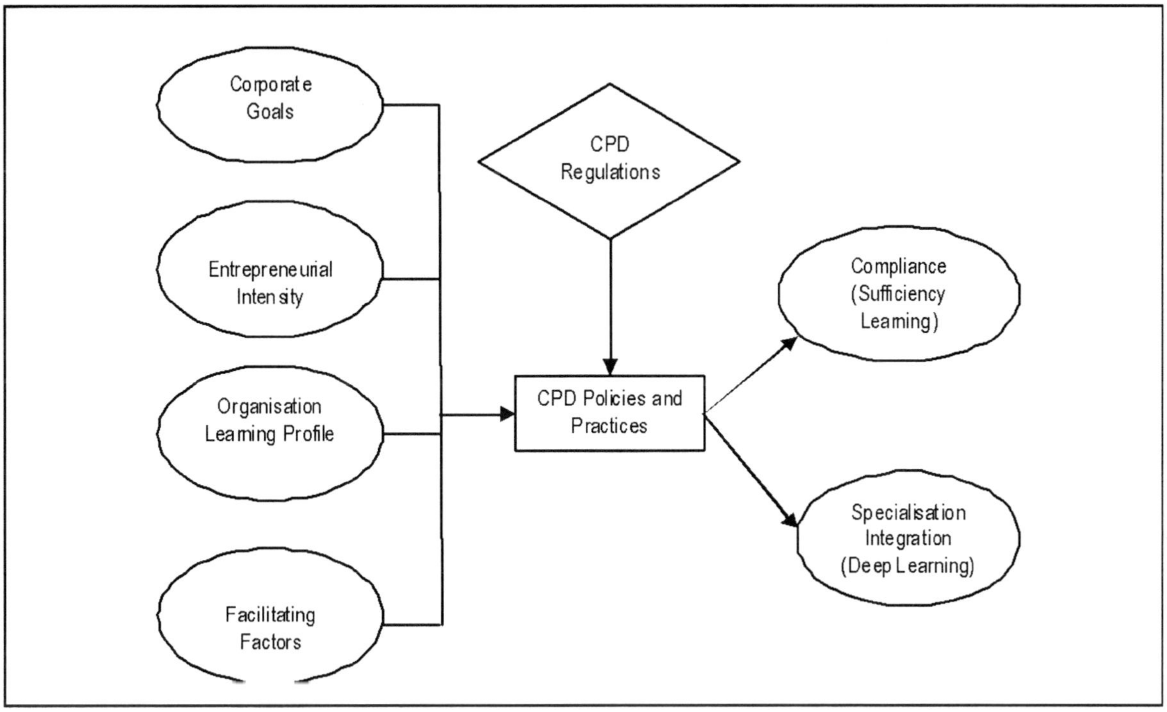

experience in order to certify that the novice professional has the required BOK specified by the association. The need for CPD programs within associations is discussed in Murphy and Calway (2008) and research areas were identified for later study;

2. **Compliance:** associations develop policies specifying compliance requirements; and systems to monitor CPD in order to ensure maintenance of sufficiency knowledge. Within an association, research could validly examine the:
   - Types of CPD available and used;
   - Scope of CPD particularly the breakdown between technical and generic knowledge;
   - Pedagogies used and the relationship of these to transfer of learning; and
   - Accessibility of CPD for all members of the profession.

Shulman's (2002) table of learning provides a model to assess the impact of CPD on the ability of the professional to perform as a functioning member of a profession; and

3. **Deeper learning:** either for a professional to achieve specialist status or to move into managerial roles (integrated knowledge). This could be the focus of a study examining the:
   - Methods used to achieve deeper learning and the degree to which professionals engage in deeper learning;
   - Relationships between senior professionals and developing professionals in the development of deeper learning;
   - Pedagogies used and the relationship of these to transfer of learning; and
   - Accessibility of CPD for all members of the profession.

4. **Transfer of learning:** for both compliance and deeper learning there is an expectation that CPD effectively transfers knowledge to professionals. Within associations and organisations research could examine the effectiveness of various methods of CPD in transferring learning.

For purposes two and three, the objectives of CPD specified by associations need to be matched against the outcomes of professional development. Our study examined whether the goals, expectation and outcomes of CPD for organisations are consistent with the objectives as seen by associations and found that the:

- Goals of organisations for CPD are consistent with the claims made by associations;
- Organisations strongly believe in the value of CPD but do not set clear expectations or measure the benefits; and
- Outcomes of CPD are considered to be important both to the organisation and to professionals employed within these organisations.

CPD and Knowledge Development

Professionals face problems which can be conceptually intricate, complex and ill-structured, often varying from case to case where successful solutions are non-repeatable in whole or in part, and requiring more than the retrieval from memory of intact prescriptive knowledge. Professionals need to build knowledge by constructing meaning in different situations using knowledge developed through effective learning rather than through committing information to rote memory (Spiro, et al., 1996). Such constructed knowledge requires both domain knowledge and experience to solve each new situation (Brown & Duguid, 1991) rather than being based on competency training, where the problems have objective, repetitive solutions. It can be argued that the need to minimise risk has increased the emphasis on competency and as a result many CPD activities have been reduced to assessing competency. This is not suitable for professionals wishing to be involved in deeper, or self-directed, career learning. To deal with continually changing and evolving problems, professionals require knowledge derived from learning and experience (Gardner & Shulman, 2005; Spiro, et al., 1996). Learning at a higher level has been referred to as deep learning (Ramsden, 1992) or double loop learning (Argyris & Schön, 1978).

Melotte (1996) argues that knowledge should be seen as being explanatory, predictive and interpretive. Professionals will not reach this level of understanding and adaptability unless their understanding of theory constructs is sound and they are encouraged to transfer learning content to a workplace context. Professionals should also be capable of transferring learning to their peers and by extension to their clients.

CPD for professionals in practice should be made relevant through drawing on the contextual experience and/or the work environment of the professional. Brown and Duguid (1991) claim that separation of knowledge and practice is unsound and argue for evolution of learning through practice. We support this emphasis when considering Work-Integrated Learning as a learning philosophy for professional development.

Learning in its broadest sense has taken place when a learner can demonstrate or display that learning later. While passing an exam can demonstrate that ordinary learning has occurred – transfer learning is always at least implicitly contrastive: it assumes learning within a certain context and asks about the impact beyond that context. Human Resources and Skilled Development Canada (HRSDC) refers to the 'so what or now what' phase of the learning process. However HRSDC (2005) also states that "…abundant evidence shows that the very 'often hoped for' transfer from learning experiences does not occur" (p.1).

The level of learning achieved from WIL and from the experience specified as required for professional level membership is seldom assessed, with the exception of the traditional professions of medicine, law and also in nursing. In these professions:

- Cultures of learning have been established; and
- Experienced (qualified) specialists take on a training role in the development of new professionals and specialists.

Professionals need a commitment to lifelong learning incorporating both career development and mandatory CPD if the objectives of professionals, associations, organisations and government are to be achieved. Learning cultures influencing CPD are affected by conflicting purposes at both organisational and individual levels: firstly consumer protection, risk minimisation, and learning to obtain and maintain accreditation; and secondly career specialisation and/or integration of knowledge at an advanced level. Our study has investigated CPD within organisations for the maintenance of competency and for developing deeper, integrated or specialist knowledge. We recognise that professionals are learners at postgraduate level; therefore the pedagogies used need to be appropriate for the needs of the learners.

CPD should add to a professional's ability to apply knowledge which contributes to their organisation. Learning takes place within a working context. Organisations may have the power to decide what knowledge is to be valued and worth learning (Webster-Wright, 2006 p.25). Organisations look to professionals to maintain and update their body of knowledge. Professionals, however, are knowledge workers who have CPD needs which are not identical with those of managers, trades persons or unskilled workers. Individual professionals choose to undertake CPD as self-determined personal development to improve knowledge and skills either for continuing competency or for personal and career development.

Professionals progress through three stages of development: acquisition of knowledge; specialisation; and integration of knowledge. New graduates or novice professionals have acquired knowledge through their degree studies (formal learning). This knowledge needs to be verified and often enhanced by professional associations before the graduate is credentialed as a full member or is certified to practice as a professional. Following the awarding of credentials professionals are required to be involved in CPD.

Professionals retain their credentials for professional practice by maintaining a competency level validated by undertaking CPD. Most associations prescribe a minimum compliance level to retain professional accreditation through formal activities such as participation in award courses, non formal activities such as conferences, seminars, workshops and short courses. CPD may also involve a range of informal learning situations such as reading journals and books relevant to the profession, mentoring, action learning from work based projects, reflective discussions with colleagues, membership of committees relevant to the profession, online learning activities, etc. The choice of process and the topics is generally left to the individual and may depend on the ability of the professional to access learning and training programs. The effectiveness of learning will be demonstrated by the ability of the learner to use the material taught.

Associations should also oversee the development of knowledge of their members beyond a sufficiency level and provide appropriate recognition of this (Argy, 2006; Australian Physiotherapy Association, 2005; Phelan, 2002).

CPD Usage, Provisions and Further Study

The individual professional, the organisation employing the professional, associations and government all influence CPD and its impact:

- Individual professionals use CPD for compliance purposes i.e. to maintain a recognisable qualification and for career development motivated by financial and strategic/power factors;
- Organisations (e.g. accounting, engineering and medical practices), use CPD to: maintain competence; preserve credibility through professional recognition of staff; and manage risk through maintaining work-ready staff. Some organisations will use CPD as a tool for business development or to introduce new services. They need to provide an environment where learning is encouraged, valued and facilitated;
- Associations (e.g. CPA Australia, Engineers Australia) use CPD to:
  - credential new members,
  - advance the careers of members through maintaining a core body of knowledge,
  - develop a culture of relevant professionalism;
  - validate the professional status of members with a view to minimising risk to professional clients; and
  - develop, and sometimes credential, deeper professional knowledge.

Associations should provide learning frameworks covering specification of learning objectives, content development, content delivery, mentoring, peer support, assessment and recognition; and

- Governments see CPD as a means of increasing productivity, skilling the workforce, increasing competence and minimising risk.

At the centre of Figure 2 is the *body of knowledge* of a profession. Organisations want employees whose knowledge is certified for marketing purposes and also to provide services to clients. New graduates who are required to undertake a formal CPD program prior to credentialing should be supported and encouraged in their efforts to demonstrate that they have knowledge sufficient to practice. Research in this area could validly be undertaken.

Professional knowledge represents more than can be obtained from competency focussed training programs. Much of what is delivered in competency focussed training programs is concerned with the transfer of explicit, abstract knowledge, that specifically excludes both complexities of practice and communities of practitioners (Brown & Duguid, 1991). If we expect professionals to cope with complexities of practice we should look at ways for professionals to obtain relevant professional development. Professionals need to make a commitment to learning which extends beyond the memorisation of facts. The experiences of communities of practitioners can be drawn on to encourage and enable professionals to integrate new ideas with what they already know, if they are going to keep and use new knowledge (Boadle, 2005).

The keys to transforming information into a thorough informed and reliable understanding of the subject material (knowledge) are "*C* words such as: comparison, consequences, connections and conversation" (Davenport & Prusak, 1998 p. 6). Comparison, consequences and connections require more than explicit knowledge. Brown, Collins and Duiguid (1989 p.37) state that professionals need to be equipped to, and capable of, engaging in professional conversations. Reflection and integration of work and learning are proven means for developing these traits in students (Calway & Murphy, 2007a).

It is not enough for educators to assume that *transfer of learning* occurs as a result of education and training. Whatever is learned will not necessarily be retained or remembered over time, or used in appropriate situations (Doyle, 2002). Melotte (1996 n.p.) argues for "better linkages between planning research, education and train-

Figure 2. Professional associations' knowledge development models

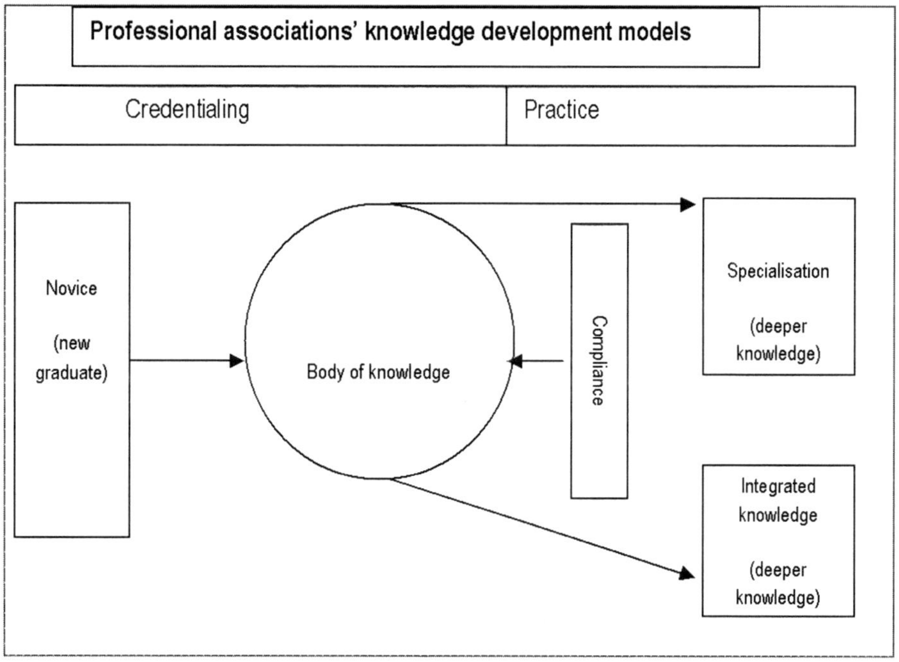

ing, competency standards, quality assurance and continuing professional development… Improving higher education and the professions requires a new interpretation of professional education and training and professional practice." Transfer of learning is said to occur when learning in one context enhances a related performance in another context. Beach (1999:103), in extending the theory of transfer of learning argued that "learning, development, and education are inherently cultural as well as personal enterprises, and, by extension, so is the phenomenon of transfer".

Shulman's (2002) table of learning is appropriate for professionals' CPD as it incorporates recognition that:

- CPD is a continuing process: CPD is not learning which has a definable start and a finite conclusion; and
- Culture is an essential component of learning.

Shulman's approach is consistent with Beach's (1999) view of learning as a metaphor of transitions, consequential and context driven, rather than static transfers providing opportunities for professional growth.

Lifelong learning at both an individual and professional level involves cultural and motivational commitment. Career development is that element of a professionals working life which makes a professional work ready and maintains and expands employability and marketability (Connor & MacFarlane, 2007).

CPD for Competency, Compliance, Credentialing

Competent is defined as:

- "Adequate; sufficient; properly or legally qualified; capable". (Johnston, 1983),
- "The ability to do a particular thing" and the alternative words suggested in the thesaurus are "able, adept, capable, effective,

> efficient, handy, practical, proficient, qualified, skilful, skilled, trained". (Knight, 2005)

The concept of competency is complex and may be defined differently by various stakeholders. "In much current usage the notion of competence has been whittled down to the ability to undertake specific tasks; it has been largely stripped of its social, moral and intellectual qualities". (M. K. Smith, 2005 n.p.) It is possible to mount a case that competency signifies an assessed minimum compliance level and this is driven by a need to minimise risk. The assessed compliance level does not necessarily guarantee that particular tasks can be performed. Some professional associations suggest that competence implies a broader focus. (ACCED-I, 1994)

Competencies provide a means of objectively measuring an individual's knowledge and skills to assess their ability to perform predetermined tasks for which set solutions exist. Assessment may relate to how one performs a given task. It implies that the person being measured or *certified* has knowledge, skill and experience in relevant technical and business areas. There may also be an expectation that the person possesses certain personal qualities and organisational skills necessary for the task, however it is unlikely that this is tested in most competency assessments (Morris, 1999).

The life span of a professional's body of knowledge is limited as knowledge develops through technical, legal, conceptual and/or social change. The Committee for Economic Development Australia's study on lifelong learning found that "the skills needed for work throughout a lifetime are diversifying. Jobs which once changed little in the course of a worker's career are now changing every few years. Qualifications achieved at age 20 are more and more likely to be out of date by the time you turn 50". (Dimopoulos & Walker, 2005 n.p.) CPD is necessary so that professionals keep up to date within their profession. This can involve professionals needing either new knowledge and/or supplantive knowledge (Atherton, 1999; James, 2000).

Professions Australia (2006), the peak body for professional associations, states that ideally the requirements for qualification should be competency based to ensure competent and safe practice consistent with an overriding objective to protect the public through the maintenance of professional standards. Technical competence usually prescribed by most professional bodies for graduates represents only a small portion of the expertise and proficiency expected by the public. Dearing (1997) advocates a concept of '*Graduateness*' which should encompass knowledge, understanding, dispositions, attitudes and values.

Associations' credentialing processes should have an auxiliary aim to maintain and improve the standards of professional practice. CPD programs with this aim should develop learning cultures which encourage both continual growth of knowledge and the professional's ability to apply that knowledge in contemporary situations. Our research considered generally the extent to which organisations use communities of practice (CoPS), mentoring and other group activities to encourage and promote the growth of knowledge of employees. This can be contrasted with the impact which compliance requirements have on the directions, focus and depth of learning of an individual through CPD.

On an individual basis and/or at a corporate level, meeting compliance regulations may be felt to be all that is required for a professional. There is a risk that compliance requirements overwhelm the learning aims of both the association and the organisation. As a consequence, professionals may not adequately maintain their body of knowledge or develop new or deeper knowledge. The ability of organisations to provide adequate professional services to clients may be compromised. The longer term viability of the organisation may not be sustainable.

In Australia for example, the adoption of the concept of key competencies has dominated discussions on attributes and skills. The DEST report (2002) recognised that technical competencies should not be the sole determinant of education policy and that emerging skills and knowledge need to be addressed.

Competencies have achieved prominence at the same time that the Quality movement was being promoted in business and education with the establishment of standards across many areas of business operations particularly in the area of product manufacturing (Standards Australia, 2007). The vocational training sector in particular found that standards could be developed for the training of individuals to carry out routine tasks. This emphasis on standards has extended into business practice through the development of quality models such as TCO, ITIL, CMM, CoBiT, Six Sigma, Malcolm Baldridge award and Scorecards (Anthes, 2004) and increased regulatory requirements such as Sarbanes Oxley, Basel II Accord, etc (Berinato, 2004).

Within organisations compliance; sufficiency; or conformance represents a minimum obligation regarding performance. Re-thinking, re-learning, and re-examining are more significant concerns for organisations and individuals who need to progress beyond the status quo and to achieve excellence. In order to progress, organisations need to promote the development and application of relevant knowledge. Davenport and Prusak (1998) emphasise the significance of the development of knowledge which originates and is applied in the minds of the knower through a mix of experiences for evaluating and incorporating new understanding and information.

IBM (2004) nominated responsiveness, variability, resiliency and focus as characteristics that businesses and employees need. From this, we argue that sufficiency training (particularly training which is specifically technical) does not adequately meet the needs of industry for the CPD of IT practitioners. IT is often seen as a profession which is multi-disciplinary and it is possible that IT requires a higher need for integrative knowledge than other professions. This would be an area for further research.

CPD, Workplaces and WIL

Associations can use the phases of Shulman's table of learning: *Engagement and Motivation; Knowledge and Understanding; Performance and Action; Reflection and Critique; Judgment and Design; and Commitment and Identity* (2002 p.2) to develop cultures and learning environments designed for effective professional development. Shulman's phases do not following a strict sequential pattern or have a beginning and an end. He believes that commitment and identity is most likely to be followed by new engagement and motivation as professionals engage in lifelong learning and continually re commit to learning in order to maintain levels of practical knowledge or to extend their knowledge base for further career development. We state that unless learning is put into context, professionals will not obtain the knowledge skills and understanding specified required in the UK by Dearing (1997), in Australia by West (1997), DEST (2002) and in the Bologna Process (Kohler, 2004).

The level of learning professionals obtain through CPD is difficult to measure as the benefits are often long-term. Learning effectiveness will be demonstrated by the ability of the learner to use the material taught; available; and/or accessible to the learner. Professional knowledge is gained incrementally resulting in benefits often occurring some time after learning has taken place. There are also instances when CPD is often required in a 'just-in-time' setting or framework as the professional needs new knowledge, skill and/or understanding to deal with a current situation, problem or opportunity.

In just-in-time learning, a professional seeks out material to research an issue of current concern. The depth of learning necessary is at the discretion of individual professionals. The learners' ability to transfer knowledge to practical situations determines the success. Material needs to be easily available and accessible and appropriately detailed to cater for a variety of users. The material should be accurate, reliable and kept up to date. Web based material may be found quickly but without expert appraisal and endorsement its credibility and relevance may be questionable. The Knowledge Transfer and Exchange site of the Canadian Health Services Research Foundation (2007) is an exemplar of knowledge transfer between researchers and decision makers. Our research has shown that organisations value the ability to get quick access to expert information for problem solving.

Academics should play a role in the development by their associations of:

- learning programs and facilities to maintain the sufficiency level of knowledge of practitioners at a professional level, as new knowledge within a profession evolves over time; and
- advancing professional's knowledge.

Learning in context, where the learning can be related to practical situations, is well established as a means of achieving deeper learning (Brown, et al., 1989). The case for WIL as the appropriate learning philosophy for professionals has been addressed in an earlier work (Murphy & Calway, 2007a). WIL incorporates hands-on work experience and learning in real-world settings that assume a level of explicit knowledge/skill on the part of the learner and the exchange of tacit knowledge/skill from the real-world to the learner. Work-Integrated Learning is a philosophy of learning which is not restricted to a single learning theory.

Doyle (2002) highlighted the value of linking learning to real workplace problems and situations. Applying theory to real life situations results in deeper learning as learners grow in confidence and are able not only to obtain clarity about the actual learning topics, but also to identify the future applications of that learning. HRSDC (2005) referred to Mosel (1957) who identified three conditions for transfer: content must be applicable to the job; the trainee must learn the content; and the trainee must be motivated to change job behaviour by applying what was learned.

Much of the literature on work-based learning emphasises training and the development of competencies. Competency based training assesses a trainee's ability to carry out defined tasks in environments where problems have been solved in the past with an emphasis on efficient and reliable performance. Where learning or training takes place in the workplace, there is a perception that educational institutions and industry structure the learning experiences in workplaces in a way that sustains existing practice in the workplace (Billett, 2001). Learning philosophies are needed which are directed at improving the occupational standards of professionals and accentuate growth of knowledge for professionals.

WIL has been found to improve teamwork, capacity to learn, initiative, communication skills oral, adaptability and interpersonal skills (Young, 1997). Educators involved in CPD within associations can deepen understanding and wisdom of professionals through appropriate learning models (Bellinger, 2004) which include developing an understanding of learning in the workplace (e.g. Andresen, et al., 2000; Brown, et al., 1989; Clark, 2000; D. Smith, 2006; M. K. Smith, 1996b, 2001)

The philosophy of WIL includes the imperatives of workforce readiness; a professional development culture (both individual and workforce related); international relevance; life-long learning; knowledge transference; human and social potential (Calway & Murphy, 2007). This philosophy is consistent with the objectives relating to advanced level learning for professionals. WIL can develop relevant professional knowledge which will provide society with greater expertise. An emphasis on WIL in the pedagogy will create a richer learning environment enhancing the professional's performance and career. It is both feasible and desirable to design CPD programs which incorporate WIL.

Key Findings

Our study, which is on-going, examined the value to professional organisations of CPD using SME practices as a case study. Maintaining and developing the professional's BoK within organisations is critical to these organisations. The impact of organisations in the CPD of professional staff was a focus of the study.

Our research suggests that CPD within organisations is strongly supported and valued within the organisations we studied. Comments included: "essential activity", "vital part of our organisation"; "increased skills reflect on client's perception and the professionalism of the firm"; "CPD is a big part of what we do"; and "Invaluable – it is a requirement but it is a necessity for the business – couldn't exist without it".

Result Summary to the Research Questions

1. Do the aspirations, culture and practices of a professional organisation impact the way CPD is conducted and consequently the learning resulting from CPD?
   a. Is there a relationship between the entrepreneurial intensity (Ireland, et al., 2006) of professional organisations and the Organisational Learning Profile (DiBella & Nevis, 1998)?

Organisations surveyed on the whole typically have 'low entrepreneurial intensity' with low numbers of new:

- Services introduced, significantly revised, or improved by the enterprise; and
- Methods or operational processes implemented by the enterprise.

There was strong evidence to support:

- An emphasis on continuous improvement in methods of production and/or service delivery;
- Cautious, pragmatic, step-at-a-time, adjustments to problems;
- Steady growth and stability as primary concerns; and
- A top management philosophy that emphasises proven products and services, and the avoidance of heavy development costs.

And little evidence to support

- Large, bold decisions despite uncertainties of the outcomes;
- Rapid growth as the dominant goal;
   b. What Facilitating Factors for learning (DiBella & Nevis, 1998) are in operation within professional organisations?

The most significant Facilitating Factors were:

- Learning Confidence: experience in learning from successes, mistakes, and specific events; experience in trying new things; belief that all groups can learn
- Learning Resources: the organisation is committed to providing high quality resources for learning.
- Shared Vision: values and vision are shared within a team or within the organisation
- Leadership: leaders are personally and actively involved in learning initiatives and in ensuring that a learning environment is maintained.
- Trusting Relationships: there is trust and open communication between colleagues and when team members respect the need for mutual growth and development.

- Multiple Advocates: staff at all levels are allowed/encouraged to look at and propose ways to develop new knowledge. Anyone can advocate change. Anyone can be a champion. Multiple advocates or champions exist.
- Learning Enjoyment: celebrating learning achievements and creating an atmosphere where humour and fun are part of the process of acquiring new knowledge.
  c. Does the Organisational Learning Profile (DiBella & Nevis, 1998) combined with the Facilitating Factors impact on the level of learning achieved by individual professionals?

The Organisation Learning Profile attempts to assess how organisations go about acquiring and disseminating knowledge and the approach taken to learning within the organisation. The major distinguishing features were that:

- Learning is directed to obtaining knowledge to improve what is already known or being done (Incremental).
- Learning to obtain knowledge that challenges the assumptions about what is known or done (Transformative) had little support.

Other parts of the learning profile where there is a significant difference are in the:

- Content-Process Focus: the organisation prefers to gain knowledge which relates to the to the definition of the services (Content) offered as opposed to the processes which might underlie or support those services (Process)
- Learning activities are directed at knowledge being obtained more by groups than by individuals; and
- The organisation prefers to develop new knowledge from sources which are more likely to be external.

Smaller differences in responses may suggest that

- Informal learning is more the mode of learning used;
- Knowledge is seen as more personal than public (or organisational);
- Learning is more likely to occur through generating and sharing knowledge and skills through actions or practices (Experiential) than from cognitive knowledge development.

Respondents were equally divided in their views as to whether the emphasis on learning was on meeting short-term needs or for long-term use.

2. What are the goals for CPD which motivate professional organisations? Is there a relationship between these goals and the level of learning achieved? What value does the professional organisation place on professional development?

Contextually questions which follow from this include:

a. Do professional organisations assess the value of CPD to the organisation? If so, how is this done?

All believe that CPD is valuable and critical to the firm. None attempt to value CPD. The benefits were likely to be long-term and therefore difficult (and even undesirable) to quantify. One measure used is the number of Professional Indemnity claims. The following comment represents the views of many enterprises: "*Value is based on perceptions of clients. If we do a good job then client is happy with professionalism and value for money, professionalism of service then business will grow through referrals from clients*". No organisation had a full measure of the cost of CPD.

b. How do professional organisations demonstrate to professionals within the practice the organisation's need for CPD and the value the organisation places on CPD?

Senior staff engage in CPD activities, in particular, internal staff development meetings. All organisations allocate time for CPD and where appropriate will support attendance at external CPD. Staff surveys indicated strong support for the fact that CPD is adequately resourced.

3. What Informal Learning is conducted within professional organisations and how does the effectiveness of this compare with more formal learning?

In the context of this research, informal learning models all involve interactions between professionals which may take place within either an organisation or a professional community. Learning may become more relevant and effective when the organisation employing a professional, actively provides an environment which enriches learning.

The only formal learning undertaken within responding organisations is for staff undertaking programs for admission to relevant associations. Non-formal learning (e.g. CPD courses, workshops) is used to a diminishing extent with queries on their value, relevance and timing. ATO briefings were highly valued as relevant and practical. Interaction with staff, other practices, and key bodies was a major source of knowledge.

- The availability and levels of mentoring (Brown, et al., 1989; Brown & Duguid, 1991; Friedman & Phillips, 2002; Orr, 1990): Some professions (e.g. the clergy and some sections of universities) have a history of supporting the growth and focus of their members in a caring and helpful environment.

Very limited: some associations e.g. ACS, CPA Australia and Engineers Australia sponsor mentoring programs for new graduates and a small percentage of participating organisations (25%) are involved.

- The existence of apprenticeship models: Apprenticeship models support a learning focus with steps undertaken by master and apprentice at varying levels as knowledge, skills and understanding (Cross, 2007).

There was little formal adoption of this although supervision by principals to reduce Professional Indemnity claims and improve service to clients was common.

- The use of effective Communities of Practice (CoPs) enabling the professional to engage in professional conversations through participation in specialist interest groups, exchanging war stories, interactive forums etc. (Brown, et al., 1989; Collins, et al., 1991; Davenport & Prusak, 1998; Lave & Wenger, 1991; Wenger, et al., 2002). "A Community of Practice (CoP) is a special type of informal network that emerges from a desire to work more effectively or to understand work more deeply among members of a particular specialty or work group". (Sharp, 1997 n.p.)

Informal Learning is the most supported form of CPD. All organisations conduct regular staff CPD meetings to discuss current issues (hot topics). Discussion is important and includes "do we have any clients that this can relate to? How do we overcome the problem? Could we have done things differently?" Organisations send staff to association discussion groups (classical CoPs), 2nd tier firms' discussion groups etc. In all cases the person attending will report back to the staff meeting to ensure that knowledge is shared, Specialist and senior staff within each organisation ensure lessons are learned and misinterpretations avoided.

4. The opportunities for Work-Integrated Learning (WIL) and how is this supported and encouraged? WIL is itself a recognised form of Informal Learning. Situated learning is likely to provide deeper learning as the relevance for the learner is more apparent. (Billett, 2001; Boud & Solomon, 2001; Brown, et al., 1989; Calway & Murphy, 2007; Dewey, 1938; Eraut, 2004; Kolb, 1983; Lave & Wenger, 1991).

Over half the organisations have used cooperative education students (and recruited from this source) or had staff who had studied part time. One principal commented that the benefits to the organisations of WIL included "Forces firm to focus on training at lower levels and gives more junior staff exposure to supervision".

SOLUTIONS AND RECOMMENDATIONS

University staff often play an important role in professional entry programs conducted for and often by professional associations and in the setting of standards between classes of membership within those associations. Professional entry programs introduce professionals to self-directed, lifelong learning and in some cases represent young professionals' first experience of practical applied learning. If maintenance and improvement of standards is an aim of credentialing then these programs should develop learning cultures which encourage continual growth of knowledge and professionals' ability to apply that knowledge.

At an advanced level, formal programs (referred to as expansive learning) (Engeström, 2001) may be offered by professional associations to develop new specialist skills and knowledge or integrate knowledge across domains. Specialist programs are developed through specialist groups or colleges within professional associations e.g. Engineers Australia, College of Surgeons. The basic level knowledge of professionals is used as the foundation for new expertise. More than basic competency is expected. Students are required to demonstrate deep understanding of the subject matter as a prerequisite to practicing within that specialisation. Advanced credentialing is often the outcome of this type of program.

Attendance at CPD activities is often opportunistic in response to a market offering rather than as a result of career planning. Quality varies and participants look to professional associations for credibility and integrity. The pedagogy associated with many programs is questionable when training "is thought of as the transmission of explicit, abstract knowledge from the head of someone who knows to the head of someone who does not, in surroundings that specifically exclude the complexities of practice and the communities of practitioners". (Brown & Duguid, 1991 p. 47)

Professional associations are examples of CoPs. CoPs provide forums by which professionals can grow within their profession. Associations are examples of informal communities who can sustain learning through sharing expertise, practices and war stories. Some of the problems of learning provided in workplaces such as insecurity, in-breeding, narrowness and exclusion may be overcome by professional associations providing and supporting a learning culture which enables both action learning and active learning.

Associations need to build learning frameworks which cover more than just content development and delivery mechanisms. They have the capabilities to provide necessary mentoring and peer support combining these with a culture which is supportive of deep learning. Knowledge Management theory advocates the creation of CoPs and the development of learning cultures. Trust, cooperation and friendship are key components of CoPs. If learning is seen to be valued then deep learning is more likely to occur. Knowledge Management proponents advocate encouraging innovation and enhancing intellectual levels to enhance performance (e.g. P. A. C. Smith & McLaughlin, 2004).

Both associations and organisations, as CoPs, are able to make explicit, relevant tacit knowledge; model strategies for learners in authentic activity; support learners' attempts at doing the task as teachers and peers; empower the learners to continue independently (Brown, et al., 1989); and facilitate face to face and electronic communication on specific topics.

The learning pedagogies are reflected in CPD delivery of learning material. Factors which influence the depth of learning achieved include:

- Learning taxonomy used: Shulman's table of learning is an appropriate taxonomy which is relevant for adult learning at a professional level, rather than the more widely known Bloom's taxonomy;
- The support given by the professional organisation to CPD and the culture of learning which exists with the organisation. This may be demonstrated by:
    - The transfer of learning which occurs: Work-Integrated Learning is an appropriate strategy which can support a number of pedagogies. Learning in context makes the potential for effective transfer of learning greater;
    - The integration of new professionals into the profession through mentoring and/or CoPs: new professionals need to consider themselves as part of the profession and to be encouraged to engage in conversations and understand the language within the profession. Apprenticeship models are a feature of some professions accreditation processes (e.g. RACGP, Australian Medical Council). In most cases the (apprentice) master is a senior member of the professional organisation; and
    - Accessibility: to learning materials; to fellow professionals; and to CPD opportunities within the profession. This may be dependent on organisations and associations; and
- Assessment methodologies used – assessment can have the primary aim of measuring predetermined competency and the ability to memorise facts divorced from the process of learning or to develop the learning which extends the students experience and knowledge.

Associations have the responsibility to credential professionals required to engage in CPD, therefore associations have a responsibility to prepare professionals for practice which includes developing capacity for lifelong learning.

FUTURE RESEARCH

In summary our key findings have shown a strong need for practical applied learning (Work and Learning Integration) recognising that CPD within SME professional practices:

- Is an integral/essential part of doing business enabling the organisation to:
    - Comply with relevant standards and legislation; and
    - Keep up to date with change.
- Can result in enhancing knowledge levels enabling the practice to:
    - Attract and maintain clients ("clients are clever enough to detect lack of professional knowledge"; "clients need and appreciate innovative solutions"); and

   - Change the way they service clients and move from low level services in which they may compete on price to the provision of more advanced services likely to be more profitable resulting in attracting better quality staff.

For these organisations:

- CPD is not costed;
- Risk minimisation is critical.
- Learning is directed to obtaining knowledge to improve what is already known or being done (Incremental) through:
   - An emphasis on content rather than process;
   - Group learning and sharing of knowledge through actions or practices;
- informal learning is more important than formal.
- organisations facilitate learning by:
   - Generating confidence in learning;
   - Providing high quality resources for learning ;
   - Sharing values and vision within the organisation;
   - Having leaders personally and actively involved in learning initiatives and in maintaining a learning environment;
   - Creating trust and open communication between colleagues;
   - Allowing/encouraging staff at all levels are to propose ways to develop new knowledge; and
   - Celebrating learning achievements;
- CPD is driven by the need to learn and develop staff rather than compliance requirements– Professional Standards legislation has had little impact; and
- SME practices studied have low entrepreneurial intensity.

Future work should examine the findings in other case environments, models and levels of learning provided for career development by associations and assess the impact of professional standards legislation.

CONCLUSION

Organisations promote their services by stressing that their professional staff are members of the relevant professional association and by listing the certifications/credentials they hold. Practice alone is generally insufficient to ensure knowledge is effective and up to date. New knowledge required to continue to practice within a profession evolves over time as a consequence of technical, legal, conceptual and/or social change. Professionals, therefore, have a CPD requirement to obtain knowledge which reflects current practice within the profession.

The framework and culture of CPD for professionals is primarily influenced by the need to protect consumers through policies which minimise risk. Professional Standards legislation may have reinforced this emphasis. Simply maintaining the sufficiency level of knowledge held at entry into the profession is inadequate. Organisations and associations have a role to play in ensuring that learning frameworks and cultures are established to enable transfer of this learning and support the career development of professionals.

Growth in the knowledge and capability of professionals is vital to the future of organisations and the economy. It should be expected that professionals develop a deeper level of learning in order to deal with globally complex environments in which they work. Most professionals do not need CPD offerings which lead to a full academic award. CPD programs should be designed to focus on a broader holistic approach to individual and corporate professional development.

As professionals 'work-learn-work', they constantly progress through Shulman's (2002) phases of learning "Engagement and Motivation; Knowledge and Understanding; Performance and Action; Reflection and Critique; Judgment and Design; and Commitment and Identity". (p.2) They learn to recognise problems, target appropriate learning, transform their knowledge and transfer their knowledge benefiting themselves, their organisations, clients and society.

CPD can provide a WIL context for learning. Work related learning situations appear not only to add current learning for future knowledge retrieval, but also make accessible existing static knowledge for future learning. Professionals have access to real world problems. It is reasonable to suggest that CPD programs for practitioners can be developed using WIL as a base which to draw on real world problems and enable the development of deeper learning.

Effective transfer of learning for professionals is enhanced through integration of learning with the work-place. The most supported form of CPD is Informal learning, much of which is based on learning in the work-place. Organisations and the providers of CPD (whether formal, non-formal or informal learning) should examine strategies to achieve transfer of learning through the practical application of learning to the work-place.

REFERENCES

ACCED-I. (1994). *Statement of professional standards and ethics.* Retrieved May 2, 2006, from http://accedi.colostate.edu/ imis_web/staticcontent/3/ standardsandethics/standards.htm

ACER. (2001). *Graduate skills assessment: Summary report.* Canberra, Australia: Department of Education, Training and Youth Affairs.

ACS. (2007). *About us.* Retrieved April 9, 2007, from http://www.acs.org.au/index.cfm ?action=show&conID=aboutacs

Andresen, L., Boud, D., & Cohen, R. (2000). Experience-based learning. In Foley, G. (Ed.), *Understanding adult education and training* (2nd ed., pp. 225–239). Allen & Unwin.

Anthes, G. (2004). Quality model mania. *Computerworld Management*(March).

Argy, P. (2006). Professionalism is a matter of practice not just attitude. *Information Age,* 2.

Argyris, C., & Schön, D. A. (1978). *Organizational learning: A theory of action perspective*. Reading, MA: Addison-Wesley.

Atherton, J. (1999). Resistance to learning: A discussion based on participants in in-service professional training programmes. *Journal of Vocational Education and Training, 51*(1), 77–90. doi:10.1080/13636829900200070

Australian Physiotherapy Association. (2005). *The health workforce.* Melbourne, Australia: Australian Physiotherapy Association.

Beach, K. (1999). Consequential transitions: A sociocultural expedition beyond transfer in education. In Iran-Nejad, A., & Pearson, P. D. (Eds.), *Review of research in education* (*Vol. 24*, pp. 101–140). Washington, D.C.: American Educational Research Association.

Becher, T. (1996). The learning professions. *Studies in Higher Education, 21*(1), 43–55. doi:10.1080/03075079612331381447

Bellinger, G. (2004). Knowledge management-emerging perspectives. Retrieved March 5, 2007, from http://www.systems-thinking.org /kmgmt/ kmgmt.htm

Berinato, S. (2004, December 16). Are you on board with enterprise risk management? You had better be. It's the future of how businesses will be run. *CIO.*

Billett, S. (2001). *Participation and continuity at work: A critique of current workplace learning discourses. Context, power and perspective: Confronting the challenges to improving attainment in learning at work.* Paper presented at the Joint Network/SKOPE/TLRP International workshop, Northhampton.

Boadle, M. (2005). Adult learning. *ACSLearn* Retrieved April 7, 2006, from http://www.acs.openlab.net.au/ content.php?article.82

Boud, D., & Middleton, H. (2003). Learning from others at work: Communities of practice and informal learning. *Journal of Workplace Learning, 15*(5), 194–202. doi:10.1108/13665620310483895

Boud, D., & Solomon, N. E. (Eds.). (2001). *Work-based learning: A new higher education?* Buckingham, UK: SRHE and Open University Press. Boud, D., & Walker, D. (1998). Promoting reflection in professional courses: The challenge of context. *Studies in Higher Education, 23*(2), 191–206. doi:10.1080/03075079812331380384

Brown, J. S., Collins, A., & Duguid, P. (1989). Situated learning and the culture of learning. *Education Researcher, 18*(1), 32–42.

Brown, J. S., & Duguid, P. (1991). Organizational learning and communities of practice: Toward a unified view of working, learning, and innovation. *Organization Science, 2*(1), 40–57. doi:10.1287/orsc.2.1.40

Calway, B. A., & Murphy, G. A. (2007). The educational imperatives for a work-integrated learning philosophy. *Journal of Cooperative Education and Internships, 41*(2), 95–109.

CAPPE. (2006). *Professionalism, ethics and integrity systems: The promotion of professional ethical standards and the protection of clients and consumers.* Melbourne, Australia: Centre for Applied Philosophy and Public Ethics.

Chivers, G. (2006). *Informal learning by professionals in the UK.* Retrieved October 14, 2008, from www.leeds.ac.uk/medicine/ meu/lifelong06/ papers/ P_GeoffChivers.pdf

CHSRF (Canadian Health Services Research Foundation). (2007). *Knowledge transfer and exchange at work.* Retrieved February 28, 2007, from http://www.chsrf.ca/ home_e.php

Clark, D. (1999). *Andragogy.* Retrieved May 15, 2006, from http://www.nwlink.com/~donclark/ hrd/history/andragogy.html

Clark, D. (2000). *John Dewey (1867-1949).* Retrieved April 23, 2006, from http://www.nwlink.com/~donclark/ hrd/history/history.html

Collins, A., Brown, J. S., & Holum, A. (1991). Cognitive apprenticeship: Making thinking visible. *American Educator*(Winter).

Connor, H., & MacFarlane, K. (2007). *Work Related Learning (WRL) in HE-a scoping study.* Glasgow, UK: Caledonian University.

Cross, J. (2007). *Informal learning.* John Wiley & Sons.

Dadds, M. (2000). Continuing professional development: Nurturing the expert within. In Solar, J., Craft, A., & Burgess, H. (Eds.), *Teacher development: Exploring our own practice.* London, UK: Paul Chapman Publishing.

Daley, B. J. (2001). Learning and professional practice: A study of four professions. *Adult Education Quarterly, 52*(39), 39–54.

Davenport, T. H., & Prusak, L. (1998). *Working knowledge.* Boston, MA: Harvard Business School.

Dearing, R. (1997). *Report of the National Committee into Higher Education (Dearing report).* London, UK: NCIHE.

DEST. (2002). *Striving For quality: Learning, teaching and scholarship*. Canberra, Australia: Department of Education, Science and Training.

Dewey, J. (1938). *Experience and education*. New York, NY: Macmillan.

DiBella, A. J., & Nevis, E. C. (1998). *How organizations learn: An integrated strategy for building learning capacity*. San Francisco, CA: Jossey-Bass.

Dimopoulos, N., & Walker, D. (2005). Lifelong learning challenges for educators, governments, firms and individuals. *CEDA Australian Chief Executive*(August).

Doyle, D. (2002). *Putting learning to work: The distance learner and transfer of learning.* Retrieved March 23, 2010, from http://www.aare.edu.au/ 03pap/doy03366.pdf

Engeström, Y. (2001). Expansive learning at work: Toward an activity theoretical reconceptualization. *Journal of Education and Work*, *14*(1), 133–156.

Eraut, M. (1994). *Developing professional knowledge and competence*. London, UK: Falmer Press.

Eraut, M. (2004). Informal learning in the workplace. *Studies in Continuing Education*, *26*(2), 247–274. doi:10.1080/158037042000225245

Friedman, A., Davis, K., Durkin, C., & Phillips, M. (2000). *Continuing professional development in the UK: Policies and programmes*. Bristol, UK: PARN.

Friedman, A., & Mason, J. (2007). *Distinguishing Australian professional bodies*. Bristol, UK: PARN.

Friedman, A., & Phillips, M. (2001). *Leaping the CPD hurdle: A study of the barriers and drivers to participation in continuing professional development.* Paper presented at the British Educational Research Association, Leeds.

Friedman, A., & Phillips, M. (2002). The role of mentoring in the CPD programmes of professional associations. *International Journal of Lifelong Learning*, *21*(5), 269–284.

Friedman, A., & Phillips, M. (2004). Continuing professional development: Developing a vision. *Journal of Education and Work*, *17*(3), 361–376. doi:10.1080/1363908042000267432

Gardner, H., & Shulman, L. S. (2005). The professions in America today: Crucial but fragile. *Daedalus*, *134*(3), 13–18. doi:10.1162/0011526054622132

HRSDC. (2005). *Transfer of learning planning workplace education programs: HRSDC*. Government of Canada.

IBM. (2004). *Services science: A new academic discipline.* Retrieved February 22, 2010, from http://www.almaden.ibm.com /asr/resources/facsummit.pdf

Ireland, R. D., Kuratko, D. F., & Morris, M. H. (2006). A health audit for corporate entrepreneurship: Innovation at all levels: Part I. *The Journal of Business Strategy*, *27*(1), 10–17. doi:10.1108/02756660610640137

James, P. (2000). *Lessons in learning the lore of the library: Theories and strategies for reader education.* Paper presented at the CAVAL Reader Education Conference: Librarians behaving badly - taking reader education to the people, Melbourne.

Johnston, G. (Ed.). (1983). *The Australian pocket Oxford dictionary* (5th ed.). Melbourne, Australia: Oxford University Press.

Knight, A. (2005). *Australian integrated compact dictionary & thesaurus*. Melbourne, Australia: Oxford University Press.

Knowles, M. (1975). *Self-directed learning-a guide for learners and teachers*. New York, NY: Association Press.

Kohler, J. (2004). *The Bologna process and employabilty: The impact of employability on curricular development.* Paper presented at the A Key Objective of Academic Studies and for Academic Institutions, Bled, Slovenia.

Kolb, D. (1983). *Experiential learning: Experience as the source of learning and development.* Englewood Cliffs, NJ: Prentice Hall.

Lambe, P. (2002). *The autism of knowledge management.* Retrieved June 28, 2006, from http:// greenchameleon.com/ thoughtpieces/autism.pdf

Lave, J., & Wenger, E. (1991). *Situated learning: Legitimate peripheral interaction.* Cambridge, UK: Cambridge University Press. (Original publication IRL report 90-0013, Palo Alto, CA.: Institute for Research on Learning) (1990).

Marden, B. (2003). High aims for professional standards legislation. *Law Institute Journal, 77*(11), 28.

Marton, F., & Säljö, R. (1997). Approaches to learning. In Marton, F., Hounsell, D., & Entwistle, N. (Eds.), *The experience of learning* (2nd ed.). Edinburgh, UK: Scottish Academic Press.

Melotte, B. (1996). *The role of qualitative research in identifying praxis and hermeneutics for professional planning education in Australia. Different approaches: Theory and practice in higher education.* Paper presented at the HERDSA Conference, Perth, Western Australia.

Morris, P. W. G. (1999). *Body building.* Retrieved April 2, 2007, from http://www.pmforum.org/ library/ papers/1999/crmpbody.pdf

Murphy, G. A., & Calway, B. A. (2007a). *Education for professionals through work-integrated learning.* Paper presented at the AARE 2006 Conference, Adelaide.

Murphy, G. A., & Calway, B. A. (2007b). *Professional standards legislation and supporting learning cultures.* Paper presented at the HERDSA 2007, Adelaide.

Murphy, G. A., & Calway, B. A. (2008). Deep learning for professional development of professionals. *Australian Journal of Adult Learning, 48*(3), 424–444.

Orr, J. (1990). Sharing knowledge, celebrating identity: War stories and community memory in a service culture. In Middleton, D., & Edwards, D. (Eds.), *Collective remembering: Memory in society.* Beverly Hills, CA: Sage Publications.

Perkins, D. N., & Salomon, G. (1992). Transfer of learning. In Tuijnman, A. (Ed.), *International encyclopedia of education* (2nd ed.). Oxford, UK: Pergamon Press.

Phelan, P. (2002). The medical colleges: Issues at the turn of the century. *eMJA -The Medical Journal of Australia.*

Professions Australia. (2006). *Blueprint for national registration of the professions.* Retrieved June 12, 2010, from http://www.professions.com.au/ natreg.html

PSC. (2006). Professional Standards Council consultation paper: *Approval of schemes under the professional standards legislation for associations with firms only as members* Retrieved March 5, 2007, from http://www.lawlink.nsw.gov.au/ lawlink/professional_standards_council /psc_ll.nsf/vwFiles/PSC%20 Consultation%20 Firm%20Only_23 _Nov_06.doc/$file/PSC%20 Consultation%20Firm%20Only_23_Nov_06.doc

PSC. (2007). *Application guidelines-guidelines for occupational associations to prepare an application to the Professional Standards Council for the approval or renewal of a scheme in New South Wales.* Retrieved May 21, 2007, from http://www.lawlink.nsw.gov.au/ lawlink/professional_standards_council/psc_ll.nsf/vwFiles/Application%20Guidelines%20 (NSW).doc/$file/Application% 20Guidelines%20(NSW).doc

Ramsden, P. (1992). *Learning to teach in higher education.* London, UK: Routledge. doi:10.4324/9780203413937

Ramsden, P. (2003). *Learning to teach in higher education* (2nd ed.). London, UK: Taylor and Francis, Inc.

Rothwell, A., & Herbert, I. (2007). Accounting professionals and CPD: Attitudes and engagement-some survey evidence. *Research in Post-Compulsory Education, 12*(1), 121–138. doi:10.1080/13596740601155587

Scottish Executive Health Department Management Executive. (1999). *Learning together.* Retrieved May 2, 2010, from http://www.scotland.gov.uk/ health/mel/mellt-00.asp

Sharp, J. (1997). *Communities of practice: A review of the literature.* Retrieved March 6, 2007, from http://www.nickols.us/

Shields, J. (2004). *Corporate learning strategies: Politics, the new order and teachers work.* Paper presented at the International Education Research Conference. Retrieved from www.aare.edu.au/04pap/ shi04532.pdf

Shulman, L. (2002). *Making differences: A table of learning.* Retrieved June 14, 2010, from http://www.carnegiefoundation.org/ elibrary/making-differences- table-learning

Smith, D. (2006). Measuring the impact of informal learning. Retrieved December 12, 2006, from http://www.trainingzone.co.uk/ cgi-bin/item.cgi?id=163069

Smith, M. K. (1996b, 1999). *Praxis, the encyclopedia of informal education.* Retrieved May 2, 2006, from http://www.infed.org/ biblio/b-praxis.htm

Smith, M. K. (2001a). David A. Kolb on experiential learning. In *The encyclopedia of informal education.* Retrieved June 16, 2010, from http://www.infed.org/biblio/ b- explrn.htm

Smith, M. K. (2005). Competence and competency. In *The encyclopedia of informal education.* Retrieved June 14, 2010, from http://www.infed.org/ biblio/b-comp.htm

Smith, P. A. C., & McLaughlin, M. (2004). Knowledge management: People are important! *Journal of Knowledge Management Practice.*

Spiro, R. J., Feltovich, P. J., & Jacobson, M. J. (1996). Cognitive flexibility, constructivism, and hypertext, 2nd ed. Retrieved April 3, 2006, from http://phoenix.sce.fct.unl.pt/ simposio/ Rand_Spiro.htm

Standards Australia. (2007). *What is a standard?* Retrieved March 12, 2010, from http://standards.org.au/ DevelopingStandards/ WhatisaStandard.aspx

Thalheimer, W. (2007). *Measuring learning results: Creating fair and valid assessments by considering findings from fundamental learning research.* Retrieved May 30, 2007, from http://www.work-learning.com/ catalog/

Webster-Wright, A. (2006). *Understanding continuing professional learning.* Unpublished doctoral dissertation, University of Queensland, Brisbane.

Wenger, E., McDermott, R., & Snyder, W. M. (2002). Seven principles for cultivating communities of practice. *HBSWK.* Retrieved from http://askmecorp.com/pdf/ 7Principles_CoP.pdf

West, R. (1997). *Learning for life: Review of higher education financing and policy.* Canberra, Australia: Review Committee on Higher Education Financing and Policy, Department of Employment, Education, Training and Youth Affairs.

Wilson, J. W., & Lyons, E. H. (1961). *Work-study college programs-appraisal and report of the study of cooperative education.* New York, NY: Harper & Brothers.

Young, J. (1997). *Comparative international study of co-operative education employers: The value of employing co-op students.* Cape Town, South Africa: World Association of Cooperative Education.

ADDITIONAL READING

Bransford, J. D., Brown, A. L., & Cocking, R. R. (1999). *How People Learn: Brain, Mind, Experience, and School*. Washington, DC: NATIONAL ACADEMY PRESS.

Brown, J. S., & Gray, E. S. (2003). *Creating a Learning Culture: Strategy, Practice, and Technology*. Palo Alto, California: Cambridge University Press.

Candy, P. C. (1991). *What is Self-Directed Learning Self-Direction For Lifelong Learning* (pp. 5–23). San Francisco: Jossey-Bass.

Ericsson, K. A., Prietula, M. J., & Cokely, E. T. (2007). The Making of an Expert. *Harvard Business Review Online,* (July-August), 6. Retrieved from <http://harvardbusinessonline.hbsp. harvard.edu/hbsp/hbr/articles/ article.jsp?OPERATION_TYPE= CHECK_COOKIE&referer=/ hbsp/hbr/articles/article.jsp& productId=R0707J&TRUE=TRUE&reason=sessionAuthenticated&FALSE=FALSE&ml_subscriber=true &_requestid=4583&ml _action=get-article&ml_ issueid=BR0707&articleID=R0707J&pageNumber=1>

Jones, E. A., Voorhees, R. A., & Paulson, K. (2002). *Defining and Assessing Learning: Exploring Competency-Based Initiatives*. Washington: Council of the National Postsecondary Education Cooperative (NPEC), U.S. Department of Education National Center for Education Statistics.

Kwakman, K. (2003). Professional learning throughout the career. *International Journal of Human Resources Development and Management, 3*(2), 180. doi:10.1504/IJHRDM.2003.002419

Leonard, D., & Swap, W. (2005). *Deep smarts: how to cultivate and transfer enduring business wisdom*. Boston: Harvard Business School Publishing.

Patrick, C.-J., Peach, D., Pocknee, C., Webb, F., Fletcher, M., & Pretto, G. (2008). *The WIL [Work Integrated Learning] report: A national scoping study [Australian Learning and Teaching Council (ALTC) Final report]*. Brisbane: Queensland University of Technology.

Poell, R. F., & Van der Krogt, F. J. (2003). Learning-Program Creation in Work Organizations. *Human Resource Development Review, 2*, 252. doi:10.1177/1534484303256996

Senge, P. M. (1990). *The Fifth Discipline. The art and practice of the learning organization*. London: Random House.

KEY TERMS AND DEFINITIONS

Communities of Practice: "are groups of people who share a passion for something that they know how to do, and who interact regularly in order to learn how to do it better". (Wenger, 2004 p.2)

Competency: is defined as "a combination of skills, abilities, and knowledge needed to perform a specific task". (Jones, et al., 2002 p. 8)

Compliance: Professional associations specify CPD as voluntary, obligatory or compulsory within their regulations for membership. When CPD is obligatory or compulsory the regulations will specify minimum requirements for professionals to maintain their credentials (Friedman, et al., 2009).

Continuing Professional Development (CPD): "The systematic maintenance, improvement and broadening of knowledge and skills, and the development of personal qualities necessary for the execution of professional duties throughout working life". (Professional Development Partnership (The), 2008 n.p.)

Deep Learning: 'learning that promotes the development of conditionalized knowledge and metacognition through communities of inquiry".

(Weigel, 2002, p. 5) It implies that learning is related to existing knowledge and experience using patterns and underlying principles and the checking of evidence in relation to conclusions and recommendations.

Informal Learning: "refers to learning that results from daily work-related, social, family, hobby or leisure activities (e.g. the acquisition of interpersonal skills developed through the experience of working as a sales representative)."(Misko, et al., 2007 p.6)

Knowledge Development: is required by professionals for career development and to maintain and/or increase the body of knowledge. The growth in knowledge will involve acquiring new knowledge which may be incremental, transformative, or supplantive (Atherton, 1999) and can be achieved through Formal, Non-formal and/or Informal Learning (Eraut, 2004).

Professionals: "A profession is a disciplined group of individuals who adhere to ethical standards and who hold themselves out as, and are accepted by the public as possessing special knowledge and skills in a widely recognised body of learning derived from research, education and training at a high level, and who are prepared to apply this knowledge and exercise these skills in the interest of others. It is inherent in the definition of a profession that a code of ethics governs the activities of each profession. Such codes require behaviour and practice beyond the personal moral obligations of an individual. They define and demand high standards of behaviour in respect to the services provided to the public and in dealing with professional colleagues. Further, these codes are enforced by the profession and are acknowledged and accepted by the community." (Professions Australia, 2007 n.p.)

Work-Integrated Learning (WIL): "... educational activities that integrate theoretical learning with its application in the workplace.... and should provide a meaningful experience of the workplace application that is intentional, organised and recognised by the institution, in order to secure learning outcomes for the student that are both transferable and applied". (Griffith University, 2006 n.p.) Work-integrated-learning provides a context for learning and in workplace situations the outcomes may not result in formal recognition but in an increased ability to perform in the workplace.

Chapter 3
APOSDLE – learn@work:
Firsthand Experiences and Lessons Learned

Stefanie Lindstaedt
Graz University of Technology, Austria

Conny Christl
Innovation Service Network GmbH, Austria

ABSTRACT

This chapter presents a domain-independent computational environment which supports work-integrated learning at the professional workplace. The Advanced Process-Oriented Self-Directed Learning Environment (APOSDLE) provides learning support during the execution of work tasks (instead of beforehand), within the work environment of the user (instead of within a separate learning system), and repurposes content which was not originally intended for learning (instead of relying on the expensive manual creation of learning material). Since this definition of work-integrated learning might differ from other definitions employed within this book, a short summary of the theoretical background is provided. Along the example of the company Innovation Service Network (ISN), a network of SME's, a rich and practical description of the deployment and usage of APOSDLE is given. The chapter provides the reader with firsthand experiences and discusses efforts and lessons learned, backed up with experiences gained in two other application settings, namely EADS in France and a Chamber of Commerce and industry in Germany.

INTRODUCTION

Modern businesses operate in a rapidly changing environment. Continuous learning is an essential ingredient to stay competitive in such environments. A typical problem within many companies is the gap or mismatch between the skills required during daily work and the knowledge obtained through formal training activities (e.g. seminars). Even though this gap has been identified by many, seminars and courses still constitute the overwhelming part of knowledge transfer op-

DOI: 10.4018/978-1-60960-547-6.ch003

portunities offered to employees. Haskell (1998) informs that in 1998, US$ 70 billion were spent on formal training and Back et al., (2001) state that in 2000, US$ 78 billion were spent on corporate training and continuing education. On the other hand, studies have shown that in today's economy only a small amount of knowledge that is actually applied to job activities (learning transfer) comes from formal training (Eraut & Hirsh, 2007; Haskell, 1998). On average, people only transfer less than 30% of what is being learned in formal training to the professional workplace in a way that enhances performance. This is independent of the kind and quality of the courses taught, but mainly depends on not considering work environment needs during and after formal training efforts (Robinson, 2003). Overall, 80-90% of what employees know of their job, they know from informal learning (Raybould, 2002). Initiatives aiming at enhancing knowledge transfer of formal training try to answer the question: "How much does the learner know after engaging in the formal training?" Instead, as suggested by the above numbers, the question which should be asked is: "To which extent can the learner apply the newly acquired skills to his/her work tasks?" (Lindstaedt, 2008). However, much of what we know today is based on research in educational settings (schools and universities) or in formal workplace training. Much less research has been conducted in informal workplace learning settings.

This chapter introduces a definition of work-integrated learning (WIL) which might differ in several aspects from the other perspectives discussed in this book. It advocates to take the strong intertwinement of working and learning practices seriously and to design learning support mechanisms so that people can learn during the execution of work tasks (instead of learning first and then applying) (Lindstaedt & Mayer 2006). We first sketch briefly the theoretical foundations of our WIL definition and identify a number of key requirements which result from our understanding of WIL and which heavily influence the way professional learning environments should be designed. We then introduce the APOSDLE[1] environment which offers a wide range of informal learning support. The main part of this chapter is then dedicated to describe our experiences during deployment, training, and use of APOSDLE within the Innovation Service Network (ISN). ISN is a network of small consultancy companies which specialize on innovation management. We show in detail which preparatory steps and modeling activities are needed in order to create an innovation management specific installation of APOSDLE and which company specific goals could be achieved with this approach. We conclude the chapter with the main results of our three months workplace evaluation and give an outlook about lessons learned and future research in the field of WIL.

WORK-INTEGRATED LEARNING

Building on theories of workplace learning such as Eraut (2007) and Colley et.al (2002) we conceptualize learning as a dimension of knowledge work which varies in focus (from focus on work performance to focus on learn performance), time available for learning, and the extension of learning guidance required. This learning dimension of knowledge work describes a continuum of learning practices which starts at one side with brief questions and task related to informal learning (work processes with learning as a by-product), and extends at the other side to more formalized learning processes (learning processes at or near the workplace). This continuum emphasizes that support for learning must enable a knowledge worker to seamlessly switch from one learning practice to another as time and other context factors permit or demand.

Research on supporting workplace learning and lifelong learning so far has focused predominantly on the formal side of this spectrum, specifically on course design applicable for the workplace

Figure 1. Learning dimension of knowledge work

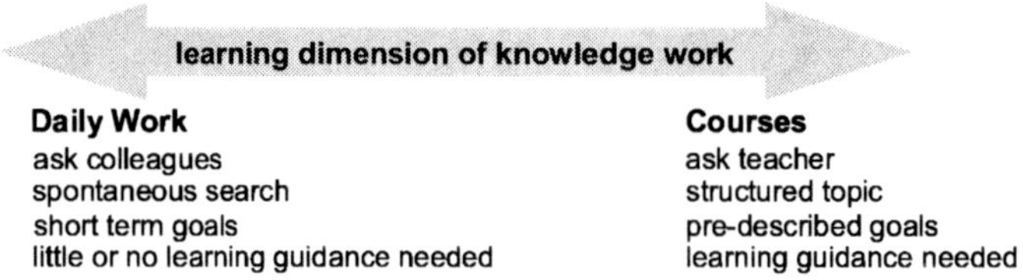

and blended-learning. In contrast, the focus of our work is on the informal side of the spectrum, specifically covering work processes with learning as a by-product and learning activities located within work processes. In order to refer to this type of learning practices, we use the term work-integrated learning (WIL). By using this term we emphasize that we investigate learning at the workplace such that it is truly integrated in current work processes, practices, and environments and makes use of existing resources (e.g., reports, project results) as well as humans (e.g., peers, communities). In general, learning during work refers to an advancement of knowledge and skills of the (knowledge) worker. Eraut (1998) defines personal knowledge and capability as "what individual persons bring to situations that enables them to think, interact and perform". WIL is relatively brief and unstructured (in terms of learning objectives, learning time, or learning support). The main aim of WIL activities is to enhance task performance. From the learner's perspective, WIL is spontaneous and/or unintentional. Learning in this case is a by-product of the time spent at the workplace (Colley et al., 2002). This conceptualization enables a shift from the training perspective of the organization to the learning perspective of the individual.

CHALLENGES FOR SUPPORTING WIL

Currently, most computational support for knowledge workers does not take the learning dimension of knowledge work (see Figure 1) into account. Instead, software companies (and research) have focused on the two extreme ends of the spectrum: dedicated e-learning systems address the deliberate learning processes (often not at all near the workplace), while performance support and workflow systems have focused on supporting the individual work tasks and processes only. These systems support two supposedly distinct roles a knowledge worker can play at the workplace: the role of the worker who performs tasks and creates value within business processes on the one hand, and the role of the learner who sits down for prolonged periods of time to learn about a specific domain or topic at the other. In order to support WIL an environment has to enable people to switch continuously and seamlessly between the two roles, thus they have to be enabled to navigate the whole learning dimension of knowledge work.

Based on two workplace learning studies (Kooken et al., 2007) and an in-depth requirements engineering process (Konkou et al., 2008) which involved four companies interested in supporting WIL within their organizations three key challenges which a WIL environment needs to support were identified:

- **Real time learning:** WIL support should make knowledge workers aware of and support them throughout learning opportunities relevant to the current work task. WIL support needs to be adapted to a user's work context and experiences, and should be short, and easy to apply.
- **Real knowledge resources:** WIL support should dynamically provide and make aware of available knowledge resources

(both human as well as material) within the organization. By providing 'real' resources the effort for learning transfer is reduced and the likelihood for offering opportunities to learn on different trajectories is increased.

- **Real computational environment:** WIL support should be provided through a variety of tools and services which are integrated seamlessly within the user's desktop and allow one-point access to relevant back-end systems of the organization. These tools and services need to be inconspicuous, tightly integrated, and easy to use. They must support the knowledge worker in effortlessly switching between varieties of learning practices.

Concerning real time learning our participatory design activities and workplace learning studies (Kooken et al., 2007; Konkou et al.; 2008, Dotan et al., 2009) suggest that learners need different support and guidance within different work situations. Specifically, it became apparent that learning guidance is needed on varying levels of informality. The learning guidance discussed here is applicable to WIL situations and covers the more informal side of the learning dimension of knowledge work. Clearly the spectrum of learning guidance could be extended to much more formal learning guidance approaches such as predefined learning modules and enforced learning processes. However, this is beyond the scope of our work and this chapter.

THE APOSDLE ENVIRONMENT

The Advanced Process-Oriented Self-Directed Learning Environment (APOSDLE) was a 48 months research and development integrated project partially supported by the European Community under the Information Society Technologies (IST) priority of the 6th framework program for R&D. APOSDLE represents a software framework which implements the requirements for WIL support elaborated above. It is a domain-independent framework which provides a variety of learning guidance functionality enabling the user to navigate the whole learning dimension of knowledge work. In order to deploy APOSDLE within a workplace environment the domain-specific knowledge has to be provided in form of a number of knowledge models (see Section 5) and learning content. This process of deploying APOSDLE at the company ISN is detailed below in Section 6. Since APOSDLE is a highly complex environment we can only provide a short overview of the functionality here. As this overview is rather short and abstract, you will find a short scenario on how APOSDLE can be used within ISN at the end of this section. For a more detailed discussion on the different learning guidance functionalities please refer to Lindstaedt et al., (2010). Lindstaedt et al. (2008) provides the technical perspective on APOSDLE.

The APOSDLE environment presents itself as desktop application (client-server architecture) which is designed to allow for high degrees of flexibility and work-integrated support. The APOSDLE main application (see Figure 2) includes a number of tabs. The *Suggest* tab proactively recommends document snippets (parts of text or multi-media documents), topics, tasks and persons based on the current context of the user. The user context is represented by the current task or topic the user is working on, and identified by a daemon that continuously monitors the user interactions in the background (while observing privacy settings). *Search* provides the user with full text search capability. *Browse* allows the user to browse through the underlying knowledge models in the knowledge base (see Figure 5) and thus exposes them for learning purposes. In *MyAPOSDLE* ad-hoc learning paths can be created. Users can simply compose a list of topics which should be covered by the learning path. When a learner later executes the learning path snippets

Figure 2. APOSDLE main application

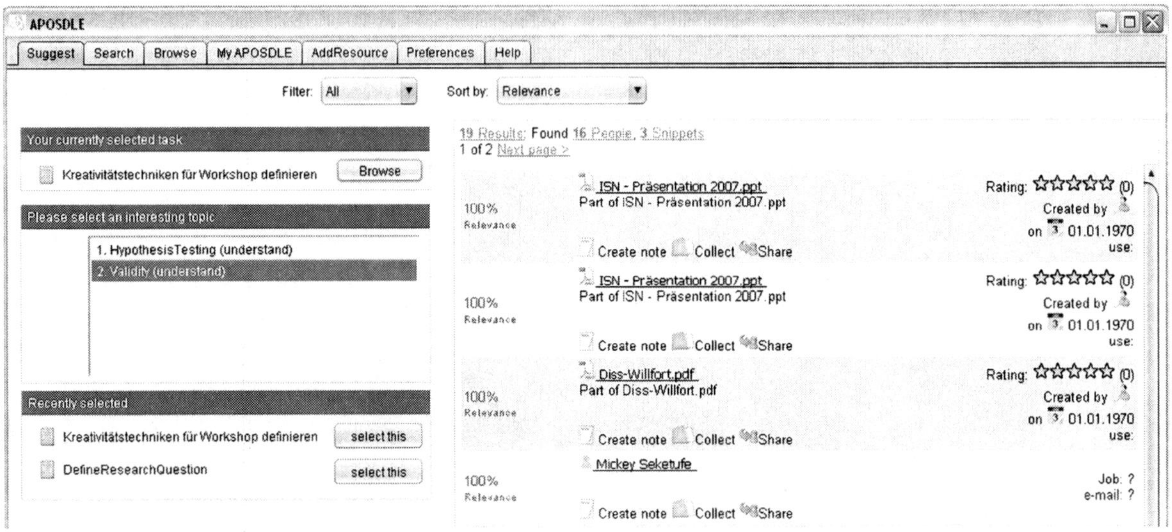

are automatically recommended for each topic. This allows quickly creating a 'learning module' for e.g. new employees. Finally, the *Preferences* tab keeps track of the work experiences a user has made by storing a usage history. This history lists all tasks and topics the user has dealt with in interactions with the APOSDLE environment. This can be used to reflect about the current state of knowledge and possible further learning goals, and is also used for the intelligent suggestions APOSDLE makes.

When a resource is selected, a resource viewer displays documents and multimedia content, and highlights all parts of the document which were found to be relevant (see Figure 3). This resource viewer also includes annotation functionality, which lets users create snippets of resources by annotating parts of documents with topics from a knowledge model. Within the *Learn more* tab these snippets of the resources are accompanied by some learning hints which motivate the user to reflect about the snippet the user has read.

Cooperation processes in APOSDLE are supported by a cooperation wizard which guides cooperation partners through a cooperation process; from a request, to the actual cooperation, to a reflection of the cooperation (see Figure 4). Different standard cooperation tools (such as skype, telephone, email, text messaging, audio/video messaging and cooperative authoring) are available and are integrated in APOSDLE via a cooperation framework (Musielak et al., 2008). The cooperation event is contextualized by allowing all cooperation partners to share their context information (such as current tasks, learning goals and resources viewed). Completed cooperation events can be stored in a cooperative authoring space and used for reflecting on these events.

In order to provide a better understanding of how APOSDLE can be used a short usage scenario is provided here:

Sandra is working for a company providing consulting on innovation management. This morning she had a meeting with a costumer from the automotive industry who wants to start a new project. Back in the office she opens her text program and starts writing an offer. APOSDLE automatically identifies (via the monitoring daemon) that she is performing the task "Writing an offer" (match

Figure 3. APOSDLE Reader, Annotation Tool and Learn more

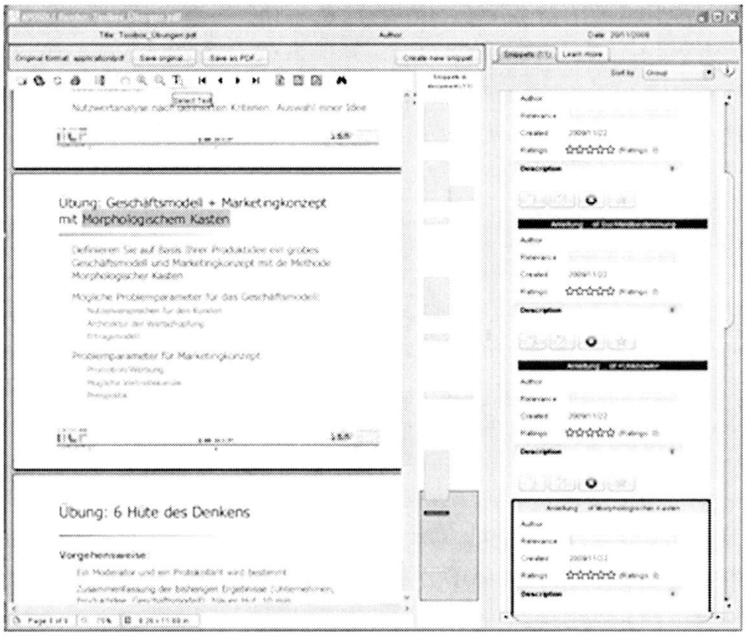

Figure 4. APOSDLE Cooperation Wizard

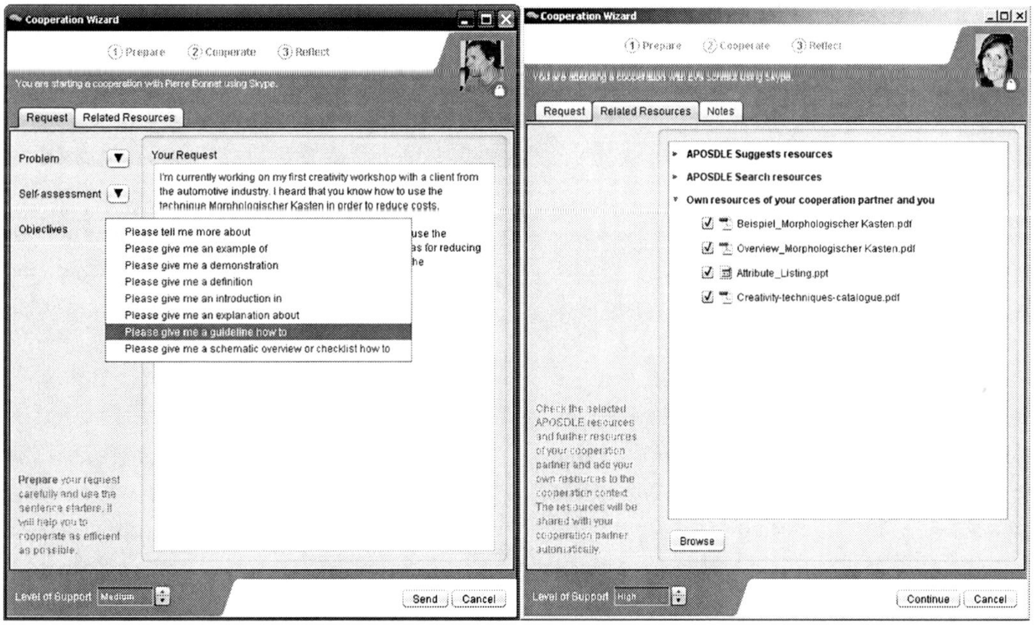

with underlying knowledge base) and suggests her similar documents from other projects (Suggests tab, Figure 2).

A few days later she is preparing the slides for the first workshop with this costumer. A colleague told her that he favors the 'value management' technique and she enters this topic into APOSDLE.

She opens the first result (Suggest tab) and a document opens indicating her exactly a paragraph (APOSDLE Reader, Figure 3) from a university script explaining the technique step by step. As she wants to know more about 'value management' she also has a look on the proposed learning hints. One learning hint shows her all other moderation techniques related to 'value management' and she clicks on one.

Besides other results APOSDLE suggests her to contact a colleague from Slovenia (Suggest tab). By clicking on his name the APOSDLE cooperation wizard opens (Figure 4) and she can structure her request very specifically. APOSDLE automatically suggests different tools to contact him. Since APOSDLE knows that he is online, skype is ranked first. She decides to contact him and the cooperation initiates automatically and provides her colleague with information on Sandra's current task (contextualized communication). Once the problem is solved Sandra adds notes to her request so that it can be used by other people to solve similar problems in the future.

THE 'INTELLIGENCE' WITHIN APOSDLE

In order to create a WIL environment that addresses the identified key challenges (see Section 3), this work-integrated learning paradigm puts one requirement on technology in the center of attention: Flexibility. It is therefore not desirable to create a system specifically matched to one enterprise and one domain. The developed software system must be as generic as possible. Deployment of the system in a new organization or in a new domain must not require substantial software changes. However, clearly domain-specific knowledge must exist in some form. The system must know about different users, their skills and learning goals and about the tasks users perform. Further the system needs knowledge about the available digital resources (text documents, multimedia documents etc.) and in which situation users might benefit from these (Christl et al., 2008). Within APOSDLE, such a domain-independent and flexible approach is achieved by employing a hybrid approach: Coarse grained semantic models (of the topics, the tasks, and the skills) and an initial set of manual annotations are complemented with a suite of soft computing approaches (such as machine learning, statistical analysis, and heuristics). This hybrid approach combines the advantages of semantic models (meaningful terminology, possibilities for inferences) with the advantages of soft computing methods which perform well in the face of uncertainty (Lindstaedt, 2008). We discuss both sides of APOSDLE's 'intelligence' below.

On the one side we rely on semantic models which have to be provided for an application in a specific domain. This has the advantage that the software as such is generic and can be tailored to any suitable domain by exchanging the models. The APOSDLE system operates on three types of models which are tightly interconnected. These are; the *topic model*, which is an ontology describing the learning domain, the *task model*, that describes the work tasks that need to be executed, and finally, the *skill model* provides the link between topics and tasks in order to realize work-integrated learning. An integrated modeling methodology (see Section 6) was created that supports the process of creating the three models in an interactive manner involving domain experts and knowledge engineers. On a technical level, services give access to the models as if they were one single knowledge structure, even though they were written using very different conceptual and formal representations. This is achieved through a unifying schema specifically developed for the purpose of supporting work-integrated learning. The main idea is that starting from the context-sensitive situation of a user, which includes his/her current task, APOSDLE is able to determine the task and topic of the user and then use the information contained in the APOSDLE Knowledge

Figure 5. The APOSDLE Knowledge Base and its relation to other components

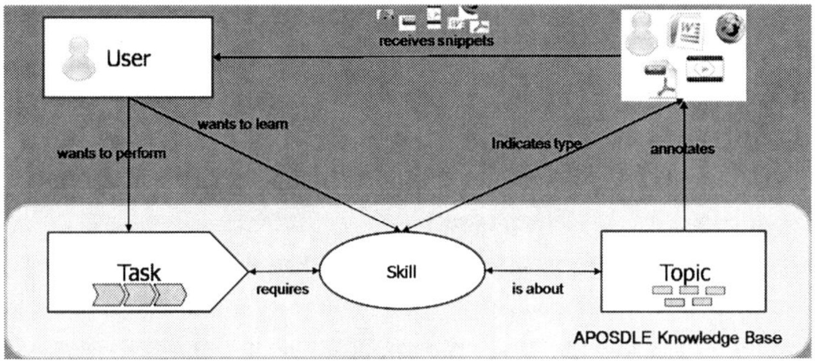

Base (Figure 5) to select appropriate resources, so called snippets (Scheir et al., 2007). A snippet is defined in APOSDLE as a document, or part thereof, which is annotated with a topic from the topic model and a predefined type triggered from the skill model. This type assigns how the respective topic is used within the resource (e.g. definition, example, explanation, presentation etc.). Based on an initial training set of manually created snippets, snippets can be created automatically.

On the other side APOSDLE makes use of a number of innovative services which employ soft computing approaches. Knowledge discovery and extraction algorithms provide similarity measures for textual documents, discover important topics and types of certain material, and extract textual information from multi-media sources. These services support retrieval of documents and also automatic annotation. An associative retrieval service performs context-based retrieval based both on textual content of documents as well as semantic metadata attached. This service can perform exact queries (such as for snippets of the type 'example') as well as fuzzy associative queries (such as expanding a query for a topic to include similar topics) (Scheir et al., 2007). A user profile service realizes a set of functionalities for maintaining and analyzing user-related information. It also provides inferences based on the usage history, such as ranking topic or snippet suggestions according to the user's learning history and ranking cooperation partners based on the history of past cooperation's (Lindstaedt et al., 2009). This data is also provided in a history view as part of the APOSDLE main application, which allows users to access change and delete any of the collected information. Security and privacy services ensure that users are authenticated, organizational and personal privacy policies applied, and access rights strictly obeyed.

FIRST HAND EXPERIENCES: DEPLOYING AND USING APOSDLE

Based on the theoretical and technical outline we focus now on the practical part of the WIL approach. The APOSDLE environment was introduced and evaluated within three different application domains. A large corporation represented by EADS (European Aeronautic Defence and Space Company), a public organization represented by the CCI Darmstadt (Chamber of Commerce and Industry in Germany), and a network of SME'S represented by ISN (Innovation Service Network). We selected the latter to report in detail on the issues a company has to face when deploying a modern learning environment relying on semantic models. Below we describe the company background of ISN, the domain and the motivation for the usage of the APOSDLE environment, and provide an experience report by showing how the semantic models can be created collaboratively by experts, while taking into account the constraints within real environments.

The Application Context, Challenges and Motivation

The Innovation Service Network (ISN) is a consulting and research company composed of a network of 6 SMEs in Slovenia and Austria working in the field of innovation management. The core team of ISN counts 20 innovation managers, who are supplemented by more than 40 further network-partners from universities, competence centers and service companies acting as a pool of experts. In order to stay competitive, the network partners need to continuously improve their skills in the focus field, rapidly learn aspects of other fields relevant to one customer project, communicate with other parties and apply both within the customer's situation. Since the network partners are increasingly involved in partner projects, the time for learning and improvement of skills becomes more limited. For efficient work on costumer projects it is necessary to identify e.g. similar projects, to avoid recreating already existing content. Furthermore, novices should be aware of the existing content or the people that can be asked. Although the partners are often outside the office they spend about 80% working on their computers. Usually learning during work task execution is done by using templates, guidelines, project documentation or internet resources and with few sharing effort to the collective.

At the time before APOSDLE deployment ISN didn't use any learning system or knowledge management system. Experts usually ask their colleagues, in order to find what they need to know - help is mainly provided via email or telephone. Due to time constraints, documentation work is rarely done. However, for efficient work it is essential to have a good overview over existing projects or the skills of other partners. The initial motivation for implementing APOSDLE at ISN is therefore to create a common knowledge base which integrates resources from different repositories and backend systems for all knowledge workers in order to improve knowledge acquisition, reuse, and sharing.

Deploying and Introducing APOSDLE

Deploying APOSDLE to a specific domain involves the creation of the required semantic models and identifying relevant content from the organization's memory (such as project descriptions, offers, reports, etc.). Despite the recent advances of formal modeling and semantic web, it is obvious that most of the organizations interested in using the APOSDLE environment neither have formal models already available, nor all the skills needed to develop them (Christl et al., 2008). Therefore, as part of the APOSDLE project we have developed an Integrated Modeling Methodology (IMM)(Ghidini et al., 2009) as a series of steps, techniques and tools to support the construction of the semantic models depicted in Figure 5. Along the example of the company ISN we below enrich the general description of the four phases of the IMM with practical examples.

The set of modeling tools contains:

- **MoKi:** The MOdeling wiKI – a wiki-based tool which supports the creation of the domain and task model (Ghidini et-al. 2009);
- **TACT:** The Task-Learning Goal Mappings Tool – a JAVA based tool which supports the creation of the learning goal model;
- **Validation Tools:** Some automatic checks and an ontology questionnaire – which support the revision and validation of the models (Pammer et al., 2009).

Phase 0: Scope & Boundaries. The first phase aims at determining the scope and boundaries of the application domain. Through questionnaires and workshops the main tasks and learning needs of the respective company are identified and resources which may be relevant for the chosen learning domain are collected. The key aspect of this phase is to support companies to identify a learning domain which is appropriate for the WIL approach taken by APOSDLE.

Implementation at ISN. Based on the outcome of a creativity workshop the learning domain was defined on project processing and project management in the innovation management domain. In order to refine this decision, all employees of ISN were asked to sketch during one week their most frequent tasks, the methods they apply or typical problems that arise. This very rough information was then structured in mind maps and further refined in the next steps.

Phase 1: Knowledge Acquisition. The goal of this phase is the acquisition of the knowledge, both from persons and digital resources, which is relevant for the 'intelligence' within APOSDLE. The knowledge elicitation phase was supported by different well known state-of-the-art techniques like interviews, card sorting, laddering, and concept/step/chapter listing are used, while the extraction of knowledge from digital resources is based on algorithms and tools for term extraction. The key aspect of this phase is twofold: First, to effectively gain relevant informal knowledge from domain experts, who are typically rarely available and scarcely motivated towards modeling; second, to collect knowledge in digital form that can be reused for modeling.

Implementation at ISN. Based on approximately 50 representative documents reflecting the innovation management domain (e.g. offers, scripts describing methods in innovation management or costumer presentations) a first list of relevant topics was extracted by using the term extraction tool within MoKi. Parallel to this activity, further information for the semantic models was gained through workshops with domain experts. For example the chapter listing technique was applied within one workshop; during one hour, a domain expert imagined writing a book about the ISN domain and listed the titles of the chapters, e.g. he labeled the first chapter of the book "strategic innovation management", and defined four subchapters: "trends", "how to define new markets", "defining core competencies", "use of external knowledge". These topics were then added to the list of topics. Two further workshops with domain experts were conducted and the results (preliminary tasks and topics) were complemented. The laddering technique was applied to identify more specific structures. For instance, the topic "method", brought up subtopics like "brainstorming techniques", or "knowledge management techniques". The card sorting technique turned out to be useful for the identification of relations between tasks and topics. Form this phase resulted a broad and good list of tasks and topics in MoKi.

Phase 2: Modeling the topics and tasks. The result of this phase is a complete formal description of the topic and task models in MoKi (Figure 6) as part of the APOSDLE 'intelligence': the topic model contains all topics about the specific application domain, while the task model concerns the activities and tasks a user can perform in the organization. Additionally the interdependencies between topics and tasks needed for the learning goal model are defined. Automatic validation checks allow validating the models and possibly revising by using guidelines for manual revision.

Implementation at ISN. First, for each topic and task in MoKi a verbal description (most of these descriptions were obtained in the Phase 1) was added. Second, hierarchical relations (e.g. *is a* and *is part of*) were identified for topics where appropriate. For example, the topic "Brainstorming" *is a* "Intuitive-creative method" and *is part of* the topic "Idea discovery". Finally the topic and task model were proofed for completeness and consistency by a domain expert and exported from the MoKi in OWL (Web Ontology Language).

Phase 3: Modeling of Learning Goals. The Task And Competency Tool (TACT) (Figure 7) provides a formal specification of the learning goal model. Starting from the initial alignment between

Figure 6. Screenshot Moki

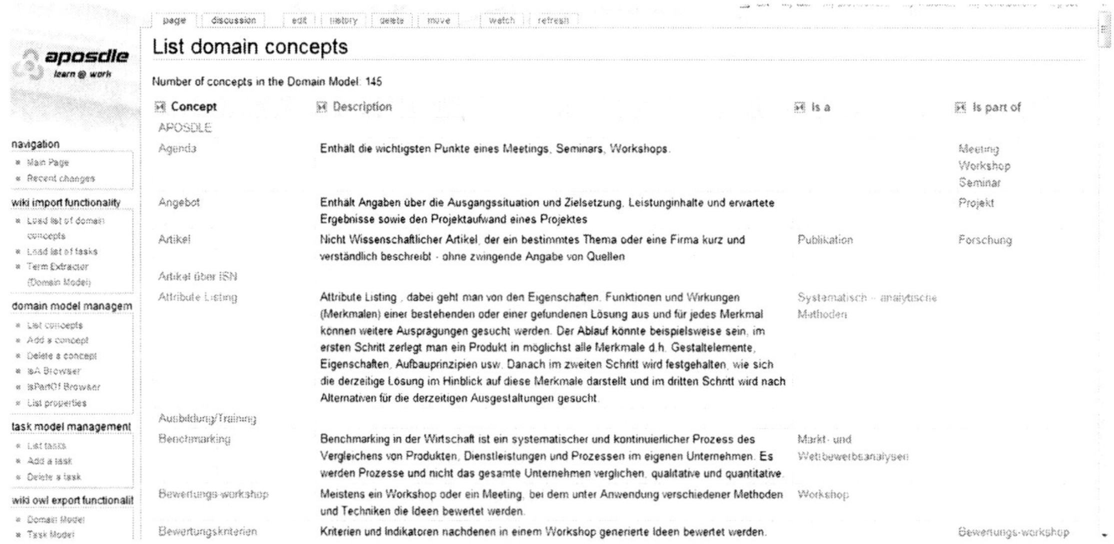

Figure 7. TACT: The Task-Learning Goal Mappings Tool

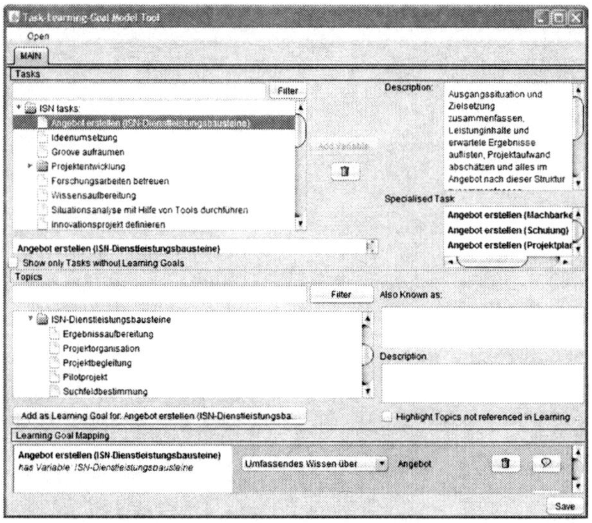

topics and tasks (Phase 2), the tool allows connecting topics and tasks through predefined skill types like "basic knowledge about", "profound knowledge of", "know how to apply" and "know how to produce". After validating the learning goal model, the entire APOSDLE Knowledge Base is ready to be plugged into the APOSDLE environment (Ley et al., 2010).

Implementation at ISN. The formal descriptions exported from MoKi were imported to the TACT Tool and topics and tasks were connected through skill types. E.g. the task "Writing an offer" was connected with the skills type "know how to produce" and the topic "offer". After connecting all tasks to topics, the final file for APOSDLE was ready.

Figure 8. Efforts needed to create the semantic models within APOSDLE

Company	CCI	EADS	ISN
Phase 0	80	24	70
Phase 1	110	130	156
Phase 2	30	134	41
Modelling of learning goals	80	9	20
Validation and Revision II	4	N/A	8
Total (hours)	304	297	295

Overall, ISN modeled 133 topics and 141 tasks in MoKi. Connecting the tasks and topics in TACT resulted in 248 learning goals. At ISN, creating the semantic models for APOSDLE required approximately 295 hours from the first creativity workshop until the final models. The numbers in Figure 8 are intended to give an impression about the time needed for modeling at the other application partner sites. However, modeling time cannot be directly compared for different models as the effort needed depends on a variety of factors, such as the scope of the domain or the availability of already existing process descriptions.

Please note, that a substantial part of these modeling efforts were spent in phases 0 and 1 of the IMM. In these phases considerable work is needed to acquire the domain knowledge from the domain experts. This is typically omitted when modeling efforts are discussed. Also, the modeling efforts reported here have to be seen in comparison to typically required work for the deployment of learning management and workflow systems. Often, for such systems the creation of domain, task and skill models is also necessary. In addition, some require much more complicated prerequisite relations (which are automatically computed in the APOSDLE case) and other models. Summarizing, it can be stated that good guidelines and tools for supporting the modeling process are crucial and have a significant impact on the resources needed for modeling.

Introducing APOSDLE to a Company

As already touched upon above, one of the challenges for the APOSDLE WIL approach is to rely on the 'real' resources of the companies' backend repositories. In order to provide the user with fitting snippets an initial set of resources has to be annotated manually. For each of the 133 topics at least two snippets were produced, i.e. parts of documents were 'tagged' with the respective topic. This training set of annotations allows the 'intelligent' services to create snippets automatically. As the quality of the environment depends on the quantity and quality of snippets, manual and automatic annotation is combined. The idea is that users create, change, and rate these annotations collaboratively over time.

A very important aspect when introducing a new environment is to figure out how to get the knowledge workers to actually use the tool. Introducing a new technology into a 'real world' context is always a challenge, and depends on the interplay of the organization, the people, and the technology. It is crucial to create a good first impression and to avoid user frustration. For preparing the users for the three moths evaluation at ISN and the other application sites we developed a threefold

Figure 9. APOSDLE Soft Goals for the Evaluation

	Evaluation Goals	Achievement
Learning	Aware of learning material	Yes
	Learning time planned and managed	No
	Aware of possible learning situations	To some extent
	Existing knowledge improved	Yes
	High quality learning material provided	Yes
Working	APOSDLE learning help task completion	Yes
	Learning material relevant to current task	Yes
	Unconnected disturbances minimized	Yes
Collaborating	Identification of experts and peers improved	Yes
	Experts accurately sorted according to relevance	Yes
	Knowledge/document sharing	Yes

training approach: First, the user's awareness and motivation is raised within the company through different social solutions several weeks before the introduction, like a poster in the meeting room or coffee brakes dedicated to APOSDLE. Second, all users get a very first short and easy introduction to APOSDLE. The focus is more on creating an interest in APOSDLE rather than explaining all functionalities. The missing functionalities are then shown step by step in the following weeks. And third, a comprehensive help section is implemented into the environment. Through short explanations, video tutorials and contact possibilities users have the possibility to speed up on their own, and directly ask questions to a support team. Additionally further organizational adjustments like APOSDLE discussions in team meetings or a bi-weekly APOSDLE lunch for all test users support the motivation.

Impressions of Using APOSDLE in a Real-World Context

APOSDLE was deployed, introduced and evaluated at ISN as well as in two other application settings, namely EADS in France and CCI in Germany. We conclude by highlighting the final workplace evaluation of our WIL approach, which focused on evaluating the extent to which the main objectives (supporting learning, working and collaborating) of the project have been satisfied by the APOSDLE solution.

The workplace evaluation took about three months and involved nineteen persons from three companies. A multi-method data collection approach was followed using a questionnaire, interviews, log data, user diaries kept while working with APOSDLE, and site visits. Based on a previous developed model of stakeholders and their goals regarding APOSDLE, we derived a subset of goals specific enough to make measurements possible (Figure 9). These goals were prioritized by the application partners and indicated whether the goal was achieved or not.

Within ISN five employees used APOSDLE during the testing period. For evaluation purposes, every user was asked to report positive and negative experiences in an online diary during the evaluation. In addition to the help section this was another way to react for example on technical problems quickly. ISN employees reported their experiences in 119 blog entries. For this section we picked out the most important user expectations and requirements gained during the deployment phase and aligned them with selected comments and experiences.

As can be concluded based on the analysis of the requirements and the blog entries at ISN, it is important to make experts aware of resources they can use for their current task in order to avoid recreating existing content. This expectation was covered by *APOSDLE Suggests.* The user is made aware of snippets (pieces of resources) and knowledgeable persons that might fit to the current work task or topic.

"APOSDLE detected that I was preparing a bill and showed me an interesting checklist for creating a bill which was completely new for me und really helpful. great:-)" [User diary] O2.P1, ISN

"APOSDLE found an ISN moderation guideline I've never seen before. I'll use it preparing my next workshop" [User diary]O2.P3, ISN

At ISN every partner is specialized in a certain topic. As experts are rarely in the office or working in different sites the improvement of communication within the network as well as the identification of experts that could be asked in problem situation, states another requirement. APOSDLE creates a user profile for each user which evolves over time through interaction with the environment. Whenever a user is an expert in a certain topic, APOSDLE recommends this expert to other users. Users can communicate through the *Cooperation Wizard,* which enables real-time coaching and supports specifying a request in a structured way.

"The cooperation assistant (..) is a very useful tool (...) which invites you to self reflect on your problem and helps others to solve your problem quickly. (...)."[User diary] O2.P4, ISN

The tab "Preferences" shows to me in which topics I am a beginner, an advanced user and an expert. I have to say that I was quite surprised about the topics I'm am an expert in (APOSDLE listed the following topics: open innovation, bionics, Neurovation and idea finding) - and this is true![User Diary] O2.P2, ISN

Like in many other companies, experts at ISN are heavily involved in projects thus limiting their time to help new employees to get up to speed. As the semantic models are visible in APOSDLE users can browse through the domain and explore relationships between topics and tasks. The Browse tab provides an interactive overview and navigation over all processes within the company and users can switch between a tree map view, a process view and a relationship view. The My APOSDLE tab supports learning by following learning paths and allows creating a quick learning course for e.g. new employees that want to get an overview over frequently used creativity techniques.

"I like Aposdle, because it gives me the feeling that there is someone else who is learning with me" [User diary] O2.P4, ISN

"APOSDLE gives me hints or induces me to think about topics that could be relevant for learning, for instance the topics "Value Management" or "Trend analysis"" [User diary] O2.P2, ISN

We can conclude that our domain-independent approach to providing varying degrees of learning guidance within workplace situations worked. Relying on existing knowledge resources and repurposing them for learning proved to be effective and cost efficient. Crucial for this is the availability of good modeling tools, experienced modelers, and high quality annotations of snippets.

CONCLUSION AND FUTURE WORK

In this chapter we gave a broad overview of our understanding of WIL and the domain-independent computational environment APOSDLE which supports work-integrated learning at the professional workplace. Along an example we provided a rich and practical description of the application of methodologies and tools to integrate and finally use the APOSDLE solution within a company. We reported on the results gained in the workplace evaluation over a period of three months which highlights the strengths and weaknesses of the APOSDLE prototype. Our aim was to give a firsthand experience report and provide lessons learned in the field of technology enhanced work-integrated learning. The APOSDLE environment[2] is available under an open source license in order to allow interested companies and research institutions to further investigate the APOSDLE WIL approach.

The current APOSDLE environment is based on static models. Knowledge workers in consultancies often require up-to-date diverse information which is difficult to predict and model in advance. ISN experts commented their most frequent knowledge needs related to new topics and projects the company was involved in. That knowledge was beyond the core business domain that was modeled in APOSDLE. Consequently, it can be said that especially for domains with dynamic nature it is important to allow adding new topics and tasks during the system usage. It also can be worth to explore, whether the models can be build up from scratch and evolve over time through the system usage, like e.g. corporate wikis.

The main conclusion of our combined evaluation efforts is that the APOSDLE solution has proven very useful for learners in highly-specialized domains in which much of the knowledge to be learned is documented. In those circumstances, APOSDLE delivered an effective work-based learning solution that enabled relatively inexperienced knowledge workers to efficiently improve their knowledge in various ways. However, it proved less effective in broad customer-driven domains where knowledge was shared to a large extent in person, can be retrieved also in the internet, or is typically not documented. This finding seems to suggest that APOSDLE needs to incorporate other kinds of communication and collaboration approaches which are oriented more towards collaborative knowledge construction instead of only focusing on question-answer interactions.

ACKNOWLEDGMENT

APOSDLE was partially funded under the FP6 of the European Commission within the IST work program 2004 (FP6-IST-2004-027023). The Know-Center is funded within the Austrian COMET Program - Competence Centers for Excellent Technologies - under the auspices of the Austrian Ministry of Transport, Innovation and Technology, the Austrian Ministry of Economics and Labor and by the State of Styria. COMET is managed by the Austrian Research Promotion Agency FFG.

REFERENCES

Back, A., Bendel, O., & Stoller-Schai, D. (2001). *E-learning in Unternehmen. Grundlagen - Strategien- Methoden - Technologien.* Zurich, Switzerland: Orell Flüssli.

Christl, C., Ghidini, C., Guss, J., Lindstaedt, S., Pammer, V., & Rospocher, M. ... Serafini, L.(2008). Deploying Semantic Web technologies for work integrated learning in industry. A comparison: SME vs. large sized company. In A. P. Sheth, S. Staab, M. Dean, M. Paolucci, D. Maynard, T. Finin, K. Thirunarayan (Eds.), *The Semantic Web - ISWC 2008. Proceedings of ISWS 2009.* (LNCS 5318), (pp. 709–722). Berlin, Germany: Springer.

Colley, H., Hodkinson, P., & Malcolm, J. (2002). *Non-formal learning: Mapping the conceptual terrain. A consultation report*. University of Leeds Lifelong Learning Institute. Leeds. Retrieved August 18, 2008, from http://www.infed.org/archives/ e-texts/colley_informal_learning.htm

Dotan, A., Maiden, N., Lichtner, V., & Germanovich, L. (2009). *Designing with only four people in mind? - A case study of using personas to redesign a work integrated learning support system*. In. INTERACT 2009 12th IFIP TC13 Conference in Human-Computer Interaction. Uppsala, Sweden.

Eraut, M. (1998). Concepts of competence. *Journal of Interprofessional Care*, *12*(2), 127–139. doi:10.3109/13561829809014100

Eraut, M., & Hirsh, W. (2007). *The significance of workplace learning for individuals, groups and organisations. (SKOPE Monograph 9)*. Oxford University, Department of Economics.

Ghidini, C., Kump, B., Lindstaedt, S., Mahbub, N., Pammer, V., Rospocher, M., & Serafini, L. (2009). MoKi: The enterprise modelling wiki. In Sure, Y., & Domingue, J. (Eds.), *The Semantic Web: Research and applications. (LNCS 5554)* (pp. 831–835). Berlin, Germany: Springer. doi:10.1007/978-3-642-02121-3_65

Ghidini, C., Rospocher, M., Serafini, L., Kump, B., Pammer, V., Faatz, A., & Guss, J. (2007). *Integrated Modelling Methodology (IMM)*. version 1. Retrieved from http://www.aposdle.tugraz.at /media/multimedia/¯les/integrated modelling methodology version 1

Haskell, R. E. (1998). *Reengineering corporate training. Intellectual capital and transfer of learning*. Westport, CO: Quorum Books.

Kooken, J., Ley, T., & de Hoog, R. (2007). How do people learn at the workplace. Investigating four workplace learning assumptions. In E. Duval, R. Klamma & M. Wolpers (Eds.), *Creating new learning experiences on a global scale* (pp. 158–171). Second European Conference on Technoloy Enhanced Learning, EC-TEL 2007, Crete, Greece, September 2007. Berlin, Germany: Springer.

Kounkou, A., Lichtner, V., Dotan, A., & Maiden, N. (2008). *APOSDLE use scenarios and requirements for 3rd prototype*, Deliverable D6.4. Retrieved June 2010, from http://www.aposdle.tugraz.at/media/multimedia/files/ aposdle_use_scenarios_requirements _for_3rd_prototypes

Kump, B. (2008). A validation framework for formal models in adaptive technology-enhanced work-integrated learning. In W. Nejdl, J. Kay, P. Pu, & E. Herder (Eds.), *Adaptive hypermedia and adaptive Web-based systems,* (pp. 416-420). 5th International Conference, AH 2008, Hannover, Germany, July 29 - August 1, 2008. Berlin, Germany: Springer.

Lave, J., & Wenger, E. (1991). *Situated learning – legitimate peripheral participation*. Cambridge, UK: Cambridge University Press.

Leemkuil, H., de Hoog, R., Lindstaedt, S., Ley, T., Scheir, P., Ulbrich, A., et al. Pammer, V.(2007). *Conceptual framework & architecture version 1*. Retrieved August 1, 2009, from http://www.aposdle.tugraz.at/ media/multimedia/¯les/conceptual framework architecture version 1

Ley, T., Kump, B., & Albert, D. (2010). A methodology for eliciting, modelling, and evaluating expert knowledge for an adaptive work-integrated learning system. *International Journal of Human-Computer Studies*, *68*(4), 185–208. doi:10.1016/j.ijhcs.2009.12.001

Lindstaed, S., Beham, G., Aehnelt, M., Hambach, S., Musielak, M., Kooken, J., et al. (2009). *The APOSDLE approach to self-directed work-integrated learning*. Retrieved April 1, 2009 from http://www.aposdle.tugraz.at/ media/multimedia/ files/ the_aposdle_approach_to_self_ directed_ work_integrated_learning

Lindstaedt, S., Beham, G., Kump, B., & Ley, T. (2009). Getting to know your user–unobtrusive user model maintenance within work-integrated learning environments. In Cress, U., Dimitrova, V., & Specht, M. (Eds.), *Learning in the synergy of multiple disciplines* (pp. 73–87). Heidelberg, Germany: Springer. doi:10.1007/978-3-642-04636-0_9

Lindstaedt, S. N., Kump, B., Beham, G., Pammer, V., & Ley, T. dotan, A., & de Hoog, R. (2010). *Providing varying degrees of guidance for work-integrated learning.* To appear in Proceedings of ECTEL 2010, Lecture Notes in Computer Science. Heidleberg, Germany: Springer

Lindstaedt, S. N., Ley, T., Scheir, P., & Ulbrich, A. (2008). Applying scruffy methods to enable work-integrated learning. *The European Journal of the Informatics Professional*, *9*(3), 44–50.

Lindstaedt, S. N., & Mayer, H. (2006). A story-board of the APOSDLE vision. In W. Neidl & K. Tochtermann (Eds.). *Proceedings of the 1st European Conference on Technology-enhanced Learning* (LNCS 4227). (pp.628–633). Heidelberg, Germany: Springer.

Musielak, M., Hambach, S., & Christl, C. (2008). *APOSDLE contextualized cooperation*. In: ACM SIGCHI: ACM Conference on Computer Supported Cooperative Work 2008. Electronic Proceedings: CSCW 08. New York, NY: ACM Press.

Pammer, V., Kump, B., Ghidini, C., Rospocher, M., Serafini, L., & Lindstaedt, S. N. (2009). *Revision support for modeling tasks, topics and skills*. In A. Paschke, H. Weigand, W. Behrendt, K. Tochtermann, & T. Pellegrini, (Eds.), *Proceedings of I-SEMANTICS '09, 5th International Conference on Semantic Systems*, (pp. 501-508).

Raybould, B. (2002). *Performance support engineering part one: Key concepts*. Ariel PSE Technology.

Robinson, D. G. (2003). *Skill and performance: They are not equal*. Apartment Professional Magazine.

Scheir, P., Ghidini, C., & Lindstaedt, S. N. (2007). Improving search on the semantic desktop using associative retrieval techniques. *Proceedings of I-MEDIA 2007 and I-SEMANTICS 2007*, Graz, Austria, September 5-7, 2007, (pp. 221-228).

ADDITIONAL READING

Albert, D., Hockemeyer, C., & Wesiak, G. (2002). Current trends in e-Learning based on knowledge space theory and cognitive psychology. *Psychologische Beiträge*, *4*(44), 478–494.

APOSDLE Consortium. (2006).*Use Cases and Application Requirements 1* (First Prototype), Project Deliverable D6.2, Retrieved on 4, 2008. http://www.aposdle.org/ media/multimedia/files/ use_cases_applica tion_requirements_1_first_ prototypeS

Brusilovsky, P., & Nijhavan, H. (2002). *A framework for adaptive e-Learning based on distributed re-usable learning activities.* In Driscoll,M. & Reeves, T.C.(Eds.). Proceedings of the World Conference on e-Learning, E-Learn 2002. AACE, Montreal, pp.154–161.

Collin, K. (2006). Connecting work and learning: design engineers' learning at work. *Journal of Workplace Learning*, *18*(7/8), 403–413. doi:10.1108/13665620610692971

Cristani, M., & Cuel, R. (2005). A survey on ontology creation methodologies. *International Journal on Semantic Web and Information Systems*, *1*(2), 49–69. doi:10.4018/jswis.2005040103

Doignon, J., & Falmagne, J. (1985). Spaces for the assessment of knowledge. *International Journal of Man-Machine Studies*, *23*, 175–196. doi:10.1016/S0020-7373(85)80031-6

Eraut, M. (2004). Informal Learning in the Workplace. *Studies in Continuing Education*, *26*(2), 247–274. doi:10.1080/158037042000225245

Fernández-López, M., Gómez-Pérez, A., & Juristo, N. (1997). *Methontology: from ontological art towards ontological engineering*. In: Proceedings of the Symposium on Ontology Engineering of AAAI.

Fischer, G. (2001). User modeling in human–computer interaction. *User Modeling and User-Adapted Interaction*, *11*(1–2), 65–86. doi:10.1023/A:1011145532042

Fox, M. S., & Grüninger, M. (1998). Enterprise modeling. *AI Magazine*, *19*(3), 109–121.

Holman, D., Epitropaki, O., & Fernie, S. (2001). Short research note. Understanding learning strategies in the workplace: A factor analytic investigation. *Journal of Occupational and Organizational Psychology*, *74*, 675–681. doi:10.1348/096317901167587

Kelloway, E. K., & Barling, J. (2000). Work as organizational behaviour. *International Journal of Management Reviews*, *2*(3), 287–304. doi:10.1111/1468-2370.00042

Korossy, K. (1997). Extending the theory of knowledge spaces: A competence-performance approach. *Zeitschrift fur Psychologie mit Zeitschrift fur Angewandte Psychologie*, *205*, 53–82.

Ley, T. (2006). *Organizational Competency Management-A Competence–Performance Approach*. Aachen: Shaker.

Ley, T., & Albert, D. (2003). *Identifying* employee competencies in dynamic work domains: methodological considerations and a case study. *Journal of Universal Computer Science.*, *9*(12), 1500–1518.

Lindstaedt, S. N., de Hoog, R., & Aehnelt, M. (2009). *APOSDLE: Contextualized Collaborative Knowledge Work Support*. Supplementary Proceedings of the 11th European Conference on Computer Supported Cooperative Work, Demos, Videos, Posters, Vienna, Austria, 7 - 11 September 2009, 7-8

Montalvo, F. T., & Torres, M. C. G. (2004). Self-regulated learning: current and future directions. *Electronic Journal of Research in Educational Psychology*, *2*(1), 1–34.

Myers, M. (1999). Investigating information systems with ethnographic research. *Commun. AIS*, *2*(4), 1.

Olgren, C. (2000). Learning Strategies for Learning Technologies. In Burge, E. J. (Ed.), *The Strategic Use of Learning Technologies. New Directions for Adult and Continuing Education, no. 88* (pp. 7–16). San Francisco: Jossey-Bass.

Pinto, H., Tempich, C., & Sure, Y. (2004). *DILIGENT: Towards a fine-grained methodology for Distributed, Loosely-controlled and evolved.* In G Engingeering of oNTologies. In: Proceedings of the 16th European Conference on Artificial Intelligence (ECAI 2004), Valencia, Spain (August 22nd - 27th 2004)

Rath, A. S., Devaurs, D., & Lindstaedt, S. N. (2009).*Contextualized Knowledge Services for Personalized Learner Support* Fourth European Conference on Technology Enhanced Learning. Demo Paper. Lecture Notes in Computer Science. Springer.

Schmidt, A. (2004). *Context-steered learning: the learning in process approach.* In Proceedings of the IEEE Conference on Advanced Learning Technologies (ICALT04). IEEE Computer Society, Washington DC, pp.684–686.

Studer, R., Benjamins, V. R., & Fensel, D. (1998). Knowledge engineering: principles and methods. *Data & Knowledge Engineering, 25*, 161–197. doi:10.1016/S0169-023X(97)00056-6

Sure, Y., Gómez-Pérez, A., Daelemans, W., Reinberger, M., Guarino, N., & Noy, N. F. (2004). Why evaluate ontology technologies? Because it works! *IEEE Intelligent Systems, 19*(4), 74–81. doi:10.1109/MIS.2004.37

KEY TERMS AND DEFINITIONS

APOSDLE: APOSDLE was a 48 months research and development integrated project partially supported by the European Community under the Information Society Technologies (IST) priority of the 6th framework programme for R&D (contract no. IST-027023). The EU project APOSDLE developed a software platform and tools to support learn@work: Learn within the context of the immediate work and within the current work environment.

Integrated Modeling Methodology (IMM): A methodology and tools to support the collaborative creation of a topic model (domain concepts), a task model (process description) and a learning goal model needed for the APOSDLE environment.

Knowledge Worker: Employee of an organization whose essential operational and value creating tasks rely on knowledge as their critical work resource. Knowledge workers are referred to as workers throughout this proposal.

MoKi: The Modelling wiKi.: MoKi is a tool that supports the creation of articulated enterprise models through structured wiki pages. MoKi enables heterogeneous teams of experts, with different knowledge engineering skills, to actively collaborate by inserting knowledge; transforming knowledge; and revising knowledge at different degrees of formality. The active collaboration is guaranteed by an automatic translation between formal and informal specifications of the different experts, and by facilitating an integrated construction of the different parts of the integrated model.

Self-Directed Learning (SDL): Self-directed exploration and application of knowledge by learners with the purpose of advancement in a learning domain. SDL largely relies on the learners' own initiatives and creativity. SDL often appears in working contexts where learners need to pursuit their learning goals within time and resource restrictions of their work. Necessary pre-requisites for SDL are: learner's having been empowered to take responsibility for their own learning attempts, means that allow them to carry out various learning activities (self-directed literature research, collaboration in groups etc.), and access to various learning resources.

(User) Context: A knowledge worker's context is comprised of many aspects, such as the step within the business process the knowledge worker is currently working in, the knowledge worker's progress in education and learning, short- to medium-term goals with regard to executing the job at hand, and long-term goals with regard to corporate and personal advancement.

Work-Integrated Learning: We coined the term work-integrated learning to refer to all different types of learning during work. We want to emphasize that workplace learning is truly integrated in current work processes and practices and makes use of existing resources. It is relatively brief and unstructured (in terms of learning objectives, learning time or learning support), where the main aim of the activities is to enhance task performance. From the learner's perspective, workplace learning is spontaneous and/or unintentional. Learning in this case is a by-product of the time spent at the workplace (Colley et al.,, 2002).

ENDNOTES

1 Advanced Process-Oriented Self-Directed Learning Environment developed within the APOSDLE project co-funded by the EU; please refer to www.aposdle.org for more details

2 An APOSDLE demo version as well as the open source software can be downloaded on www.aposdle.org

Chapter 4
Industrial Training in Engineering Education in Spain

Urbano Dominguez
Universidad de Valladolid, Spain

Jesus Magdaleno
Universidad de Valladolid, Spain

ABSTRACT

Practical training in companies has been recognized for many years as an important component of the education of new engineering graduates all over the world. The format used to provide this education to students varies widely not only across national boundaries, but also within each country. This chapter deals first with the state of industrial training in engineering education in Spain, both in the old engineering degrees and in the new ones, following the European higher education area requirements, which are now in the process of introduction. An analysis is also carried out on the evaluation and assessment of industrial training when this activity is a part of first cycle engineering curricula, and the role played by the tutor is discussed. Finally, some weak points of industrial training in Spanish curricula are discussed, as well as some proposals to overcome that situation and to move towards a global approach of industrial training in engineering education.

INTRODUCTION

Among the components of innovative curricula in Higher Education courses, practical stages at companies are considered all over the world as one of the most wanted requirements sought by employers. This is especially important for engineering graduates wishing to compete in a global labour market.

These practical stages may take different forms and the way in which they are incorporated to curricula varies widely not only from country to country but also within the same national boundaries. These different forms can be in general referred to as work integrated learning (WIL). In this chapter we consider Industrial Training (IT)

DOI: 10.4018/978-1-60960-547-6.ch004

as a form of WIL very common in Engineering Education. As for the geographical scope of this analysis it will be centred in Spain, with the due reference to the situation in other European countries involved in the development of the European Higher Education Area (EHEA).

IT for engineering students can be approached in three different ways. In one of them, there is a formal allocation of quite large stage periods in companies, lasting up to one or sometimes more semesters, with formal recognition in the curricula of engineering graduates. In the other, practical internship appears as an optional subject, and usually it has short stage periods and it has assigned a low number of credits. Finally, in some study programmes IT is not considered at all, and students engage in it only to improve their curricula vita (CV) and their job prospects.

The contents and nature of the practical stage may be very different, even in the cases in which IT is compulsory. In some cases that period may cover a full semester and there are countries providing public economic support to the student while at the industry. Also, some universities provide students with internships while others, the responsibility rests with the students.

It was with the design of curricula adapted to the EHEA, when different voices were raised asking for including work placements on a compulsory basis in Spanish undergraduate engineering programmes (Domínguez & Magdaleno, 2003; Cámara et al., 2009).

In this chapter a review is carried out of the structure of the new Degrees in Spain and how IT is considered in them. Some proposals are discussed on the methods to improve the evaluation of students and the outcomes of work integrated learning experiences, with a special reference to Mechanical Engineering Degrees in Spain. Finally, there is a discussion on how to improve the present situation of IT in Spanish Engineering curricula.

INDUSTRIAL TRAINING IN ENGINEERING EDUCATION

Practical IT activities can be defined as periods of engineering education outside the University geographical space oriented towards providing the students with knowledge, competences not easily obtained in class-rooms, and carried out while they provide some services to the industries implied in the educational process.

"Practical" and "industrial" terms must be understood here in a broad sense. Practical activities mean those educational tasks that are carried out without requiring more theoretical knowledge over that already gained by the student at the moment when the practical stage is initiated. On the other hand, industrial refers here to any organised human group implied in producing goods or supplying services. In this sense, the term industry includes public or private manufacturing or services firms but also public administrations, co- operatives, trade unions, non-governmental organizations (NGOs), foundations and other collectives.

Work placements play a fundamental role in the education of 21st century engineers who have to act on a global and competitive environment. Many authors involved in engineering education have been working on this issue for many years, and some contributions to this topic can be seen elsewhere (Ahrens, 2000; Domínguez & Magdaleno, 2003; El- Sayed, 2001; Greve et al., 2009; Keleher et al., 2007; Maki, 2002; Patkó et al., 2009; Roelofsen, 2009; Schenck & Boots, 2001; Welters & Van de Wetering, 2009).

Engineering Core Curricula requirements have been discussed in the framework of the Socrates Thematic Network Enhancing Engineering Education in Europe (E4) (Heitmann, 2003). Within the personal requirements for Bachelor Engineering Degrees, those related with the practicum are:

- Function efficiently in project groups and teamwork.

- Understand the interaction process between people working in teams, and be able to adapt himself to the requirements of his working environment.
- Display an understanding of the influence of engineering activity on all life and the environment, and demonstrate a high moral and ethical approach to engineering tasks.
- Apply his learning ability to undertake appropriate further training of a professional or academic nature.
- Show an appreciation of the uncertainty, ambiguity and limitations of knowledge.

Another document on output standards, produced by The Engineering Professor´ s Council of Great Britain (EPC, 2000), also lists several of the social skills among the most- valued attributes of engineering graduates according to the opinion of the employers.

Some of these of abilities can be more easily acquired through student placements in companies rather than by formal University education even in the cases when it incorporates innovations such as project- based learning.

Benefits of IT can be increased by adding an intercultural component as in the case of internships carried out in foreign countries. A good example of this is the Leonardo programme for EU students about or just after graduating who follow a period of training at different jobs. In the case of engineering students this is carried out mainly at factories. However, this programme only benefits a rather small number of students, far less than in the case of the Socrates programme, which has proved decisive in the configuration of the EHEA.

Industrial internships can provide the students with a real and direct knowledge of their future work and also constitute the best way to put into practice the knowledge acquired at the University. Their knowledge will be applied to specific projects and programmes at the company they are in. In those internships the students are trained in tasks such as the integration in working teams, fulfilment of specific objectives within predetermined deadlines, assumption of personal responsibilities, contact with business administration, etc., topics which are not properly covered by formal academic education.

Some of the main attributes assigned to work placements in higher education curricula are:

- It is a part of the student educational process and it must be assessed with reference to its contents, methodology and results.
- It must be fully integrated in the curricula.
- It must report benefits to all those involved: universities, industries and students.
- It usually implies activities not easily carried out at University centres.

If these are the common features to be requested to work placements, many of the activities receiving that name today are not such since they do not fulfil one or several of the requirements. Very often they do not take into account the first and the second points. Sometimes the only part receiving a clear benefit is the industry while it is harder to find out the returns to the university or even to the student, who is supposed to be the main actor in the process. This kind of malfunctioning can be found especially when the University concentrates its attention to the administrative part of the process but it does not take into account its educational nature, and when the academic tutor does not play in it any significant role.

For a good running of these activities, some incentives should be established for each of the implied agents:

a. **Industries.** If there are not enough offers of places for students from industry, as is the common case in Spain, establishing adequate stimulus for the later should be a priority. According to the philosophy and operative principles of industries the incentives should be mainly of a type financial or fiscal. Tax rebates may be as efficient in promoting industrial placements for students as direct money supplied to industries from public funds. On the other hand, in some public administrations, co- operatives and small companies, technical assistance provided by the student on training could be better valued than any kind of economic return.

b. **Institutions.** Those Universities, Engineering Schools and Departments that offer proper and well organized programmes receive a bonus in the quality of the education they offer to students (see below the section on IT and quality in education). Usually the companies have to pay a fee to educational Institutions (even in the cases in which the student receives no salary) to compensate them for operational, insurance and administrative expenses. In some programmes for industrial placements offered by the governments, a financial compensation for the each student is also granted to Universities. Since the development of new initiatives in this issue is a costly task, Universities should receive from governments some kind of financial assistance to encourage those initiatives.

c. **Students.** The first incentive for students will be to have a proper recognition in their CV of the time spent in industrial placements. Extending this period longer than the minimum required in the curricula could be rewarded by some additional credits as those for free optional subjects. In the curricula that are being changed now, 10% of the total number of European Credit Transfer System (ECTS) can be applied to free optional subjects. However, adding more credits to IT should not compromise the possibilities of the students for incorporating non- engineering subjects to the educational process. However the main return for the students will be the educational turnover they receive from IT, if it is properly managed. So the Universities should specially watch the educational aspects of the process to verify that students are really under training and are not used as cheap hand labour instead.

d. **Lecturers.** Probably this is the aspect requiring a stronger action for improvement. In some situations the academic tutor plays a purely nominal and passive role. In these cases there is not a proper follow up and evaluation of the educational side of the process and the results tend to be very poor. Once adequate procedures have been established for actively implying lecturers in these activities, the time and effort devoted by them should be considered with the same criteria applied to other academic activities. To evaluate the work carried out by academic tutors, it is necessary to consider all the time required for the different activities. First, the time spent by the tutor to contact industries and external tutors. Moreover, the follow up of the students´ progress students tutored, including going to the site where they are placed, also require some dedication of the tutors. Finally, to check and assess students´ reports and to mark them also requires tutor´s additional time.

ENGINEERING EDUCATION IN SPAIN: AN OVERVIEW

General Framework of Higher Education in Spain

Higher education in Spain is carried out at Universities, structured in Faculties, Higher Schools and University Schools, most of which are being converted into Faculties or Higher Schools as the new curricula according to Bologna's requirements are being implemented. There are 75 Universities, and around one third of them are privately owned, some by Church organisations.

Universities are autonomous institutions, and higher education matters are under the regional governments' rule although there is a common framework defined by the central government.

The number of Higher Education students in Spain according to the data provided by the Ministry of Education (ME, 2009), is over 1,300,000 in the academic year 2009- 10. From these, around 350,000 are enrolled in Engineering Courses.

Trying to change the system in operation for decades, the central government issued in 1987 a Royal Decree (RD, 1987) on "Common regulations for University Degrees valid all over the Spanish territory". In the previous system almost all the subjects were compulsory and the study programmes were substantially equal for all the Faculties/ Schools issuing the same Degree at different Universities in Spain (Domínguez, 2002). Under the regulations that have been in force up to now, some flexibility in the curricula were introduced once the common core, the so called trunk subjects covering around 60% of the credits, was guaranteed. One credit corresponded to 10 lecture hours or equivalent practical/ laboratory workload. Each University was allowed to organise the rest of the study programme and a part of it is to be defined by the student as optional. The Universities should submit their curricula to the Universities Council for approval.

Some new regulations to put the Spanish system in line with European Higher Education Area (EHEA) have been approved by a Royal Decree issued in October 2007 (RD, 2007). That Decree rules the general and specific competences which are required for a Programme of Study to be verified, and to have official validity all over Spain.

A list of the competences sought for first cycle Degrees is also included in that text. First, the students must show a proper knowledge and understanding of their study field, and they must also know how to apply their knowledge in the defence of arguments and in the solution of problems in their speciality. Moreover, they must be able to transmit information, ideas, problems and solutions both for specialized and not specialized interlocutors. Finally, graduates have to be able to engage autonomously in life- long learning.

The RD (2007) establishes a minimum of 240 credits for all first cycle Degrees which are referred to as "Graduates". The text speaks of European credits, or ECTS, as defined in the RD (2003). One ECTS unit is equivalent more or less to 25 hours of students′ working time RD (2003). In the Royal Decree (RD, 2007), 60 of those credits are allocated to basic education in one of the five different branches of knowledge considered (Architecture and Engineering in our case). In this document is also contemplated up to a maximum of 30 credits for external practices, which may include practical training in companies.

Specific Regulations for Engineering Degrees

In Spain, many University Degrees related within the framework of the so called regulated professions, as Engineering, grant the Graduates direct access to the professional market and allow them to join the appropriate Engineering Council without any other requirement. With reference to Engineering, up to now the name of the Degrees was the same as the one of the professions and the Engineering Council involved. Namely the Technical Engineering Councils, for first- cycle Degrees, and the Engineering Council, for integrated 5 year Degrees.

For the new Degrees their name cannot be the same as the pertinent profession and Council. As a consequence, the Ministry of Education has established the competences required to new graduates for being entitled to act on the fields of the already existing Engineering Professions. The particular cases we are going to deal with in this work are the Graduates in the field of Industrial Engineering, which is formed by some of the classical engineering studies (Mechanics, Electricity, Chemistry, Textile, Electronics, and Industrial Management). Among them, and for the sake of clarity, we will refer in particular cases to the Graduate in Mechanical Engineering.

Two Ministerial Orders have been issued, one for Industrial Technical Engineering (First, Bachelor or Graduate level), and the other for Industrial Engineering (Master level) (Order, 2009a, and Order, 2009b, respectively). The Orders establish the requirements for new Graduates for them to be able to engage in professional activities with the same rights of the existing Technical Engineer and Engineer Degrees. Order (2009a) defines conditions to be fulfilled in the new Industrial Engineering Degrees, in addition to those established by the RD (2007).

The Order (2009a) considers several blocks of subjects and it allocates a minimum number of credits for each of them for Technical Industrial Engineers courses. There is a first block of 60 ECTS of basic subjects, common to all Engineering and Architecture Degrees. A second block, with another 60 ECTS, is devoted for subjects that are common to all branches of industrial engineering (as indicated above). Another 48 ECTS are allocated to each specific technology and, finally, 12 ECTS are assigned to the Final Project.

For each one of these blocks there is defined a list of competences to be acquired by graduates, with specific emphasis in the scientific and technical knowledge involved. In some respects the list of items is very detailed, and there are 11 descriptive paragraphs in the common industrial block, with 60 ECTS, and another 8 items in each specific technology (48 ECTS). Since the Order (2009a) defines up to 180 ECTS of a Degree which has 240, then only 60 ECTS are left for each University to be distributed among other compulsory and optional subjects.

In the academic year 2008- 09 only a few engineering new programmes were offered and others added in 2009- 10. While it is envisaged that only new study programmes will be offered in 2010- 11, many of them are still in the process of being officially approved.

A common feature of the study programmes approved so far is the reduction in the number of optional ECTS with reference to the existing ones. This restricts very much the ability of the students to shape their own curricula, which is the general trend in EHEA. Moreover, it limits the credits available for IT when it is included among the optional subjects.

ENGINEERING IT IN SPAIN WITHIN THE FRAMEWORK OF THE EUROPEAN HIGHER EDUCATION AREA

The situation in Europe varies widely from country to country. Compulsory IT to obtain an Engineering Degree in first or second cycles is required in about half the European countries. In general, compulsory practical training activities are more common in Northern and Central European countries that in the South (Domínguez & Magdaleno, 2003). This is a requirement for graduation and is independent from other practical periods which might be required of graduates in some countries by Engineering Councils before granting them membership.

According to a survey carried out by the Spanish National Accreditation Agency (ANECA, 2007), around 40% of the Spanish Engineering Students have carried out IT stages before graduating. This percentage goes up to 70% when considering all the University students. This puts Spain slightly over Czech Republic, United Kingdom and Italy, who are at the lower end among European Universities with reference to this parameter in European University. At the other end of the scale are The Netherlands, Germany, and Finland, having rates of IT between 100 and 90%.

On the other hand, when work placement is compulsory the contents and nature of the practical stage may be very different. In some cases that period may cover a full semester and there are countries providing public economic support to the students while they are at the industry. Also some Universities provide students with practical placements while in others these should look by themselves where to carry out them.

In the recently approved curricula of Graduates in Mechanical Engineering, practical training is considered in different ways or possibly not at all. A total of 18 curricula have been analyzed. Only in three Universities (17%) that matter has to be taken prior to graduation. In others (more than 50%) It appears listed as an optional subject. Finally, in the rest it is not clear if this activity is considered or not, although the number of credits available for optional subjects is so low that most probably IT will be not included in them. With reference to academic recognition of IT in new study programmes, usually it has allocated 6 ECTS, although in some cases it may go up to 12 ECTS.

In the new industrial engineering curricula approved so far, IT will have a very limited recognition, if any, in terms of ECTS. On the other hand, as the number of ECTS available in those curricula for complementary activities or free option subjects will be very small, it will be very difficult to find in this way recognition for internships and other practical activities. In this respect, the change to new curricula according to Bologna will mean in Spain less opportunities than before for those activities. This is a weakness of the change to EHEA, which needs to be corrected.

Most of the Universities have an office in charge of students´ placements in companies, either on its own as an educational activity addressed to the students, or in connection with employment both for undergraduates and postgraduates. In any case they have a regulation of the kind of activities that could be included in IT. The regulation usually has the form of a normative that also includes the model of University-Company Agreement, which specifies the general framework.

Only in some special cases all three parties involved in these activities receive economic compensation from the Government. There is a type of those practical stages, usually called Alternating Education ("Formación en Alternancia" in Spanish), ruled by the Regional government (Magdaleno et al., 2002). They are usually carried out during the summer holidays, and they involve a rather high number of hours of practical training. However, only a handful of students can benefit from it in every yearly call.

Students may or may not receive payment from companies, depending on the specific agreement for the practical stage. According to ANECA (2006), Universities usually charge a fee to the companies for expenses for each student stage.

An interesting experience of university- industry cooperation in education has taken place in Universities of Castile and Leon since 2001. It is the programme PAVEs (from its Spanish designation), financed by the Regional government. In this programme, financial support is granted to allow some members of the staff of companies be engaged as teachers by the Universities (Magdaleno & Domíngez, 2006). The agreement between the University and the company concerned establishes what members will be engaged as lecturers for a period of one year. The compromise is to teach students for six hours per week during an academic course, although it is usually extended on a yearly basis.

One of the aims of the PAVEs programme is to send students to factories where they can learn on and about that environment by in- situ lectures. In that way they deal with real practical situations and they work with some equipment that cannot be found at the Faculties.

Students feel in general very satisfied with the experience. In some cases this opens the door for them to carry their Final Projects in the company involved, and sometimes they become part of its staff once they have graduated.

As for the academic point of view, PAVEs work under the direct supervision of the lecturer in charge of the subject of the study programme to which the activity is linked. Usually the assessment of this practical activity is a part of the final mark in the subject concerned for every student in the course group involved.

We consider that Universities should include practical training in companies in their Engineering study programmes as a compulsory subject. However, the reviewed carried out on the situation of IT in Spain shows that the case is very different, and in most of the Universities it appears as an optional subject, usually with a rather low workload recognition (6 ECTS) as an average.

INDUSTRIAL TRAINING AND QUALITY IN EDUCATION

IT can have a very positive influence on the quality of engineering education when it is properly carried out and adequately supervised as it has been pointed out by Domínguez & Magdaleno (2003), Orrel (2004) and Zabalza (2000). In a survey of innovation in Engineering Degrees addressed to teachers (Domínguez et al., 2006), these were asked about the importance of different factors on competences in new Engineering degrees. Over a maximum score of 5 points, they value IT with 4.22, followed by multi- disciplinarily, team working and others.

Practical IT activities allow for:

- Approaching students to real industrial life, providing them with knowledge on group dynamics, management styles, and business culture.
- Providing the students with experience on how to act in working communities and gaining social skills, necessary for their future work as engineers.
- Relating student´ s academic knowledge to practical problems. In that way they learn how to apply theoretical education to real problems solving. On the other hand, and especially in engineering education, the students can find in industries practical resources and machines in operation which could not be easily found at many Universities.
- Connecting industries and universities by providing new links, developing common activities and acquiring common responsibilities in the educational process. Through these contacts Universities get a better insight of industry´ s needs and they can tune curricula to improve the quality of the education provided to the students.

To get the best for the student in work placements these should be incorporated at the right stage in the educational process leading to the Engineering Degree. On the other hand, the formative aims associated to them have to be related with the rest of processes taking place within the University walls.

To have this system working properly at improving the quality of engineering education is necessary to incorporate an academic tutor with specific duties. Some authors remark the lack of appropriate reward for academic tutors implied in IT (Huerta et al., 2009).

Teachers tutoring these activities can also benefit from them in several forms. They are in close contact with professional and industrial environments when they programme and tutor work placements for their students. From this contact they can have new information to be incorporated in their lectures to put them more on line with industrial life. As work placement is one of the many ways of active learning, they can extend this pedagogical approach in their day to day activity. Finally, they may receive inputs from students involved in work placements, not only on professional but also on personal side, when they are tutoring students´ experiences in the practicum.

A review of several aspects of quality in work placements has been carried out by Orrel (2004). It analyzes a programme on the Practicum of students at Flinders University. Among another factors, the practical stages of students are considered a mark of quality for the Universities and for the study programmes in which they are included.

The perception of students' involved in IT has been used by Osman et al (2008) to measure the quality of its educational experience. Surveys carried before and after the training programme took place show a clear improvement in their competences such as communication and attitudes. On the other hand, students ranked more than 90% the benefits for career guidance, qualification and job prospects.

Since it has a positive influence on the educational process, some kind of added value should be attached to the activity when assessing the quality. The Spanish Ministry of Education considered as an index of quality in higher education the rate of the number of credits implied on work placements with that activity over the total number of credits allocated to any activity of a practical nature. On the other hand, contacts between University and industry through internships activities are also recommended in some integrated proposals of engineering education (Ahrens, 2000).

The Spanish National Accreditation Agency (ANECA, 2007) has carried out a survey on this matter in Spanish Universities. In its Annex D, it analyzes the correlation of IT, and another 10 items, with the development of competences in higher education. The former appears highly correlated with the ability of working in teams. It also shows high correlation with other competences, such as ability for coordinating activities, development of leadership, capability for negotiation, and ability for detecting new opportunities. The role of IT in the development of competences, with the specific reference to engineering students, has been analyzed elsewhere (Monyarch, 2005).

THE EVALUATION AND ASSESSMENT OF INDUSTRIAL TRAINING

The system built up by most of the European Universities, not only in Spain but in many other countries, takes care of the administrative aspects of the process, which is managed by student's counselling or employment offices. The regulations are pretty clear in establishing the conditions of the University-industry relationship, and usually the business runs pretty well in this respect within the established framework. Usually they also list the rights and obligations of the students involved. However, what is not so clear in most of those regulations is the kind of obligation acquired by the academic tutors.

The rate of success of IT programmes is usually measured by the number of students and companies involved each year and their annual rates of growth. There is an appropriate body at the Universities for the evaluation of the activities offered by industries to students and the way in which they are carried out. This body should guarantee a real educational benefit to the students involved. However, only in some cases there are references to the academic aspects in the control of these activities.

However, we consider academic recognition of IT as one of the main indicators of success, if it has to act as a part of the educational process. To measure that, Universities should go beyond the evaluation stage and they have to assess the student's performance with reference to well defined academic criteria. Each University has to care for them in order to evaluate IT of students with similar guarantees to those existing for the rest of academic subjects.

The assessment system has to take into account two main points:

- Active role of the academic tutor as responsible for the feedback of results into the educational system.
- Continuous contact between academic and industrial tutors during the period of the stage of students.

As for the assessment of these activities, there are not clear rules similar to those used to evaluate other subjects directly taught by Universities.

Very often a kind of pass mark is granted without a quantitative evaluation graded according to the results obtained by each student. If IT is to be extended and generalised in the European Union the assessment system should be carefully designed to guarantee that proper results are achieved by the student in the training period.

A discussion on the issue of assessment by using external as well as internal references, which can be applied to IT, is presented in Maki (2002). Some of the points there outlined can be taken as a first step for defining an assessment system useful for that kind of activities.

There are some data on IT in ANECA (2006). According to this report, in a survey of over 41 Spanish Universities the main goal sought with those activities by Universities is to contribute to the education and practical experience of their students, with a score of 77 over 136 points. By contrast, academic recognition is marked with only 18 points, which is less than a fourth of the main criteria.

The evaluation of IT is also a component of the procedures for quality assessment to be considered by the different Spanish Accreditation Agencies (there are one National and 17 Regional Agencies in Spain). According to a document circulated by the Spanish Network of University Quality Assessment (REACU after the acronym in Spanish), external practices have to be considered just the same as any other subject in the study programme (REACU, 2009). They must have adequate planning and the students should be assessed following academic criteria. For REACU, the indicator of quality is defined by the rate of students successfully finishing the practices over the total number who applied for them.

FINAL REMARKS

In these pages we have reviewed different aspects of IT in undergraduate engineering education to conclude that, in spite of including the word "training" in its name, it has a truly educational function and has a positive influence on the students in order to develop some competences in addition to those acquired at the University premises.

In the process of educating Engineers for a global labour market, we propose a consensus among national governments, Engineering Councils, Universities Administrations and Accreditation bodies, to consider IT as a requirement to be formally considered in study programmes.

Moreover, Universities, that usually pay attention to the administrative aspects, should also encourage its academic evaluation and provide methods and tools for the assessment of students engaged in practical stages. In this respect the consideration of the academic part of work training requires, as analyzed previously, a proper recognition of the academic tutors and the adequate reward of the time and efforts they have devoted to this task, as it is done with their other academic duties. This point is considered to have the highest importance to increase the quality both in IT activities and in the Engineering education.

As mentioned above, IT is not receiving in new Spanish curricula a treatment according to its relevance in engineering education. In our opinion, several things can be proposed in order to overcome this weakness of new Spanish curricula, and the incorporation of IT to Engineering study programmes on a compulsory basis. First, a proper discussion on work placements should be carried out among those responsible for engineering study programmes. They also should evaluate benefit/cost rates and the possible positive impact that it could have on those rates. After that, and if the influence of practical training appears positive, they should work for a consensus for including these activities in all curricula for Engineering graduates. That could lead to including IT in the study programmes already approved and those on the way to be approved. Placements should be an optional subject in all Engineering degrees with an academic recognition of at least 12 ECT.

Once this move has been accomplished and after reviewing its impact on the competences of new Engineering graduates, our proposal goes a step forward. That will be the modification of curricula to include in them IT as a compulsory subject. This will put these curricula on a more professional track as it happens, for instance, in Medicine, Nursery, and Education among other Spanish First Cycle Engineering Degrees. If these steps are given, Spanish Engineering curricula will be on line with existing First Cycle degrees all over the world with reference to practical training in industries.

REFERENCES

Ahrens, C. (2000). European Civil/ Construction Engineering Management (ECEM/ ECM) as an example for integrated European engineering curriculum. *Proceedings of the International Conference of SEFI*. Paris.

ANECA. (2006). *Criterios e indicadores de calidad para la autoevaluación de las unidades de empleo que gestionan las universidades*. Agencia Nacional Para la Evaluación de la Calidad. Retrieved February 01, 2010, from http://www.aneca.es/media/ 158226/estu_oil_fase3b_ criteriosindicadores.pdf

ANECA. (2007). *Reflex. Informe ejecutivo. El profesional flexible en la sociedad del conocimiento*. Agencia Nacional Para la Evaluación de la Calidad. Retrieved February 01, 2010, from http://www.aneca.es/media/ 158162/informeejecutivoaneca_jornadasreflexv20.pdf

Cámara, J. M., Sánchez, P. L., & Represa, C. (2009). Las prácticas en empresa en el marco del EEES. *Proceedings of the XVII Congreso Universitario de Innovación Educativa en las Enseñanzas Técnicas*. Valencia.

Domínguez, U. (2002). Spanish engineering curricula. Moving towards the European higher education space? *Proceedings of 30th SEFI Annual Conference,* Florence.

Domínguez, U., & Magdaleno, J. (2003). Industrial training in education. Towards an international approach in engineering curricula. *Proceedings of the International Conference of SEFI*. Porto.

Domínguez, U., Martin, M. A., García, J. M., & Magdaleno, J. (2006). Engineering education in Spain. The teacher's point of view on innovation. In U. Domínguez (Ed.), *Innovative teaching and learning in engineering education* (pp. 219-236). Valladolid, Spain: Escuela Universitaria Politécnica. El- Sayed, J. (2001). Industrial- academic integration takes learning out of the classroom. *Proceedings of the SEFI Annual Conference*. Copenhagen

EPC. (2000). *The EPC engineering graduate output standard*. Interim Report of the EPC Output Standards Project. The Engineering Professor's Council. Warwick. Retrieved February 24, 2010, from http://www.epc.ac.uk/publications / standards/index.php

Greve, D., Schonk, P., & van der Aa, P. (2009). Matching, monitoring and assessing learning outcomes of students in practical assignments. *Proceedings of 37th SEFI Annual Conference,* Rotterdam.

Heitmann, G. (2003). Employability through innovative curricula. In Borri, C., & Maffili, F. (Eds.), *E4 thematic network* (*Vol. C*). Firenze, Italy: Firenze University Press.

Huerta, M. M., Portela, J. M., Díaz, J. E., & Pastor, A. (2009). Seguimiento de los alumnos en las Prácticas de Empresa. *Proceedings of the XVII Congreso Universitario de Innovación Educativa en las Enseñanzas Técnicas*. Valencia.

Keleher, P., Moxham, L., & Shakespeare, P. (2007). Work integrated learning: A comparative analysis of an Australian and UK experience. In *Proceedings 15th World Conference on Cooperative Education, Transcending Boundaries: Work-Integrated Learning for the Global Economy*. Suntec International Convention & Exhibition Centre. Singapore.

M. E. (2009). *Datos y cifras del sistema universitario Español*. Curso 2009/2010. Ministerio de Educación. Secretaría General Técnica. Retrieved February 24, 2010, from http://www.educacion.es/educacion/universidades/ estadisticas-informes/datos-generales.html

Magdaleno, J., & Domínguez, U. (2006). Industrial training in engineering curricula. Proposals for spanish degrees. In Domínguez, U. (Ed.), *Innovative teaching and learning in engineering education* (pp. 145–153). Valladolid, Spain: Escuela Universitaria Politécnica.

Magdaleno, J., Domínguez, U., & González, J. (2002). La integración obligatoria de las prácticas en empresas en los planes de estudio de ingeniería. *Proceedings of the X Congreso Universitario de Innovación Educativa en las Enseñanzas Técnicas*. Valencia.

Maki, P. (2002). *Using multiple assessment methods to explore student learning and development inside and outside of the classroom*. NASPA's Net Results. Retrieved January 10, 2010, from http://virtual2.yosemite.cc.ca.us /mjcinstruction/CAI/Summer2005Institute/ TrainingMaterialsUsed/NASPA's%20NetResults%20Using %20Multiple%20Assessment% 20Methods.doc

Monyarch, J. (2005). La adquisición de competencias de los ingenieros a través de las prácticas tuteladas a la empresa. *Proceedings of the XIII Congreso Universitario de Innovación Educativa en las Enseñanzas Técnicas*. Maspalomas.

Order. (2009a). *Orden CIN/351/2009, de 9 de febrero, por la que se establecen los requisitos para la verificación de los títulos universitarios oficiales que habiliten para el ejercicio de la profesión de ingeniero técnico industrial*. Retrieved February 24, 2010, from http://www.boe.es/boe/dias/ 2009/02/20/pdfs/ BOE-A-2009-2893.pdf

Order. (2009b). *Orden CIN/311/2009, de 9 de febrero, por la que se establecen los requisitos para la verificación de los títulos universitarios oficiales que habiliten para el ejercicio de la profesión de ingeniero industrial*. Retrieved February 24, 2010, from http://www.boe.es/boe/dias/ 2009/02/18/pdfs/ BOE-A-2009-2740.pdf

Orrel, J. (2004). Work-integrated learning programmes: Management and educational quality. *Proceedings of the Australian Universities Quality Forum 2004*. AUQA Occasional Publication. Retrieved January 10, 2010, from http://www.auqa.org.au/ auqf/pastfora/2004/program/ papers/Orrell.pdf

Osman, S. A., et al. (2008). The importance of industrial training: Students' perception in civil engineering sector. *Proceedings of the 7th WSEAS Conf. on Education and Educational Technology*. Retrieved January 10, 2010, from http://pkukmweb.ukm.my/ ~upak/paper/importance.pdf

Patkó, G., Szentirmai, L., & Váradi, A. (2009). Adventures in engineering wonderland. *Proceedings of 37th SEFI Annual Conference*, Rotterdam.

R. D. (1987). *Real Decreto 1497/1987, por el que se establecen directrices generales comunes de los planes de estudio de los títulos universitarios con carácter oficial y validez en todo el territorio nacional*. Retrieved February 24, 2010, from http://www.boe.es/aeboe/ consultas/bases_datos/doc.php ?id=BOE-A-1987-27707

R. D. (2003). *Real Decreto 1125/2003, por el que se establece el sistema europeo de créditos y el sistema de calificaciones en las titulaciones universitarias de carácter oficial y validez en todo el territorio nacional.* Retrieved February 24, 2010, from http://www.boe.es/aeboe/ consultas/ bases_datos/ doc.php?id=BOE- A-2003-17643

R. D. (2007). *Real Decreto 1393/2007, por el que se establece la ordenación de las enseñanzas universitarias oficiales.* Retrieved February 24, 2010, from http://www.boe.es/aeboe/ consultas/ bases_datos/ doc.php?id=BOE -A-2007-18770

REACU. (2009). *Recomendaciones para el seguimiento de los títulos oficiales.* Red Española de Agencias de Calidad Universitaria. Retrieved February 24, 2010, from http://aplicaciones. acsucyl.com/ acsucyl/export/system/modules/ org.opencms.module.acucyl/ elements/galleries/ galeria_descargas /Seguimiento_titulaciones _REACU.pdf

Roelofsen, W. R. (2009). Engineering education: More fun with action learning. *Proceedings of 37th SEFI Annual Conference,* Rotterdam.

Schenk-Brill, D., & Boots, P. (2001). Innovative engineering. Experiences with industrial coeducation. *Proceedings of the SEFI Annual Conference.* Copenhagen.

University. (2001). *Studying in e-espace and other challenges for learning. Learning by working. University and Industry co-operation as a resource for learning.* E4 - Activity 5 & Best Symposium. Helsinki University of Technology. Helsinki.

Welters, A. T. J., & van de Wetering, H. T. J. (2009). From competence to knowledge based competence. *Proceedings of 37th SEFI Annual Conference.* Rotterdam.

Zabalza, M. A. (2000). El prácticum y las prácticas en empresas en la formación universitaria. *VIII Congreso Universitario de Innovación Educativa en las Enseñanzas Técnicas.* Santiago de Compostela.

Chapter 5
Work-Integrated Learning for Engineers in Coordination with Industries

Walter Nuninger
University of Lille, France

Jean Marie Châtelet
University of Lille, France

ABSTRACT

This chapter presents the curriculum and training process applied in the IESP department, standing for "Ingénieur d'Exploitation des Systèmes de Production" of the Ecole Polytechnique Universitaire de Lille (Polytech'Lille, F59 Villeneuve d'Ascq, France), with an emphasis upon the WIL training process for Exploitation Engineers of Production Systems. The IESP department is dedicated to lifelong learning and apprenticeship leading to a Master's degree in Engineering (French Engineer level) in the production field. It is an accredited program. This practice relies upon a background experience of 18 years with close partnerships with industry from many sectors, such as; energy, metallurgy, food industry, automotive, chemical engineering, and aeronautics. The graduates from Continuing Vocational Education and Training (CVT) that are already employed in the company improve their position. Younger graduates from Initial Vocational education and Training (IVT) that also validated their abilities are much more employable. They can work in any industrial sector dealing with engineering production having a strong technical and managerial skill base. In this chapter, the historical setting up and evolution is elaborated in the legal French education framework. The training model based on the IESP professional profile is presented. This model presents an academic curriculum with WBL that integrates a real formative work situation in the industry. The tools and methods developed all along the training process are also focused within a sustainable development policy. Finally, success and difficulties or challenges with mobility due to the globalization of the economy and innovation with respect to the economical crisis are also discussed.

DOI: 10.4018/978-1-60960-547-6.ch005

Figure 1. scheme of path to French diploma of higher education with European equivalent (LMD)

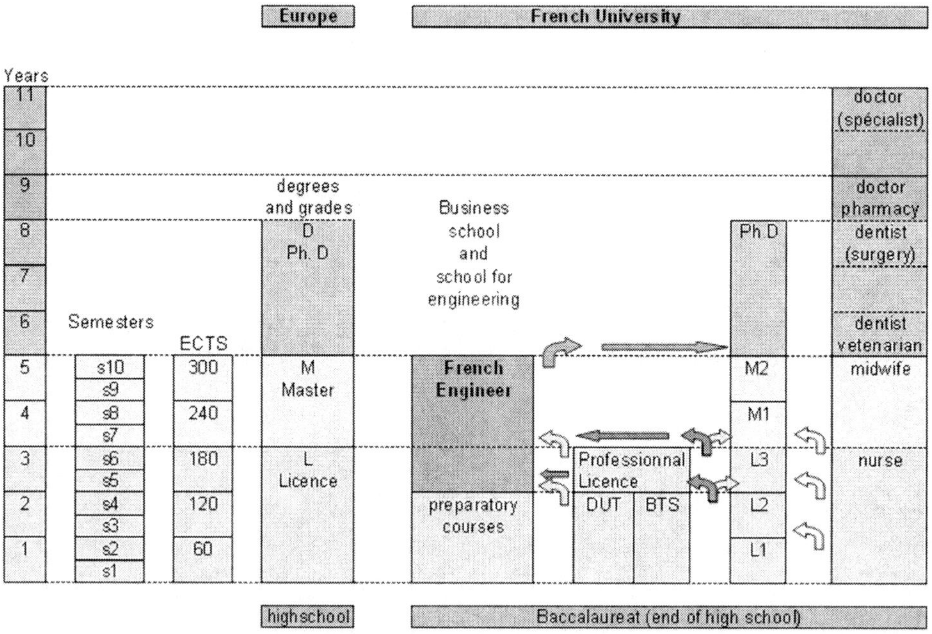

INTRODUCTION

To help readers to understand the French Education framework, it is important to explain the French higher education system and the peculiarity of the French engineer diploma (master degree program of 5 years after high school, standing for chartered engineers). Since the Bologna process (1999), the specificity of French education system was adapted to European qualification through the Master degree. The modes of access are summarized in Figure 1 with the equivalent European grades. For someone to be employed in France it is made easier if the Baccalaureat degree is passed (it is not an obligation but this level of qualification validates studies during high school that gives the basics (language, mathematics, history)). Therefore one enters university (higher education) to prepare a technical certificate in two years or a grade L. With such a bachelor degree, graduates can start their professional activities. Note that in France universities are public establishments where training fees are mostly reduced. Private institutions exist with more expensive fees for students. Several state social aids limit social discrimination. A master's degree can be obtained by further studies at university. In France, it is a requirement to undertake a minimum training period within industry that is work experience mostly dealing with practice in a real environment. The duration depends on the training but forms part of the curriculum. Figure 1, highlights the comparison between the French education system and the European systems, which is License (Bachelor), Master (Engineer), Doctorate (Ph.D) or LMD. Gateways exist between university trainings and engineering trainings. The European Credit Transfer and Accumulation System (ECTS) apply and take into account the academic workload by semester.

At the university, courses and training are conducted by teachers and researchers ("enseignant-chercheurs") mostly hired with a PhD qualification. They work within a laboratory environment engaged in research. Some of them are first chartered engineers especially when teaching in school of engineering like Polytech'Lille.

Table 1. Network, Polytech'Lille and IESP department figures in 2009.

	Polytech Network	Polytech'Lille	IESP Department	ITII Network
Birth	Creation 2000	Since 1969 (EUDIL), 1985 (IAAL), 1992 (IESP), in merged 2000	Since 1992. In partnership with ITII N-PdC since 1984	Since 1992
Structure (engineer training of higher education)	12 schools (engineer) 160 trainings	8 departments, 2 training in apprenticeship.	2 vocational training, CVT since 1992 IVT since 2004	23 training centres in France
Field	70 specialty	8	Production	50 fields
Hired engineer	48000	6056 (upon 7874 graduate)	150 (CVT) 17 IVT (IVT)	17000
Students enrolled	12000 (1000 Ph.D)	1200 (132 Ph.D)	55 CVT, 99 IVT	7000
Graduates a year	2900	350	15 CVT, 26 IVT	1700
Laboratories	80	17	See Polytech'Lille	
Other figures		120 students(preparatory) 160 trainers (searchers)	228 graduates by CVT since 1992	Link with professional unions

Courses at the university and school engineering are closely aligned to research initiatives to facilitate a better transfer of knowledge and innovation. In 2006, French law for research planning (French Government, 2006) created the French Assessment Agency for Research and Higher Education (AERES) establishing a more inter twined relationship between training and research. It's an independent administrative authority evaluating higher education institutions and organizations within a global perspective.

History of the IESP Department

The school Polytech'Lille is an Engineering College of the University of Lille (North of France). It is part of the national Polytech network composed by 12 French Engineering schools covering five main domains (mechanics, computer science, electricity and electronics, civil engineering and food processing and biotechnology) within 50 scientific and technical fields taking advantages of research context. The academic syllabus lasts 5 years (2 years after high school + 3 years) to get the French engineering diploma (Master degree as show in Figure 1). It is accredited by the Engineer Title Commission (CTI). Since 1992, our IESP department (Ingénieur d'Exploitation des Systèmes de Production) has overseen the delivery of a degree in engineering (specialty production) through Initial Vocational education and Training (IVT) and Continuing Vocational education and Training (CVT). Since 1994, the diploma is delivered in partnership with the Institut des Techniques d'Ingénieur de l'Industrie of North-Pas-de-Calais (ITII N-PdC, Institute of techniques of the engineer of the industry). Table 1 outlines the IESP Department of the school Polytech'Lille of the University of Lille and its networks.

The history of the Department and School is summarized in Figure 2. The department was created in 1992 as a response to the demand of industries. The educational project is based on a partnership with 7 industrial groups; these are; BSN, EDF-GDF, Michelin, Pechiney, Renault, Rhone Poulenc, Snecma and 2 professional unions, namely; Metallurgy (UIMM) and Chemical union (UIC). At that time, France was lacking process engineers with leadership skills. Indeed young graduates preferred to peruse careers in the areas of finance, research and development, methods and marketing sectors instead of process engineering. The aim was therefore to promote technicians with skills known in the company thanks to an

Figure 2. History of Polytech'Lille and IESP Department. Training is initiated in 1992 as an engineering school further merged with two engineer school of classic education to form Polytech'Lille of University of Lille1 (Wikipedia, 2010)

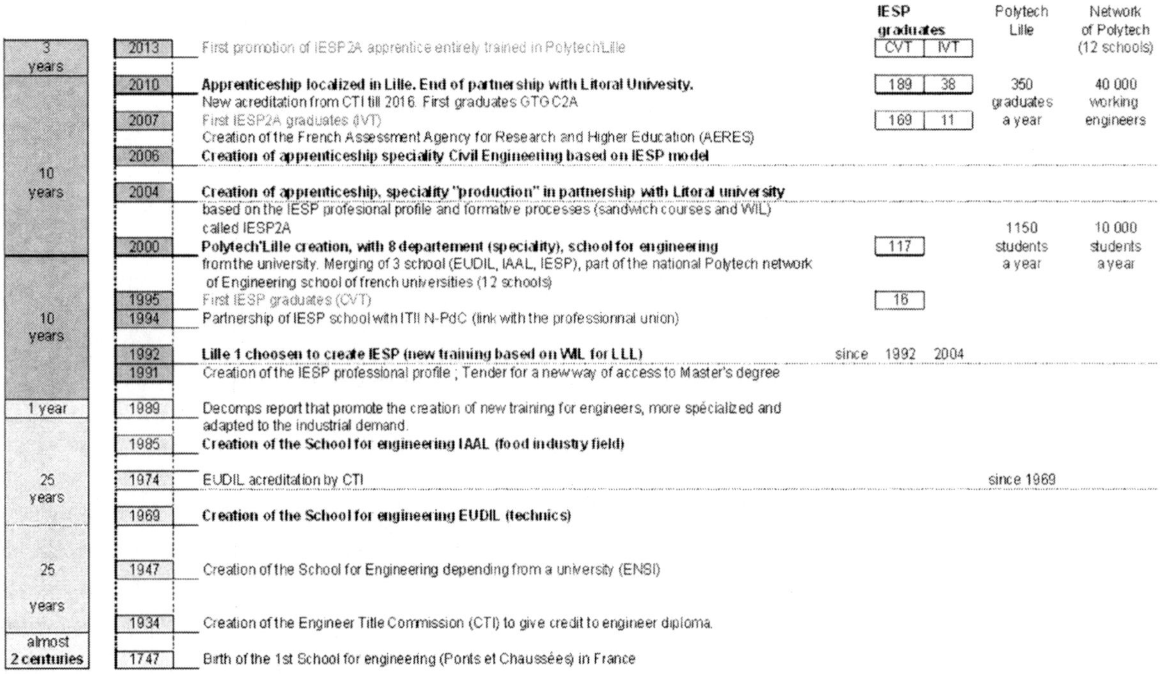

accredited diploma using a different path for learning and training, inductive approach and work integrated experience.

The CVT program was initially created for employees with professional experience of duration longer than 5 years but in fact this has exhibited 15 years industry experience as a mean value. In 2004, to help the recruitment process of our partners, an Initial Vocational Training (IVT) program was offered for young adults who more than likely did not have industry experience; typically less than 26 years old. Both training culminated in a unique model based on a professional profile of competencies for process engineers (repository of capabilities, so called "referentiel IESP") and a university syllabus with an academic curriculum and a work integrated experience (specifically designed for skill acquisition; therefore such work experience go further than just an internship period in the industry). Consequently, university curriculum includes an academic component conducted at university and an industrial component (sandwich course). The frequency of sandwich courses is designed to optimize transport costs between the company's location and the university, satisfy the industrial partners' constraints (e.g. participating in company life) and the length of academic curriculum duration (apprenticeship), in order to fulfill the ITII Charter. This approach gives priority to the development of the learner's skills. Our work based learning program is focused upon the professional profile defined by industrial partners. It is conducted as an academic program that focuses on inductive pedagogy, a work placement training period, with a shared responsibility by the university and the organization for tutoring and assessment of the learner's skill proficiency. Our strong university-industry partnership arises from our decision to establish a board of manage-

Figure 3. General design of the training model (based on formative work situation) both for IVT and VCT. A link between two interdependent realities that complement and cooperate

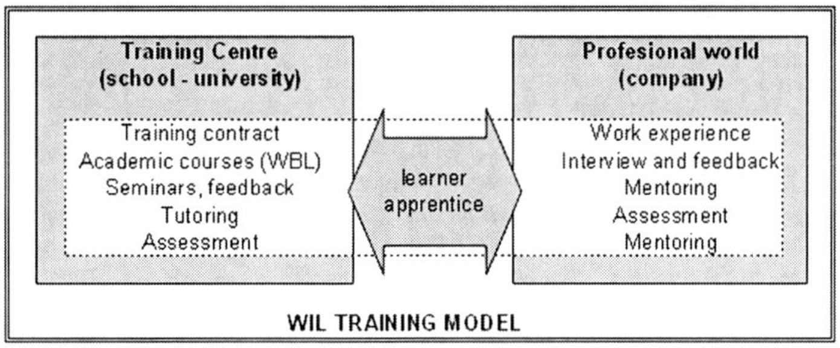

ment (with both industrial and university membership) to oversee the curriculum and because of the shared university/organization mentoring processes for work placement and the assessment of learner skills.

WORK INTEGRATED MODEL AND PROGRAM

General View of the Training Model

The Work Integrated Learning program conducted by IESP demonstrates a balance between an academic curriculum and a work integrated experience in an industrial setting; sandwich courses. Figure 3 shows each institutions role in the training process: two realities that contrast and complement, inseparable and interdependent.

Two types of student groups are targeted.

- A group with professional experience/qualifications enrolled in Continuous Vocation Training (further called IESP FC). These students have a strong knowledge and experience of professional life but with a little experience and knowledge of scientific processes and methodologies; we call them learners ("apprenants")
- The second group is composed of young adults fresh graduated as a technician or with a technical certificate (2 years after High school diploma) and who can enroll in the Initial Vocational Training program (apprenticeship). These students called apprentice ("apprentis"), typically, have a good scientific background but has a lack of knowledge and experience of a professional life.

In addition, IESP accepts so called 'atypical' candidates, including apprentices with some work experience or Bachelors degree. Consequently some individualization occurs in study plans. In CVT, the first semester of the training is dedicated to the acquisition and reactivation of learning mechanisms whereas in IVT, the focus is placed upon gaining knowledge of the industry environment.

Accreditation and Quality Charter

The core of the training is the IESP professional profile that establishes the required abilities of the production systems engineer. Any alterations to the curriculum are undertaken to meet our industrial partners' demands and needs. Through this mechanism the appropriateness, currency and relevance of the training is compared to the professional reality. Quality is also monitored and improved to achieve consistency with the ITII engineer charter (ITII, 2009); whose main requirements are given

Table 2. Training requirements from the charter (ITII, 2009) and CTI accreditation (CTI, 2010)

	Training close to the industry	Training with sandwich courses	Value management
ITII charter	- strong partnership with industry - professional profile - project management ability - sciences and technology skills - equilibrium between courses for sciences, engineer tools and professional background - both validation of academic courses and work experience in the company - focus on knowledge, know how, soft skills and ability acquisition through work experience	- sandwich courses - inductive pedagogy - problem based learning - teachers and tutors should taking into account the industrial reality - duration of training (1800 h for IVT, 1200 h for CVT) - formative situation in the company - work contract for learners - contract for apprenticeship	- individualization of training with respect to learner history and evolution - cooperative assessment of ability - assessment of the training process by the learners - participation of the training centre (school) to the managing groups and improving discussion group of partners (ITII)
	Goals of CTI	**Main requirements for CTI**	**Accreditation indicators**
CTI Audits	- development of quality in engineering education - promotion of engineering curricula and careers in France and abroad - adaptation of engineering training to globalization of economy and to the ever-growing international higher education area	- implementation of effective internal quality assurance - competence-based engineering education - international visibility - satisfy the European standards	- graduate flow and employability - match the needs of the industry field of activity - management courses - English level (independent and proficient user) - grade of trainer - industrial participant (stakeholders) - partnership - learners mobility - research and innovation - training assessment - organization (evolution, curriculum improvement)

in Table 2. Therefore there is a regular assessment of the training and benchmarking of the curriculum and where necessary updates are undertaken in order to respond to industry requirements. The institution and management practices are also reviewed. Indeed, like any accredited diploma of engineering (master level) by the Engineer Title Commission (CTI, 2010), periodical audits are conducted. The IESP program has been fully accredited since its creation and was recently renewed (2010) for 6 years which demonstrates its established quality and continued relevance (see Table 2). Audits consider all aspects of the program; curriculum design, learners' employability, training realization, school management and organization, link with research and industrial partnership, international actions and development. This assessment mechanism guides the development of the process, respecting each of the participant's goals and provides a framework with for mutual enrichment; a requirement for this kind of training in France (Crequis et al., 2010).

Role of the IESP Engineer, Production Field

About 98% of IESP graduates work as managers in industry, in large international groups or small and medium enterprises; there is a wide range of opportunities offered in any sectors (see Figure 4).

IESP Model in Details

Table 3 presents the main aspects of the curriculum whereas Table 4 gives the topics of the academic courses with duration. Training last 6 semesters for apprentices and 5 semesters for CVT as the 1st semester stands for professional qualification processed during the recruitment of the learner.

Figure 4. distribution of the company sectors where our learners and apprentices are employed

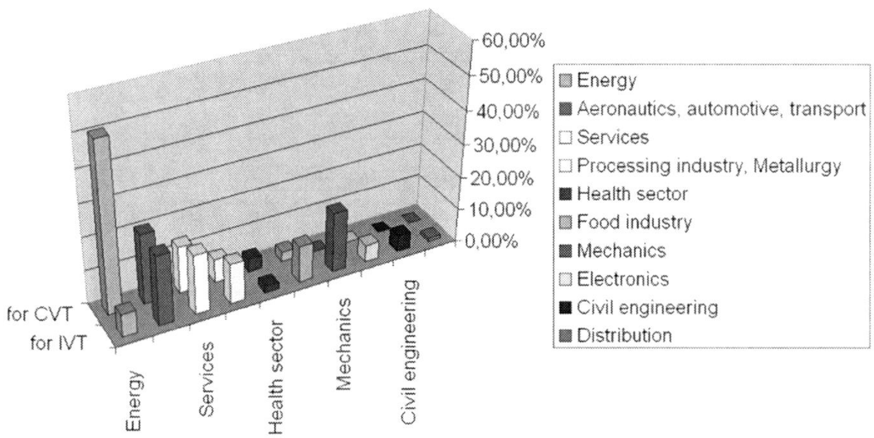

Table 3. Unique Quality Training Design (IESP, 2010)

Training	Lifelong learning (CVT)	Vocational learning (IVT)
Core of IESP training	IESP Professional profile (production field, techniques and human ability) in partnership with professional union - Sandwich courses – Inductive pedagogy - Work experience based on a formative situation – Dual tutoring (school tutor and company mentor) – Relocated seminars	
Grade required	BAC+2 (Grade L 2) as a minimum	
Grade reached	Chartered Engineer (BAC+5, Master's degree in Engineering) from Polytech'Lille	
Student	- average age of 35 year - employees (technicians, supervisors…) - professional experience longer than 3 years with personal evolution potential detected by the company	- young adults (less than 26 years' old) - no professional experience required - motivation detected within interview at school and in the company for recruitment
Legal condition	Work contract by the company (salary) Training convention with the department	Work contract for apprenticeship with the company (salary). Transport is refunded by Formasup (Crequis,2009)
Training fees	Training fees are paid by companies through the professional training plan (legal obligation in France), human resource policy or financial aids from professional union and state. Learner may have a personal participation.	Free registration fees for the apprentice. Training financing comes from legal tax for apprenticeship paid by the companies and regional aids.
Duration	Professional Qualification + 5 semesters	6 semesters
Frequency	1 week per month at school (average)	1 week at school for 3 weeks in the company
At school	1200 h (academic curriculum)	1 800 h (academic curriculum)
In the company	1800 h (formative situation)	3 600 h (formative situation)
Personal workload	2100 h (estimation based on polls)	900 h (estimation based on polls)

The curriculum is implemented through vocational training and sandwich courses. The frequency of the alternation between the two is based upon the company's needs and transport constraints. Our students operate with local and national industries (70%) within France. A one week regrouping for formal lessons reduces training costs (distance between training centre and company, easier scheduling). It is a truly co-operative educational approach between the IESP team and

Table 4. IESP curriculum for CVT (above) and IVT (below) (IESP, 2010)

Continuous Vocational Education and Training (CVT) **Learners (employees) under training agreement**		
	Academic	**Company**
Validation of the previous professional experience		
Basics Sciences Mathematics, Physics, Statistics, Thermodynamics, Chemistry, Mechanics, Materials	204 h	
Techniques of specialty Methods and tools for solving problems, Quality and value management, maintenance management, logistics operating, production management, data analysis, Industrial data processing,,	260 h	
Methods and tools for engineering Team building, professional and personal development, project management, IT Tools, data processing, Automatic, Work sociology	261 h	
Engineer professional background English, Work sociology (Knowledge of the business), Economy and Management, Writing Skills and Public Talking, Environment/Safety, Economy and Management, Law, Operations research and simulation	391 h	
Project in the company Mini-project	16 h	300 h
Formative situation (work experience) Mission of engineering level	**68 h**	**1500 h**
Initial Vocational Education and Training (IVT) **Apprentices**		
	Academic	**Company**
Techniques of specialty Methods and tools for engineering, project management	360 h	
Methods and tools for engineering Quality Environment, Workshop steering, OGP, Supply chain, experimental design, ergonomics, maintenance, Economy, industrial innovation, production system design, security	386 h	
Engineer professional background English, Inventory Management, Communication, Marketing, Negotiation, Management Control, Sociology of Organizations (work sociology), Accounting, Economics, Law, HSQE, Sustainable Development, and Human Resource Management.	674 h	
Project in the company Experimental and Reasoned Knowledge of Professional life	17 h	1200 h
Formative situation (work experience) Mission of engineering level	103 h	2400 h
	1800 h	**3600 h**
Including conferences, industrial visits and 3 relocated seminars on the following topics: and personal development; language (abroad); Health, Safety, Quality & Environmental (HSQE) and management		

the companies. The purpose is to reinforce what the learner is learning (prior, experiential, formal, non formal learning) and also to formalize and assess the knowledge, skills and competencies they acquire within the workplace. Its strength is to learn at, by and for work through the training and formal learning situations linked to industry outcomes (ability and competences, sector knowledge and organizational culture, adaptability and employability of learners and apprentices). The program covers technical knowledge, professional background and engineering project methodology in a proportion of 33% each incorporating work experience and language skill development (English).

Figure 5. Kolb learning cycle (Kolb, 1984)

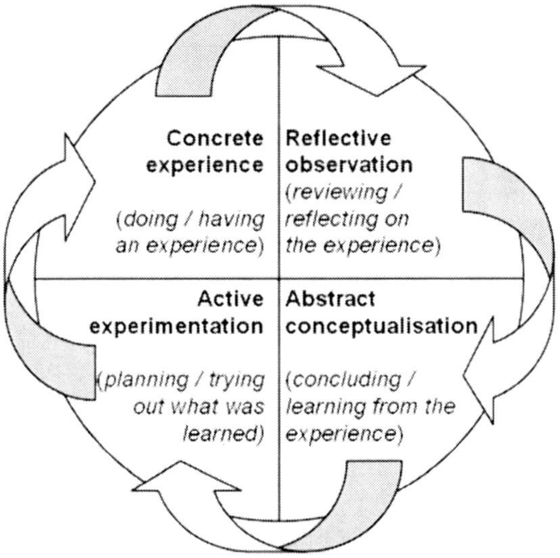

Figure 6. WBL triangle and wining-trio virtuous circle of relations

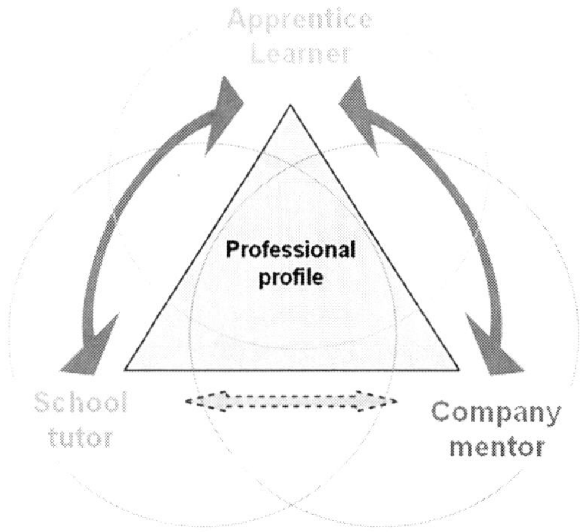

Academic knowledge is evaluated by the stakeholders (lecturers) whereas work based experiences are jointly evaluated by the industry tutor and school tutor. Benefits of work based learning are dependent upon the interactions and discourse between the Higher Education community and professional sector. Training, trainees and trainers are close to the industry (run in partnership with it, work in it or work with it). Teaching and training methods are adapted to the specific learner group, Inductive pedagogy and Problem based pedagogy are used (starting with a real life example, upon which theory is built and solutions realized). Such methods are better ways than deductive pedagogy (first learning the theory and then applying it) for a quick and effective acquisition of knowledge and skill (capacity acquisition) (Michel, 2005). Nevertheless, they are more complex to perform, including the forwards and outwards experimental learning process (Kolb, 1984) shown in Figure 5. It helps the learner to acquire knowledge and abilities leading to skill development. Errors are accepted and corrected thanks to a reflective process. Lecturers facilitate the groups of learners helping to choose and define the study topics or problems. In so doing, make it explicit and clarify the context, express the purpose, analyze and model, plan and define assessment criteria, learn and solve problems, provide a positive criticism of student's approach, present their results and proofs and transfer their experience to others. This is a formative process with assessments of competency acquisition, with not only the learner but also the lecturer, be allowed to make errors. This is not possible with summative evaluation as in inductive pedagogy.

Our Work Integrated Learning program develops learning in, at and for work. Competency development is of interest to the learner as a personal goal, to the company in developing the organization's skill base (individual and collective), and also the university in its educational mission (Demol, 2002); as shown in Figure 10. The IESP path of work integrated learning with assessment is given in Figure 7.

Examples of work situations integrated with the IESP training are outlined in Table 5. Company goals (outcome, meeting deadline) and knowledge/skill acquisitions are jointly validated by company and university. This is a philosophy central to the curriculum and it is a cooperative decision, a moral and written contract with in-

Figure 7. training and assessment steps during work formative situation

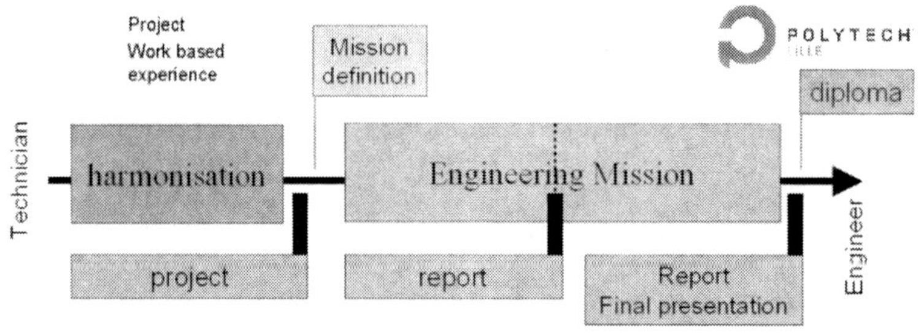

Table 5. examples of formative work situations integrated to IESP training

Field	Company	Task
Metallurgy	Aluminium Pechiney Aluval	Implementation of TPM Foundry
Energy	Bugey Nuclear Plant (EDF)	Implementation of conditional maintenance alternators
Energy	Gravelines Nuclear Plant (EDF)	Project High Ground Control define a new job
Energy	EDF GEH Garonne	Quality organization design: ISO 9001, ISO 14001
Energy	EDF LONS LE SAUNIER	Inspection and maintenance of hydroelectric
Energy	EDF-GDF Corse	Controlling Explosive Atmospheres in thermal Corsican
Transport	KEOLIS Métro de Rennes	Consolidation (update of the design an planning) of work cycle
Automotive	RENAULT Sandouville	Forward and build new production lines
Automotive	RENAULT Technical centre	Define and deploy the logistics to the international
Transport	SNCF Tech. Centre Hellemmes	Prepare the workshop production of axles competition
Aeronautics	SNECMA Services	Physical flows optimization (incoming/outgoing) for all activities
Manufacturer	SNECMA Engines	Industrialization of a new computer for combat aircraft
Aeronautics	THALES	Industrial Project Airbus-Thales
Manufacturer	TOSHIBA Neuville les Dieppe	E.O.L. end of life - reverse logistics
Food industry	DUYCK Brewery	Optimization of production (+30%)
Food industry	METAROM	Optimization of production
Manufacturer	DECATHLON	Opening a new subcontractor in foreign country (Italy)
Transport	ALSTOM	Transport Renovation of the subway of the Caire (Egypt)
Telecom	ALCATEL Cables	Industrialization of a new submarine cable

volvement of each participant (university, industry, and student) based on the "wining-trio" virtuous circle (see Figure 10). Ultimately, the learner is required to demonstrate his understanding of the task as an integrated part of the action plan and general objectives of the company stressing the contribution of his actions. The engineering aspects are highlighted by the learner through his demonstrated understandings and skills from the profile (see Table 6 and Figure 9). The success of the task is expressed in terms of achieving expected learning outcomes (measurable and surveyed), percentage of time dedicated to the task (completed or partial completion) and assigned resources (financial and human). Finally at the end of the task, the learner will show his proficiency of competencies through achievement of skill acquisition and organizational objectives.

Table 6. IESP professional profile

Field of application		Nature of results	
Product Process Installation Human Organization (structure) Environment		**Quality**	satisfaction of the customer (respect for the conditions of contract), the staff (conditions of employment), The French legislation (rules, law)
		Safety	safety of possessions, people, Information
		Profitability	increase in capital, Financial, technical, human, knowledge, competence...
		Reactivity, feed-back	in connection with time cadence schedule, lead time, time of answer...

Table 7. Balanced score card principles to allow performance measurement of the company in a near future

TARGETS		MEASUREMENTS		GOAL		PROSPECT	
STRATEGY	setting values and delays	EQUILIBRIUM	using indicators	SYSTEMIC APPROACH	Profitability Satisfaction Continuous Improvement Personal involvement Ethics Work Welfare	DEPLOYMENT	Finance Customers Process Development Sustained Development Safety-Security

Consequently, he will be considered an engineer with developed leadership skills (see Table 14). Their behavior can be characterized by the 7 steps of implementation of Balanced Score Cards (BSC) consistent with the core of the curriculum focusing on the task, strategies and the organizational needs as given in Table 7. Work Based Learning needs to take into account requirements for satisfactory outcomes for the learner, the industry and university. As a consequence indicators and targets are chosen to evaluate actions for each task (see Figure 8). Corrective measures are considered and implemented as they arise.

The professional profile is linked to the expected competencies of an Operating Engineer of Production Systems who is required to work in an ever-changing landscape, dealing with problem/project, problem/project and where necessary actions to guarantee the on-going success of the organization. Proof of this engineer attitude is given using the set; Quality, Reactivity and Feedback, Profitability and Safety (see Table 6). In addition, the engineer's responsibility covers the application fields of the professional profile; Process, Installation, Human, Organization and Environment. As a production manager and an engineer, they have to use both skills (improvement, development, innovation) and skills (energizing, anticipation and decision ability). Our professional profile was developed with the input from working groups including heads of enterprises, professionals, unions and university experts. Our curriculum design was based on this professional profile and the pedagogy was adapted considering the co-operative education constraints: 'Forming Engineer in a different way, thanks to learning in a different way'.

Mentoring in the Company and Tutoring at School, Cooperative Actions in the Wining Trio

Learner knowledge and skill acquisition in this Work Integrated Learning approach is facilitated by tutoring and mentoring by university and industry participants (see Figures 6 and 8). The role of mentors and tutors mirrors the 10 com-

Figure 8. functional links for WIL training process (design of task, cooperative assessment) (Chatelet et al., 2006)

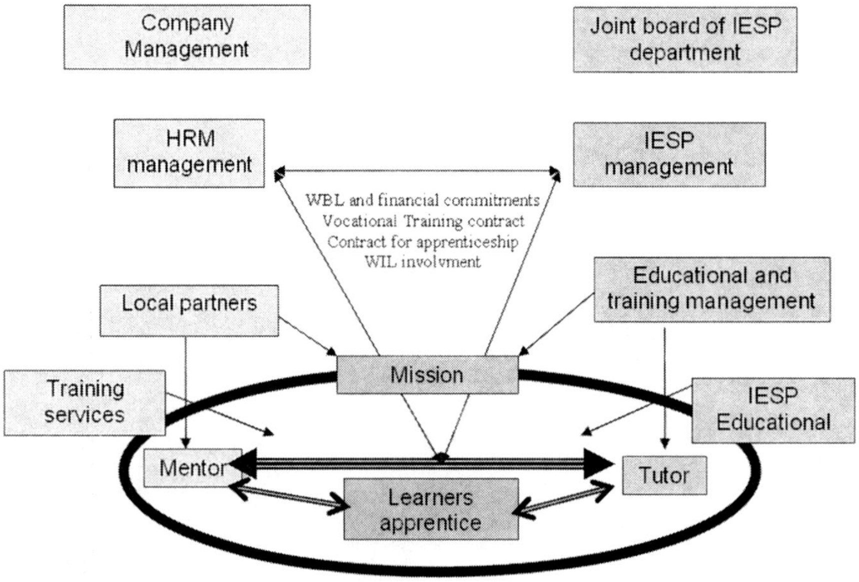

mandments of Bova et al. (1981) and are adopted by the department (see Table 8). Tutors are given specific training within the training centre. The aim being to help the definition of the task and the identification of the role of each one. In doing so, it is not to resolve all of the issues for the learner, but instead to coach and guide them on their path of personal evolution, self analysis, behaviors and acquisition of skills and abilities. Coaching should be efficient with the position of the learner in the company adapted to the objective(s). Thanks to the Work based learning triangle (see Figure 6) that is consistent with the sustainable virtuous circle of participants (Figure 10). In so doing emphasizing the development of good practice for technical, human and economic understandings. Tutors are members of the teaching staff with recognized competency in the life long learning Program of partnership. They work together in guiding (Houssaye, 2000) to help the learners to achieve their goal. Mentoring is the establishment of a personal relationship of recognized value for the purpose of professional

Figure 9. virtuous circle of "ability of IESP engineer"

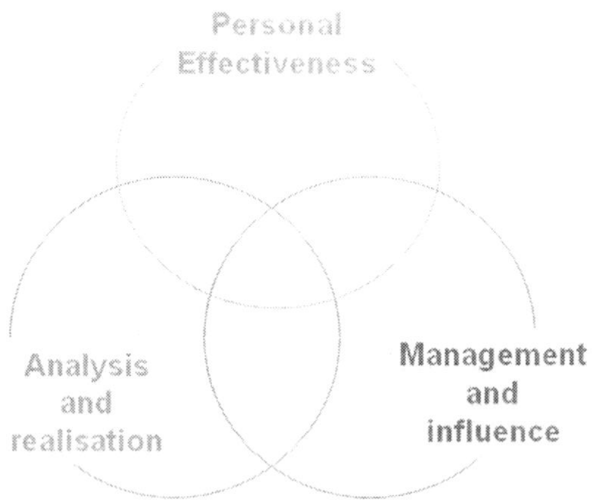

instruction and guidance mentor-protégé couple (Evenson, 1982). The tutors will assist the learner in acquiring scientific, technological, economic and social skills and knowledge. Tutors listen, encourage, give advice but do not provide solutions. The aim being to help the learner to discover his new identity as an engineer (Paul, 2004). Tutors

Figure 10. "ability transfer" virtuous circle (left) at the origin of the training process policy of development; "IESP curriculum" virtuous circle (right) leading to a sustainable personal development and value transfer in a competitive and globalize world. With respect to the sustainable virtuous circle dealing with the overlay of the 3 circles "Social", "Ecology", "Economy"; it is clear that the wining-trio of processes (recruitment, assessment, partnership) and wining-trio of actors (learner, tutor, mentor) overlay the sustainable disk (respect of both social and economy framework): the enhancement of the aim is therefore "sustainable"

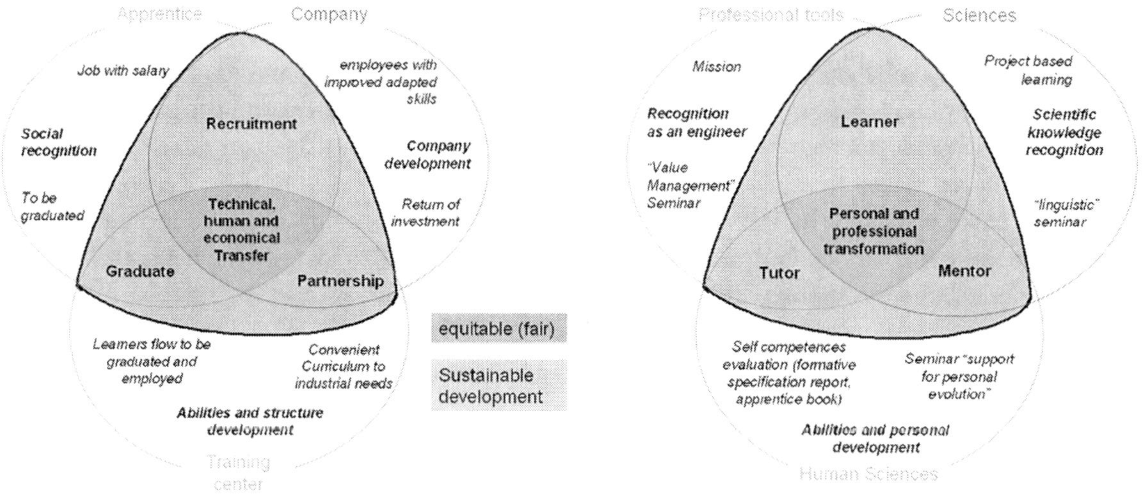

Table 8. The 10 commandments of mentor role (Bova and Phillips, 1981) and used by IESP

	One-to-one relationship: mentor-protected			Winning-trio: mentor-protected-tutor
1	Mentor-protected relationships grow out of voluntary interaction.		1	Voluntary participation
2	The Mentor-protected relationship has a life cycle: introduction; mutual trust-building; teaching of risk-taking, communication, and professional skills; transfer of professional standards; and dissolution.		2	Formalize the relationship between the academic tutor, the entreprise-tutor, the learner (the wining trio)
3	People become mentors to pass down information to the next generation.		3	Explain the different issues attempted and help to characterize the Work situation. Training journey.
4	Mentors encourage protégés in setting and attaining short- and long-term goals.		4	Interview and follow-up with one visit in the company as a minimum per year
5	Mentors guide technically and professionally. Mentors teach protégés skills necessary to survive daily experiences and promote career-scope professional development.		5	Mentor brings methodology and problem solving tools. Tutor brings specifics tools and gives access to professionals sources of information
6	Mentors protect protégés from major mistakes by limiting their exposure to responsibility.		6	Mentor and tutors set up together the most efficiency way to improve the learner competences
7	Mentors provide opportunities for protégés to observe and participate in their work.		7	Help company-tutor to establish an engagement letter in consistency with the needs of enterprise and school.
8	Mentors are role models.		8	Set up the trio **mentor-protected-tutor**
9	Mentors sponsor protected organizationally and professionally.		9	Cooperative assessment is made by mentor and tutor
10	Mentor-protected relationships end, amiably or bitterly.		10	Best practices are collected in the referee for tutors.

Table 9. The IESP Engineer roles as a leader

Interpersonal role	Informational role	Decision making role
Symbolic Leader Leader Liaison Officer (contact with the outside)	Active Observer Internal broadcaster of information Spokesman (outside the company)	Entrepreneur Regulator (Environmental disrupted) Dispatcher Resources Negotiator

should be familiar with operational tools; reference guidelines (professional profile, tutor's book of good practice), collaborative distance work platform and Learning Management Systems (i.e. Moodle), reference documents (for engagement letter, framing report and final report), problem and project based pedagogy, tools for evaluation (assessment grids for report, oral presentation, competencies), apprentice booklet (that include writing, feedback, ability evaluation, e-portfolio). Mentors gain the satisfaction of being able to transfer skills and knowledge accumulated through extensive professional practice (California State Board of Education, 1983; Krupp, 1984). The school administration provides an introduction to the rules but the mentor teaches the skills necessary to comply and cope with them (Driscoll et al., 1985). The mentor provides the learner with opportunities to develop professional competence through a cycle of observation / assessment / practice / assessment. This permits continuous communication and constant feedback to the learner. Classroom skills develop under the mentor's constant and consistent assistance. The mentor also guides the protégé though the maze of local and state administration systems which potentially influence the practices of the classroom teacher. Finally, the mentor directs the learner to professional organizations for academic and professional development. Difference between 'assessment' and 'evaluation' is clearly defined to avoid confusion (Griffin, 1984). However, as beginning lecturers they need to expand their techniques, improve their teaching skills and learn classroom management (Huffman and Leak, 1986), At the IESP it is designated that a lecturer could become a mentor after two complete cycle of tutoring; at least after 6 years. The first three years are undertaken in conjunction with a mentor who facilitates their learning to gain the necessary expertise as a new lecturer (Califonia State Department of Education, 1983; Krupp, 1984). Note that IESP has set up a referee for tutors (referee tutorat IESP) to identify the best practice linked with the IESP spirit.

Cooperative Assessment

Assessment is based on knowledge and skill acquisition and proficiency. Boolean notation ("validated" and "not validated"; stands for an insufficient level of ability) is adopted for the continuous vocational training learners (CVT) no matter what the kind of course is undertaken, including the work experience evaluation. The same system is applied for apprenticeships with respect to the work experience but academic achievement is designated by a result of a total of 20 points. In each case, both courses (academic) and work experience (company) has to be validated to pass the grade of master. An 'independent user' level in English expression/language should also be passed to graduate. Only technical competences are needed for managers, IESP engineers will have at least ten roles to play as a leader (Mintzberg, 1973) partitioned into 3 groups, namely; interpersonal, informational and decisional roles as outlined in Table 9 and Figure 9, explaining balance score card use.

For any manager or employee in the company, the activities are closely relied to 3 ability groups (Katz, 1974); personal effectiveness, analysis and

Table 10. Example of IESP general definition for "insufficient" level; applied for courses and seminars and used in the pedagogic booklet describing the courses (lessons).

Item	Contents
Courses	Duration, kind of course and training, contents of curriculum with knowledge, skills, resources
Goal	Description of the expected results (form, nature, delay) with form of assessments (conditions)
Principles of assessment definition	With respect to the goal, the trainee (or apprentice), despite the guidelines during training, the contribution of the group and complementary actions (individual or collective) and / or accompanying measures imposed at the additional prior jury did not meet expectations in the following topics of abilities, particularly in the sense that he / she:
Personal Effectiveness	- Do not meet its commitments - Has no active attitude to the difficulties - Does not show a minimum of autonomy - Do not get involved over time (active participation, investment in time...) - Has not acquired the basic skills - Not at all mobilized resources and tools available for lessons or self - Does not show down from his knowledge and against the background
Analysis and realization	- Has not been able to achieve the goal (achievement required) even with assistance and successive cropping - Failed to improve results - Has not acquired the learning process - Unable to organize work to achieve the results expected individual or collective - Does not detect mistakes - Do not rest on its laurels for making proposals or implement methods - Not arguing - Do not follow the reasoning because it focuses on details without a global vision
Management and influence	- Not an appropriate behavior (compliance with operating rules, usage and etiquette...) - Assumes no responsibility with negative consequences on the environment (group, team...) - Does not take a decision and does not explain their choices - Is not in tune with its environment (closed position) - Still stood its ground without making any effort to collect the merits of other points of view argued - Do not cooperate and do not draw benefits of teamwork (sharing of best practices, expertise, knowledge, skills, life skills...)

realization, management and influence. Personal effectiveness abilities are linked to the motivation and personal commitment to work. This includes personal organization of activities, establishing priorities and adapting resources such as time and information. Analysis relies on their problem solving abilities, operational conduction involving situation analysis, decision making and management. Their managerial abilities need to take into account how they involve their work environment, subordinates, peers and superiors and other stakeholders. The assessment criteria of IESP training fits theses concepts thanks to a considered definition of expected goals of courses and work integrated situations (see Table 10). It also relates to ability level definitions (see Tables 11 and Figure 9). All validation of academic courses for CVT relies on the general definition of each ability levels. This definition is a description of the expected goals within the 3 groups of ability. It is further clarified for each course (also for seminar, for instance the positive action in the group) to urge assessment to be rational (based on facts) even if subjective decision-making might remain (tutoring and individualization allow a better knowledge of learners). The assessment grid is available to facilitate evaluation (see Table 12). The definition of these levels is of course not exhaustive but will allow lecturing staff to assess, as objectively as possible, the engineering trainee with regard to the chosen pedagogy. The seminars are also evaluated with common rules in addition to the validation condition of the linked courses. Validation is obtained as soon as level 0 (insufficient) is overcome by the learner (level 1 to 4). If this is not the case, remedial activities are

Table 11. Assessment means at the IESP Department

Assessment	Methods used
Knowledge acquisition	based on PBL and tutoring during courses (discussion), tests and final examination
Rating	based on professional results with respect to industrial demands
Knowledge acquisition	Based on tutoring during courses (discussion) and examination
Know how acquisition	Assessment during the work experience in the company and during the PWL activities
Skill evaluation	Confrontation upon the virtuous circle of ability to acquire: personal, analytical and working, management and leadership

Table 12. Elements to be evaluated jointly with the company to asset the formative work situation

Capacity				Ability level				
Personal Effectiveness	**Analysis and realization**	**Management and influence**		0	1	2	3	4
Self-control	Analytical Thinking	Active listening						
Self confidence	Conceptualization - ability to synthesize	Customer Focus						
Personal Commitment	Curiosity - research information	Impact and Influence		0	insufficient			
Adaptability	Rigor	Political acumen and network		1	basic			
Teamwork	Initiative - Anticipation	Developing Others, transfer		2	fair			
Cultural adaptability	Innovation	Leadership		3	autonomous			
Transversality	Achievement			4	expert			

proposed by the pedagogic team to ensure success to further assessment and decided during final semester tribunals. Such assessment consists of new examinations, additional and adapted homework and new specific tutoring or lessons. Personal enrichment activities can also be asked based on identified improvement points: goal is individual evolution. In any case, the learner should pass the final evaluation and should meet their 'commitment'. Otherwise the student is deemed unsatisfactory and training is ceased, in many cases, resulting in consequences for his professional evolution in the company and completing tasks. The learner is identified as not being able to achieve outcomes for a successful passing grade at this time. Ideally the training process should, as soon as possible, identify the possibility that the learner may fail and respond accordingly and appropriately to correct the situation (remedial). The same issues may be identified at company level and that is why it is paramount that there is a close working relationship between the university and industry. Further, as a consequence it is extremely important for IESP credibility and integrity that the IESP assessment system correlates with the professional standards of competencies and their evaluation. Cooperative assessment is based on questioning, challenging perspectives and defending arguments and facts. However ultimately the guiding principle for professionalism, professional integrity and credibility remains: Is the trainee an engineer, with respect to the IESP professional profile, who is able to be employed as an engineer-leader or be recommended to your professional network of industry partners or clients?

Sustainable Development Policy to Develop our Model

In many part of the economy, there is an assumption that a complex system of determinants will

Table 13. Main steps of our work integrated model of training

	Step	Tools	Goal
1	Recruitment	Interview with candidate, Visit to the company, explanation about training process and formative requirement	Detection of potential Ensuring the involvement of the company in the process and candidate project
2	Mentoring	Tutors training, participation to final assessment jury, learner interview during the process, one visit per year in the company (CVT) and one per semester (ICT)	Explanation of each one role (coach), understanding of IESP professional profile, ensure involvement of mentor for guiding and cooperative evaluation, help the mission definition and further identification of improving points of learner, feedback, benchmarking
32	Definition of formative mission	Formative situation with both company results and ability acquisition goal with respect to the IESP professional profile	Ensure ability acquisition with right position in the company within work based experience and personal evolution, a way to improve personal and professional evolution
3	Academic curriculum	Inductive pedagogy, WBL, English courses and seminars, conferences	Adapted to the public with real needs, benchmarking, and ability to analyze problems, independent user of English, knowledge of other industrial sectors.
5	Feedback	Proofreading experience, e-portfolio, apprentice booklet, interviews with tutors	Reflexive attitude, behavior evolution, personal evolution, transfer of ability, network
6	Cooperative assessment	Ability assessment through interview and questions, courses assessments	Proof of ability acquisition (human & technical fields), both ability and knowledge evaluation

tend to a state of equilibrium. When a system lacks this tendency we use terms like virtuous circle and vicious circle (or vice versa) to describe these unstable patterns of events. Spending investment resources to achieve the goal can be seen as a virtuous circle when effort is spent in actions that can increase the learner, mentor (company) and tutor (school) satisfaction creating a corporate culture. Result will also create network loyalty and improves feedbacks and strong partnership for the benefit of all; initiating another iteration of a virtuous cycle. IESP department and training based development follows this policy of practice. Tools and training process stick the definition as shown in Figure 10. A virtual circle respecting each actors and sustainable development within a target that remain technical and human transfer of ability and values; ability defined in the professional profile. The point of consistency of each actor goal is obviously part of the sustainable development: satisfying it has an impact on social (recognition) and economic (existence) fields. Transfer of ability allows the sustainable goal. Table 10 makes the link between theses aspects and the IESP professional profile. Respecting these concepts is a tool for managing such training process and leads to quality. The IESP department therefore considers the centre of the virtuous circles as it manages the training process in a dynamic way to enhance the transfer, increase the competences. For instance relocated seminars (group cohesion), positive thinking during action and assessments to identify the improvement points (improve the self-knowledge of the learner, of the training, of the company) lead to the same performance (see Figure 10).

Main step of this sustainable process are given in Table 13. First the recruitment process which guarantees that all actors have the same "vocabulary" and pending even if goal might differs. Candidate backgrounds (grade, professional and academic history) and motivation, but also the involvement of his company with respect to the learner personal and professional project are validated (support and finance of the project of his actual (CVT) or future (IVT) employee). For any demand specific contact or visit to the company are made by members of the IESP department to present the training process, inform about the training requirement and mostly the

Table 14. Flow of student for IESP

Training (date of creation)	CVT (1992)	IVT (2004)
Groups	20 learners	26 apprentices
Rate of recruitment (accepted upon received demands)	50%	36%
Graduates (2010)	187 since 1995 (first graduate)	41 since 2007
Number of women	1 per group (not constant, 2 in 2010)	13%

ones of the formative situation of engineer level (that includes both company goals and ability acquisition goals to be validated; efficient role of mentoring and tutoring; right position of the learner in the company). Note that some of our partners (e.g. Snecma, Renault) identify in their payroll "individual potential" for whom the company designs a peculiar and complete "work experience path" (3 to 7 years) to prepare them to enter our Work Integrated Training. In case of EDF (energy field) a unique Vocation training framework was designed called proc@dre leading to grade L2 further associated with a specific employment leading to Master grade through the training ; it ends in 2012 but a new framework for technician promotion exists: Polytech'Lille-IESP is one of the 10 national selected trainings by the company. The department informs and supports the individual demands of candidates by explaining financing and company opportunity to improve its intern-skill within the IESP training and guiding company to the support services. 75% of trainings are financed by Vocational training company plan, 25% by FONGECIF (closer to individual demands but with support of the company). This rate now increases due to new French law for social modernization in 2001. The update of the social modernization act in 2010 in France will lead to main changes in the funding for training of employees with consequence upon the strategic workforce planning (GPEC). Finally job seeker can also be enrolled (1 by group in general) within reducing fees (in charge of the department depending on annual resources, state and region facilities for such public). Training of the company mentor and school tutors complete the process while the mission is designed to achieve company results and formatives ones. Respect of inductive pedagogy and project based learning during academic courses is the right process. It requires times and efficient planning of schedule. Including relocated seminars improve group cohesion and **experience transfer** through proofreading experience. Finally, the cooperative assessment is a way to alert the learner all along the process when there is a risk of failing, to help him identify the improvement axes and to urge the personal evolution thank to questioning, exchange, feedback, proofreading experience. Future engineers should develop the ability to manage industrial project in a multicultural environment. Our network of partners and graduates maintain development and consistency of training.

Feedback on learners and apprentices at the IESP

Table 14 illustrates the students flow at the IESP department. Our learners are already employees with professional experiences (Long Life Learning) or young adults entering the professional system (apprenticeship). Curriculum is different for the two groups that follow distinct courses however pedagogy and work experience are based on the same requirements. Due to the technical field (production) learners and apprentices are mostly men. Nevertheless we do have women following the training and act to improve the rate of women which seems easier for apprenticeship. With respect to CVT, rate is 1 woman by group in general whereas the rate is 13% for IVT. Public come from social promotion and they come back to school for a diploma or they enrolled apprenticeship for financing reasons (to have a salary). Apprentices generally choose this kind of training for two reasons: it is another way to learn taking advantage of the financing and work experience it implies. IVT is a way for social promotion; a reason that explains a more regional recruitment of apprentices (30% in 2009) with groups from different social classes, mostly lower social class as they did not have the same access to the university of engineering college for financial reasons. With respect to CVT, work experience has brought out the need for training to enforce the capacities: final competencies recognition through salary requires diploma. Historically, the training is a reply to industrial demands coming from national (and now international) companies. As a consequence, recruitment is mostly national (70% in 2009) and keeps on that way however we act to develop regional enrolment. Until 2008, CVT fees where directly paid by the companies but the number of personal and mostly FONGECIF financing has jumped (indirect financing by the company through professional unions) and will keep on that way due to new French law this year with the creation of the joint funds of career security: the routes of financing simply changed.

We do take into account social actions with respect to jobseekers (one by group for CVT) with financing solutions such as; decrease of fees, financing by the department, regional aids, etc. which is a hard task: indeed work experience requirement are the same so the learner has to find a company for a 3 year training period which is not easy. On the other hand it is an opportunity to go back to professional activity. Of course, skill people are required and evolution potential detected. Disabled persons can enrolled depending how they fit the recruitment process (1 per group as an average value but with only small disability (problem with writing, taking notes, dyslexia, sensory)) Tutors are more precisely pointed and technical solutions can be developed (printing lessons, recording of speech, more time during writing examination). Another interesting figure about our students is the rate of companies' size, e.g. SMEs and big firms. History gives the explanation but we try to develop access by employee from SMEs stressing on the information about the training process and formative requirement. They are part of the training process and it's an opportunity to detect their future inner-skills. Difficulty is not so much to inform and promote the training as a real way to improve the skill in the company but to find the financing and right mentor in the company: at first view the goals of each actors (students, industrial, school) might differs. Indeed, the company has a productive logic with time constraint and training process can be considered as a drawbacks and lack of time, but the work experience do make the young efficiency, they also bring new ideas, point of view, positive criticism thanks to the training because the school no more focus on the knowledge acquisition but also on skill acquisition. In the end almost 97% of the learners stay in their company when passing the diploma (100% have a higher position internally; 44% change site in the group, and 70% of jobseeker are employed) and a few start their personal project within firm creation. 100% of graduate apprentices are employed within 6 months seeking job mainly in their company or another one (see Figure 11 and Table 15).

TODAY AND IN A NEAR FUTURE

Mobility

The location of our region makes it historically sensitive to European framework. Our partners have international activities and therefore require engineers that speak English and can easily work and move to participate to international projects. Mobility should then be developed for our learners and apprentices. Many solutions are already

Figure 11. Learners (left) and apprentices (right) distribution with respect to the kind of company. Histogram of apprentice recruitment rate and learners (CVT) professional evolution rate

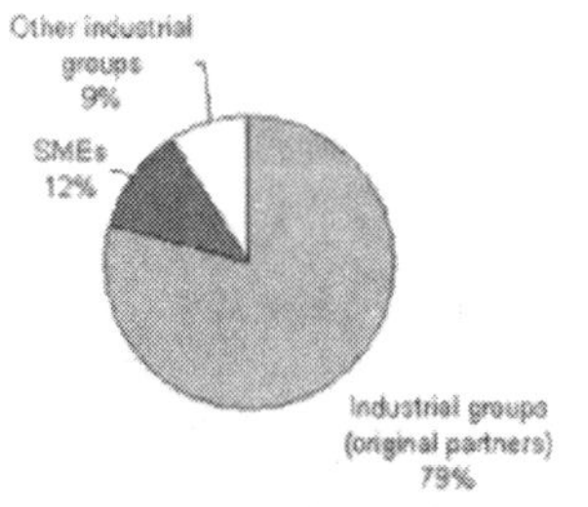

CVT recruitment has always been national and stabilises to 25% of regional

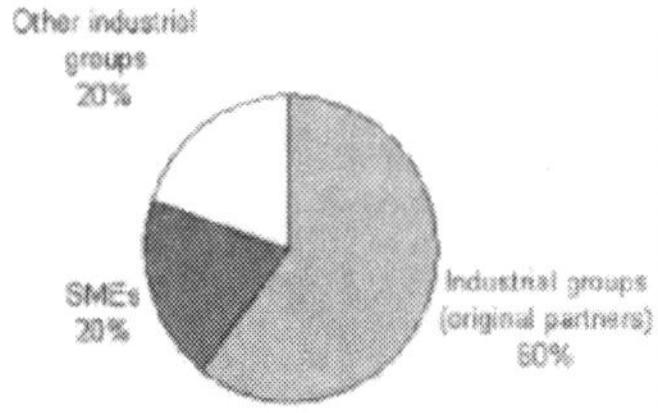

IVT regional recruitment decreased lately to 35%

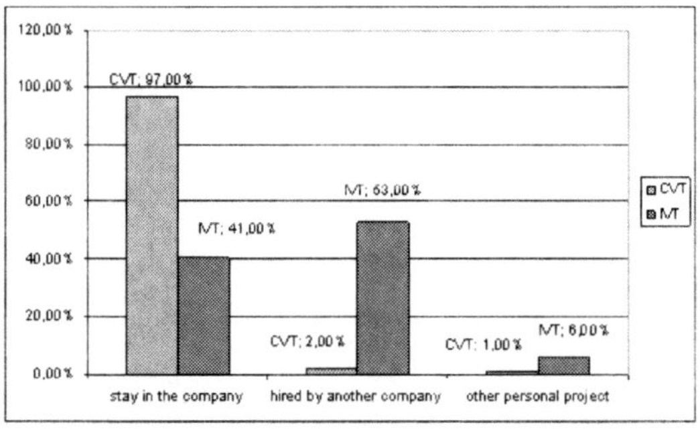

Table 15. Rate of graduates upon enrolled students at the beginning of the training through the years. Note that for apprenticeship, training can be extended if the company accepts to extend the contract; this is not the case in CVT. If English level is not passed, only part of the diploma is validated

Graduate rate upon enrolled	CVT	Apprentice (IVT)
After a 3 years training (normal duration of training)	90-95% 10-5% stop the training	62% 8-12% stop the training
Main reason for failing	English qualification Ability assessment	Mainly ability assessment in the work experience, Validation of academic courses English qualification.
Rules of validation	Assessment failing means stop of the training; except for English.	When failing, apprentice can have an extended contract if the company agrees.
After 1 more year	95-100%	73%
Main reason for failing	English qualification	English qualification
After 2 more year	100%	88%
Main reason for failing	-	Stop of training
Main reason for training stop	Training is stopped when individual (project, potential) is not adapted. In some case due to company bankrupt (economical reason)	
Remarks	Average figures based on 18 years of practices	Average figures based on only 3 promotion (not quite significant to conclude)

implemented thanks to European or regional scholarship in addition of required TOEIC level for an engineer like seminar abroad for culture knowledge and language capacity and training work period abroad. Vocational training can now be proposed to foreigners that could now work and study in France (legally students with work contract), passing a French degree. Further goal is to send our learner for some specific schedules abroad or to combine academic periods in France and industrial periods in European countries. Foreign tutors should be found and trained to our process, a way to reduce cost for mentoring and risk of failing due to distance. But it increases anyway due to sandwich courses (transport, housing, etc) and work located abroad. Foreign professional experience has now been integrated to the definition of the work formative situation in addition to the already existing seminar abroad, courses in English or conference in the curriculum. Written

English supports will also be developed. Mobility is the ability to work in a multicultural environment: IESP profile already includes it through capacity of dynamization (information gathering on the environment, culture, organization…) and for improvement (systems identification, analysis of practices, continuous observation and correction when necessary).

Innovation Awareness

Training is linked to research world but with respect to our goals we use the term innovation. Innovation can be technological improvement (relative of absolute of a new process, a new product; new methods or new organization) but also commercial novelty with new trades/markets or news commercial techniques (confers Ansoff matrix of 2 dimension: markets with respect to products: 4 directions are identified, namely; market penetration, and market opportunities, product development and diversification). Improvements don't fit the definition as it is only a marginal field change without impacting the others. Development impacts all fields but in a marginal manner as it still follows the usual policy. Innovation stands for a radical change in a field. It's a break point that will lead to changes in all fields. Due to the new economic challenges of our industrial partners, it is obvious that learners should be aware of technological survey and to the detection of innovation projects in the company. Innovation can be considered both as research and development projects for big industries and as technology transfer for small companies. Apprentices are therefore a peculiar vector for firm development and success within new and original methods (breaking point). Work based learning is a way to help small companies to detect, develop and implement new methods for optimization of products or production processes to allow higher performance, quality and security. As a consequence, specific tutoring should be available to afford more precise qualification with respect to the field of application (additive academic or conference program but also within tutoring by expert, modification of schedule to get quicker knowledge). Our educational system closed to our partners is a plus to help learner and firms to meet the right partners to finance their projects, conclude contracts and find legal support and additive skill if required. Our vocational training is of interest as the IESP profile defines abilities dealing with the leadership skill that lead to innovation (mission, strategy, and organization). The department is involved in a structure for innovation supported by ITII N-PdC to enforce the links between actors of innovation in SMEs with the apprentice main actor and part of the innovation process with an effective and worthwhile work based experience that will increase skill, competencies and autonomy but also the company intern-skill. In addition benchmarking of learners during academic period will be of interest for our groups of apprentices and the competency of our tutors. Conferences in the training dedicated to the topic are included and of course teacher-researcher inform about their research activities and future development, linking the two worlds. Actually one learner (CVT) and four apprentices are concerned by formative missions closed to laboratories. Works of the department with respect to WIL, assessment and apprenticeship is also part of the innovating process for us.

CONCLUSION

Today out of 350 graduates from Polytech'Lille 15% were work based learners. Our first promotion passed the degree in 1995 and today 228 IESP Engineers work in the industry (41 apprentices). Academic program is made in Lille with trainees from any part of France. Until today we have worked with more than 70 different companies including strong partnership with Renault, EDF, Safran, Rio Tinto and Danone. Work Based Learning education allows the same level of knowledge than any Polytech Engineer who has followed the conventional educational path but adds competencies that can be directly mobilized. As a consequence our engineers do have immediate employability and are well paid. Due to the success of our methodology and choices, the trend is to increase the number of apprentices and learners to prepare the renewal of generation. Work Based Learning is the right solution to transfer and keep individual skills in the company. Work is recognized as a process of learning (Costley, 2000; Armsby, 2006; Dewblam, 2006). For our department and team the strong partnership allows us to keep in touch with the industrial reality and keep on updating our competencies. Development of Lifelong learning process and legal possibilities for the accreditation of competences will help the social modernization. Today formative situation will introduce professional and vocational periods abroad, some courses or conferences in English, including some English written supports. Innovation has always been implicit in the missions. Today we urge our partners to identify it as such and to explicitly incorporate it in the engagement letter. Our action in supporting theses prospects will be participation in structures that favors vocational training and helps to solve financing and reducing cost of training including environmental aspects. If weakness should be stress, let's focus on transport, housing cost due to sandwich courses, tutoring cost for a real individual follow up of trainee and mission with our partners. Some courses like seminars also have a cost inherent to this kind of trainees and already optimized with financing balance. Aim is not for public school to make huge benefit but to keep on the effectiveness of training in an organized structure. Some reduction of expand can be done but mobility and sustainable development give us new challenges and solutions to innovate. Schedule is also difficult with respect to industrial constraints but financing is not only the clue. Motivation, involvement, easy contacts between actors whatever the location is and convenient pedagogical methods and tutoring and tutor training too! Actually the IESP Department is writing down the teacher best practices ("referee enseignant"). Aim is to complete the already existing "referee tutorat" for the best practices of mentoring. The core of our Work Integrated training with our identified consistent virtuous circles is really our strengths; defending the department values we keep on explaining it, trying to develop new ideas with respect to closer assessment with the companies. Feedback surveys allow us to identify problems and propose correctives actions. As a conclusion, we might say that our work based learning program is a good reply to our industrial partners that respects each other's goals within a sustainable training process and department policy. The main reason for such a policy relies in the history of the training but also on the charter we respect including strong requirement as; industrial partnership, consistency of curriculum with company needs, training process dedicated to ability acquisition, equilibrium between Sciences, Engineer tools and professional background including, professional profile with both management and technical skills, cooperative assessments of both academic courses and work experience, adapted inductive pedagogy focusing on WBL, value management of the training and the organization of it. Our model is compliant with the Tuning Dynamic Quality development Circle for Educational Structure in EUrope (Tuning Project, 2007). Within the process, the learner's skills are developed and positive thinking and analyses enhanced. As a consequence employability is improved for apprentices due to a better self knowledge and learners (CVT) can pretend to higher position in the company. The companies are our clients and we do keep on adapting academic curriculum and methods with respect to their needs and our legal constraints linked to the conditions of delivering an accredited diploma of such level taking into account our school structure. International framework is included in a natural manner in our curriculum, adapting the training to the actual issues. Training is open to foreigners (language constraints and state authorizations) and mobility encouraged and urged. Alternation is a huge constraint due to increase cost for transport (tutors) and housing abroad (learners) but solutions can be found within partnership with foreign organizations. Trainings with the same kind of process and mentor requirements, companies for work experience, finance diversification. Innovation (method or technological transfer) is also integrated with support of our school structure based on the research community. So new opportunities for company are opened based on employee hiring with both improved skills and knowledge of company sector and activity. Once again the department value sticks the training process, which is

"Ingénieur autrement, Apprendre autrement !" © (IESP, 2010)

" The IESP, learning in a different way, to become an engineer!" © (IESP, 2010).

A process based on alternation of periods at school and in the company with WBL, designed as a global vocational formative process integrated to the industry goals (WIL), respecting quality (value management) aimed at development, employee hiring enhancement, respecting all stakeholders.

REFERENCES

Armsby, P., Costley, C., & Garnett, J. (2006). The legitimacy of knowledge: A work based learning perspective of APEL. *Journal of Lifelong Education, 25*(4).

Balanced scorecard Institute. (2010). *Official website for BSC*. Retrieved from http://www.balancedscorecard.org /basics/bsc1.html

Boud, D., & Solomon, L. (Eds.). (2001). *Work based learning: A new higher education*. Buckingham, UK: SRHE Press.

Bova, B. M., & Philips, R. E. (1981). *The mentor relationship, a study of mentors and proteges in business and academia*. (ED 208 233)

California State Department of Education. (1983). *Califormia mentor teacher program, program advisory*. (ED 241 473).

Châtelet., et al. (2006). *The social and educational challenge of work based learning in European higher education and training. Results of a pilot experience. Chapter 5: Quality control*, (p. 153), (Ed. Reinhard Schmidt), Firenze.

Châtelet, J.-M., & Nuninger, W. (2011). *Quality Seminar for WIL leading to Value Management and Innovation*. XIIIème Congrès de la Société Française de Génie des Procédés, SFGP 2011.

Costley, C. (2000), The boundaries and frontiers of work based knowledge. In D. Portwood & C. Costley (Eds.), *WBL and the university: New perspectives and practices*. Birmingham, UK: SEDA Paper 109.

Crequis, A., Leroy, N., Nuninger, W., Berthoux, A., & Pauwels, J.-P. (2010). *3 partners, 3 frameworks, 3 goals: The WIL to achieve the success of all*. International Conference on Work Integrated Learning, Wace 2010, Hong Kong, 3 5 February, 5 pages.

CTI. (2010). *Presentation of the commission for accredited diploma*. Retrieved from http://www.cti-commission.fr/ spip.php?page=sommaire-en

IESP Departement. (2009). *English presentation booklet of the IESP department and training*. Polytech'Lille.

Dewblam Project. (2006), *The social and educational challenge of Work based Learning in European higher education and training: results of a pilot experience*. Ed. Reinhard Schmidt, Firenze, December 2006, 352 p.

Direction de l'information légale et administrative (French State informative center). (2005). *Apprenticeship in France, chronology*. Retrieved from http://www.vie-publique.fr/ politiques-publiques/ apprentissage-enseignement -professionnel/ chronologie/

Driscoll, A. (1985). Designing a mentor system for beginning teachers. *Journal of Staff Development, 6*(2).

Evenson, J. S. (1982). *Workplace mentoring, far West laboratory for educational research and development*. (ED 246 182).

Fernandez, A. (2008). *Les nouveaux tableaux de bord pour les managers*. 4ème édition Septembre

French Agency for Research and Higher Education Assessment. (2006). *Official website*. Retrieved from http://www.aeres-evaluation.fr

French Government. (2002). *Loi de modernisation sociale*. (Loi n°2002-73 du 17.01.2002, et décret d'application du 24.04.2002, Version consolidée au 01 juillet 2007).

French Government. (2006). *AERES establishment law*. (n° 2006-450 of April 18, 2006 followed by the decree No. 2006-1334 of November 3, 2006 version 2009, February 6 defining AERES organization and operation).

Funding Agency to advance French research (ANR). (2010). *Official website and ANR presentation*. Retrieved from http://www.agence-nationale -recherche.fr/Intl

Griffin, G. A. (1984). *Crossing the bridge: The first years of teaching*. Paper prepared for the National Commission on Excellence in Teacher Education. (ED 250 292).

Houssaye, J. (2000). *Le triangle pédagogique*. Edition Peter Lang.

Huffman, G., & Leak, S. (1986). Beginning teachers' perceptions of mentors. *Journal of Teacher Education, 37*(1), 22–25. doi:10.1177/002248718603700105

Iribarne, P. (2004). *Balanced Scorecard et Management Qualité - AFNOR*. Retrieved from http:// www.itii.fr/img/ itii_charte_2009.pdf

ITII. (2009), *ITII Engineer charter*, Retrieved from http://www.itii.fr/img/ itii_charte_2009.pdf (on line, french), Institute of Technical Engineering of Industry of North-Pas-de-Calais, created 2006 updated 2009

Kaplan, R. S., & Norton, D. P. (1992). The balanced scorecard: Measures that drive performance. *Harvard Business Review*, (Jan – Feb): 71–80.

Katz, R. L. (1974). Skills of an effective administrator. *Harvard Business Review*, 51.

Kolb, D. A. (1984). *Experimental learning experience as a source of learning and development*. New Jersey: Prentice Hall.

Krupp, J. A. (1984). *Mentor and protege perceptions of mentoring relationships in an elementary and secondary school in Connecticut*. Annual meeting of the American Educational Research Association. (ED 245 004).

Michel, M. (2005). La démarche inductive en pédagogie. *Le Portique, 9*. Retrieved from http:// leportique.revues.org/ document182.html

Mintzberg, H. (1973). *The nature of managerial work*. Harpercollins College Division.

Nuninger, W., & Chatelet, J.-M. (2011). *Quality WIL based model for IVT and CVT*. 17th World Conference on Cooperative Education and Work Integrated Learning, Wace 2011, Philadelphia, 14-17 june, 5 pages.

Nuninger, W., Chatelet, J.-M., & Abt, A. (2009). *Feedback on work based learning for engineers as a reply to industrial demands*. 16th World Conference on Cooperative Education and Work Integrated Learning, Wace 2009, Vancouver, 23-26 june, 5 pages.

Paul, M. (2004). L'accompagnement: Une posture professionnelle spécifique. In L. Schlesinger & J. Heskett (Eds.), Breaking the cycle of failure in services. *Sloan Management Review, 31*, 17-28.

Polytech-Lille. (2010). *Official site*. Retrieved from http://www.polytech-lille.fr/iesp

Tuning Project. (2007). *Tuning general brochure of Tuning Educational Structure in Europe*. European Commission, Socrates & Tempus. Retrieved from http://tuning.unideusto.org/ tuningeu/images/ stories/ template/General_Brochure _final_version.pdf

Chapter 6
Global Impact for your Institution:
International Experiential Education for Technical Students

Thomas M. Akins
Georgia Institute of Technology, USA

Debbie D. Gulick
Georgia Institute of Technology, USA

ABSTRACT

In the United States, students pursuing degrees in technical fields (engineering and computer science) are the smallest population of students who study abroad other than agriculture (only 1% of agriculture students studied abroad in 2009). According to the Institute of International Education's Open Doors Report in 2009, only 3% of engineering students studied abroad. Only 2% of computer science students studied abroad (Institute of Engineering Education, 2009). Currently the Open Doors Report does not account for students doing international experiential education so the only statistics we have are for study abroad. The world is in high demand of engineers and computer scientists, and with this global need, it is imperative that educational institutions focus on producing globally-minded and culturally competent engineers and scientists.

This chapter describes Georgia Institute of Technology's (Georgia Tech) model for producing globally competent engineers. It details two aspects that Georgia Tech thinks are vital to its success: (1) the need for institutional support and resources and (2) making international experiential education a part of an institution's culture.

DOI: 10.4018/978-1-60960-547-6.ch006

INTRODUCTION

Georgia Tech has a longstanding tradition of applying technical knowledge in a practical fashion, of blending theory with practice. The faculty and students are entrepreneurial and the campus culture promotes this mindset. A great part of that culture comes from the 98 year-old cooperative education program, which is the fourth oldest in the United States and the largest totally optional program of its kind. Among Tier-1 research universities in the country, Georgia Tech has the largest group of students formally participating in experiential education programs. Co-op began at the Institute in 1912, six years after it was started at the University of Cincinnati. It was then, and is still now, considered an academic program, reporting to the Provost, or the chief academic officer. In the beginning years, students alternated between classes and employment every two weeks. This was changed after a few years to a monthly rotation and then finally settled into a quarterly schedule after World War II. It was so popular, that the school changed its semester calendar to accommodate the co-ops and everyone was attending Georgia Tech on the quarter calendar. In 1999, however, the University System of Georgia mandated all public institutions of higher learning adopt a semester calendar. Co-op students still alternate between school and work, but on the semester schedule.

In 2002, after 90 years of continual successful operation, it became apparent that changing economics in the U. S. and different attitudes among students and employers were causing many to seek out internships as opposed to committing to a co-op program of work and study. In order to capture these students and make this a formal experiential learning program as part of Georgia Tech's academic offerings, the Cooperative Division reorganized to promote an official undergraduate internship program. These are mostly students entering their third or fourth year of study, who have not previously worked in the co-op program. The office then became known as the Division of Professional Practice (DOPP). The Co-op Program remains the mainstay of DOPP, but the internship program has grown exponentially since 2002. In the spring of 2004, due to Institute reorganization, the Graduate Co-op Program, for Master and PhD students, was moved from the Graduate Office to the DOPP. The Work Abroad Program started in 2005 as DoPP's fourth program, and as a part of the International Plan (IP) initiative, which is outlined in more details in a later section of this chapter.

WORK ABROAD PROGRAM DETAILS

The Work Abroad Program consists of undergraduate and graduate students within any field of study. Georgia Tech has Colleges of Engineering, Sciences, Architecture, Management, Computing, and Liberal Arts. Approximately 60% of undergraduate students are working toward a degree in engineering. The College of Engineering consists of the schools of: aerospace, biomedical, chemical and biomolecular, civil and environmental, electrical and computer, industrial and systems, materials science and engineering, mechanical, and polymer, textile and fiber engineering.

Typically Georgia Tech students work for one semester. All of the Work Abroad opportunities are full-time internships or co-op jobs outside the United States. The majority of the internships are paid. Students can work during the fall, spring, and/or summer semester. Georgia Tech students are rarely given academic credit as internship credit is not helpful to an engineering curriculum. One service that Georgia Tech gives to its students is enrolling them in a full-time audit course while they are working abroad. This service has numerous benefits. This class allows them to maintain full-time status; it is tuition-free, and non-credit bearing for the student. The class appears on the student's transcript as "International Internship."

This is also good for university personnel to be able to see what the student is doing and why they are not enrolled in classes for a semester. This is a unique ability that Georgia Tech has due to its long-standing co-op program. It helps increase international experiential learning opportunities to students because they do not have to pay tuition (cost savings for the student). In addition, it assists in obtaining work permits and visas as it illustrates that the student is not trying to emigrate to the country but simply taking part in an internship as part of his/her studies. This is also a good way to track students who are going abroad.

The Georgia Tech Division of Professional Practice's Work Abroad Program does not utilize an instructor or teacher. The relationship is between the host employer and the Georgia Tech student. The Division of Professional Practice's Work Abroad Program employs two full-time staff. Their roles are to develop internships around the world for students. In addition, they support students in resume reviews, one-on-one career advising, cover letter writing, and various global marketing sessions (for example, German CV Writing). Once a student obtains an international internship or co-op the staff supports the student on:

- Visa and work permit support
- International health insurance, including medical evacuation and repatriation
- One-on-one advising on working abroad
- Cross cultural pre-departure orientation
- Safety abroad orientation
- Peer mentors
- Work Abroad Handbook with many helpful resources
- Written online cross cultural and career assignments while abroad
- Returnee event
- Opportunities to talk to students about their international experience and to become a peer mentor

The Georgia Tech staff member's responsibilities are split regionally. On the Georgia Tech campus, we find that staff can excel in learning to work within a region enabling them to understand the work culture and visa and work permit rules, resulting in better advisement for students.

Peer mentors are a key aspect to the Georgia Tech Division of Professional Practice's Work Abroad Program. The concept of the peer mentor program is to provide support to students getting ready to work abroad. We believe that peer mentors help aid in the understanding of cross-cultural perspectives and ease the anxiety before working abroad. Peer mentors are Georgia Tech students who have worked abroad and after returning from their assignment volunteer to be a peer mentor to students who are getting ready to embark on a Division of Professional Practice's Work Abroad Program in the same country or region. This allows students to ask questions about dress attire, cultural questions, housing advice, etc. to a peer. By doing this, employers are relieved from numerous questions and emails, thus freeing more of their time to prepare for the intern/co-op. It also allows students to learn from their peers, which we find to be an excellent and necessary part of the learning process.

The internships and co-op jobs that students participated in during 2008-2009 vary greatly within the Division of Professional Practice's Work Abroad Program. In 2008-2009, 77% were corporate, 13% were research, 5% were non profit and 5% were government positions (see Figure 12). Due to the history of the co-op program, the Work Abroad program uses the face-to-face practicum approach. Georgia Tech firmly believes that this is the best learning method – a complete hands-on, immersive experience. This has been the approach used for almost 100 years in the cooperative education program and it has proven to be very successful. However, the challenges are having a sufficient budget for global travel for staff and to have appropriate funding to run a program that is highly labor-intensive. Consequently, it is quite fortuitous that many of the students have mentors at their work place. Yet, many others are asked to work almost completely independently. The internships and co-op jobs are mainly represented by the apprentice/mentor model and the face-to-face practicum model. The strength in both of these models is the immersive learning that takes place when one works in a foreign country. Living and working in a foreign country away from your peers and parents forces a person to truly grow as a person. The interpersonal learning that takes place on these internships is immense.

WORK ABROAD PROGRAM STATISTICS, 2008-2009

The Work Abroad Program had 149 work abroad terms by Georgia Tech students during 2008-2009. We sent 132 students abroad to work (some worked multiple semesters thus totaling 149 work terms) in 32 countries from 25 Georgia Tech majors. This was a 29% increase in student participation from the previous year. Program numbers have continued to increase annually since the program's inception. See Figure 6 in Appendix for the top five countries where students from Georgia Tech worked. Germany has remained in the number one spot every year since 2005. France fell from the top five chart this year and Switzerland made its way into the listing. India has been in the top five every year.

The Work Abroad Program works with students from all fields of study at Georgia Tech. Approximately 60% of those students are engineering, which is representative of the student body. All major schools of engineering are represented in the top five (see Figure 7, Appendix).

Seventy five percent of students who went abroad were undergraduates, 24% were graduates and 1% graduated/unknown (see Figure 14).

Full listing of charts in Appendix:

1. Top 5 Countries (Figure 6)
2. Top 5 Majors (Figure 7)
3. Male to Female Ratio (Figure 8)
4. Ethnicities of Students (Figure 9)
5. Countries 2008-2009 All (Figure 10)
6. Major Field of Study (Figure 11)
7. Types of Employers (Figure 12)
8. Year in School (Figure 13)
9. Level in School (Figure 14)
10. International Plan vs. Non-IP Students (Figure 15)
11. IP Students by Country (Figure 16)
12. 2008-2009 Enrollment by Term (Figure 17)
13. Students Working Abroad by Region (Figure 18)

MAKING INTERNATIONAL EXPERIENTIAL EDUCATION PART OF AN INSTITUTION'S CULTURE

In late 2002, Georgia Tech began its early preparation for re-accreditation with the Southern Association of Colleges and Schools (SACS). A leadership team, or council, was formed to begin the process that would culminate in the affirmation of the school's academic endeavors. The Executive Director of DOPP was a member of this council. The SACS accreditation is broken into two parts: (1)Compliance, or the statistical portion (based on factual information) which reports number of students, faculty, research, etc., and (2) a Quality Enhancement Plan (QEP), that demonstrates Georgia Tech's willingness to create new initiatives to provide more opportunities to students in order to expand and enrich their educational experiences. After many, many months of meetings, planning, discussions, and philosophizing, the Institute settled on two programs to be part of this QEP: (1) an undergraduate research option, and (2) the International Plan (IP). It is the latter of these that we focus on at this juncture.

In 2005, the International Plan officially began. The International Plan offers a challenging academic program that develops global competence within the context of each participating student's major. It is a four-year program that builds global competence by requiring students to engage in a minimum of twenty-six weeks of international experience (work, research, or study) related to their discipline, to develop a proficiency in a second language, and to take internationally oriented coursework. This experience provides students a deeper global competency than traditional international opportunities. Degree requirements are not modified but are satisfied with specialized courses and appropriate choices of elective courses, including three globally focused courses within the major area. Approximately 74% of the degree programs available at Georgia Tech participate in the International Plan option, with more majors being added every semester.

Our vision is bold ... Georgia Tech will define the technological research university of the 21st century and educate the leaders of a technologically driven world. (Clough, 2002)

In order to achieve the goal as set forth by the former president of Georgia Tech, Dr. G. Wayne Clough, graduates of the institution must have a sense of global competence, which is outlined in the following section.

Defining Global Competence

Basic global competence is characterized by a graduate's ability to:

- Demonstrate proficiency in a second language;
- Exhibit knowledge of international social, political, and economic systems;
- Assimilate easily into foreign communities and work environments; and
- Communicate with confidence in a global context.

Additional skills sets developed by working abroad are:

1. Flexibility
2. Adaptability
3. Ability to work with a diverse work team
4. Ability to think outside of the box
5. Problem solving skills

The students who elect to go into the International Plan must spend 26 weeks abroad in no more than two terms (in an effort for students to have a more immersive experience). The student can choose to study, research, or work abroad. The students can also do a combination of any two or more of these activities in order to satisfy these requirements. A model that Georgia Tech sees as the most comprehensive and immersive began as the Technical University of Munich (TUM) – Siemens Model. Students originally would spend a semester or more studying at TUM, then work as an intern with Siemens in one of their German facilities. Based on the success of that program, it has been expanded to include several German universities and many various employers. It lasts for approximately 12 months, and operates as follows. Students who elect to go on this year-long program can spend eight weeks on the Languages for Business and Technology (LBAT) Program in Germany. This program is for students who have already completed a German language course at the second year level. This program gives students a chance to improve their German language skills through study in small classes and interaction with German student tutors. After completion of the LBAT Program, students can study abroad at a variety of German universities for one semester, enrolled in German-speaking classes. Students can then work abroad for six months after their classroom studies. By the time they begin working with an employer, their German language skills are excellent, they have adapted to their host culture, and they are able to work well within their host company. This year-long program is an excellent option for students, and really shows to the faculty the value added to a degree program by participating in an international internship.

The IP leadership team found it imperative to create a work abroad program in conjunction with the International Plan. Georgia Tech has a strong culture of applying theory to practice through the long-standing co-op program and the leadership team believed that one of the strongest methods in creating globally competent graduates is through global work experience. Georgia Tech has created this culture through being a pioneering institution in cooperative education in the United States and it supports its co-op students by allowing them to be enrolled as full time students in an audit course while working with no tuition being required. Georgia Tech is also one of the first universities to start a comprehensive Work Abroad Program to support all of its students. One key characteristic in offering this opportunity for students is that the Work Abroad Program is centralized so that all students and faculty know where to go to be involved in this international effort. It also cannot be over-emphasized that any work abroad efforts must be under the auspices of the Provost, or chief academic officer, to give credibility and substance to these experiences. All of these lead to better communication and a more effective program.

This effort can be extremely successful if given the tools needed. Higher level administration must believe in the purpose of the program, and make sure it becomes "institutionalized," or a part of the campus culture. Then, program staff can come to work and do their jobs instead of coming to work justify the existence of the program. It allows staff to work easily with faculty, students, and other campus units to ensure a successful program.

THE NEED FOR INSTITUTIONAL SUPPORT

Financial

Georgia Tech made a formal commitment for a five year period to support financially and organizationally all aspects of the IP, including teaching more sections of language classes, offering a greater variety of languages, supporting global-oriented coursework in participating academic units, and providing assistance in the area of international internships, or as it has come to be known, the Work Abroad Program. A budget was set aside to provide for an International Practicum Coordinator, administrative support, and support for travel as well as other office expenses. (Subsequently, due to the success of the program, and the need for higher visibility on campus and abroad the coordinator overseeing these activities was promoted to Director, Work Abroad Program.)

Due to the International Plan, the Division of Professional Practice was able to hire a staff person to work only on international internships and co-op opportunities. In order for a program to be successful, it needs to be someone's full-time job. Often times, a university will add "international positions" to an individual who already has a full workload. This usually leads to failure of the program and extreme frustration on the part of faculty, staff, and students. Administering a work abroad program is a lot of work, especially in the beginning. Originally the International Plan funded one International Practicum Coordinator and one half time administrative staff person. After a year, Georgia Tech realized this was not adequate. At that time, a Work Abroad Coordinator was hired in lieu of the administrative support staff. This allowed for two people to be able to travel and advise students.

Essentially, an institution needs staff for advisement and job development, resources such as office space and appropriate computer hardware and software, sufficient funds for travel, and very importantly, marketing. The ability to travel must not be overlooked, as a personal presence in foreign countries not only helps communications with employers and partner institutions, but shows a deep commitment on the part of an institution's administration.

Faculty Buy-In

It cannot be over-emphasized that institutional support is crucial to any international experiential education program. Georgia Tech's long standing history and successful co-op program combined with the International Plan give a platform for the Work Abroad Program to succeed. To begin, the Work Abroad Program staff has been able to network through faculty. The International Plan Committee (IPC) is comprised of one faculty member from each participating discipline. Work Abroad staff and the Executive Director of DOPP are part of the IPC as ex-officio members and thus have had the opportunity network with academic faculty. This has proven to be incredibly helpful in job development. The IP started as a committee of faculty and Vice Provosts working together. This type of interaction and development led to a program that the faculty believe in and support. Faculty play a key role in assisting the Work Abroad Program to develop international internships. Due to the very diverse background of the faculty at Georgia Tech many of them have research connections globally. These connections are already invested in Georgia Tech and thus taking an intern is a second prong of involvement that deepens the relationship and can be very rewarding.

Faculty also have a lot of interaction with students (they are the face of Georgia Tech and academics to the students) and thus can recommend individuals to the program, advertise internships, etc. A vital aspect to a successful program is working with the academic faculty. Academics drive the institute, and faculty support along with that of the higher administration, have been keys to Georgia Tech's success.

REVIEW, ASSESSMENT, AND RESEARCH OPPORTUNITIES

The IP, and thus the Work Abroad Program, will come under review formally at the end of the 2009-2010 academic year. The formal review will look at the first five years of the program's existence.

The Division of Professional Practice Work Abroad Program's assessment is tied into Georgia Tech's QEP assessment. In addition it uses the standard Division of Professional Practice assessments. Georgia Tech Quality Enhancement Plan states "Today's graduates live and work in a highly interdependent global community. In keeping with the impact of globalization on business practices and world economies, Georgia Tech seeks to globalize its educational experiences by preparing students for success and leadership roles in business, government, and academic careers with advanced communication and technical skills developed within an international context."

The objective of this initiative is to increase the number of undergraduate students who graduate with global competence in the international practice of their major. Global competence is the product of both international studies and experiences designed to instill a deep and multi-faceted understanding of global relations, intercultural differences, and international disciplinary practices. Among the desired outcomes of the initiative is to achieve Georgia Tech's strategic goal of "..**having 50 percent of its undergraduate students graduate with an international experience by 2010** (Clough 2005)."

In addition, the QEP states that "To facilitate assessment of the initiative's impact, the four student learning outcomes and one program objective are further defined as follows. Graduates exhibit:

- Second Language Proficiency
  - Communicate in a second language. Students choosing to use English during their international residency experience must attain the ACTFL proficiency - "Intermediate-Mid" and those choosing to use their second language during their international experience must attain the ACTFL proficiency of "Intermediate-High."
- Comparative Global Knowledge
  - Demonstrate knowledge about their culture within a global and comparative context;
  - Demonstrate knowledge of global issues, processes, trends, and systems;
  - Demonstrate knowledge of at least one other culture, nation, or region, such as beliefs, values, perspectives, practices, and products.
- Intercultural Assimilation
  - Readily use second language skills and/or knowledge of other cultures to extend their access to information, experiences, and understanding;
  - Convey an appreciation for different cultures in terms of language, art, history, etc.;
  - Interact comfortably with persons in a different cultural environment and be able to seek out further international or intercultural opportunities.
- Global Disciplinary Practice
  - Use cultural frames of reference and alternate perspectives to think critically and solve problems within the discipline in the context of at least one other culture, nation, or region;
  - Collaborate professionally with persons of different cultures and function effectively in multicultural work environments."

Pre and Post Cross Cultural Assessments

In 2005, the International Practicum Coordinator worked with the Director of Assessment and the Director of Study Abroad to develop a pre and post cross cultural assessment. This assessment is designed to address measuring a person's global knowledge, global disciplinary knowledge and ability to assimilate. Many of the learning outcomes are related to assessment criteria used by the Accreditation Board for Engineering and Technology (ABET). The study and work abroad surveys were also designed to be able to compare and contrast study and work abroad outcomes. These surveys focus on abilities to communicate, lead, work in cross functional teams, function in a different culture, and many more other skill sets. It is a self-assessment that the student rate's their ability before their internship and after their internship. Every student that works abroad must complete a pre and post assessment (see Chapter Appendix for forms).

Division of Professional Practice Assessments

All programs in the Division of Professional Practice have the following assessment tools that students must complete during their work term:

1. Assignment record – the student and their supervisor detail the student's tasks and goals for the length of their internship/co-op Supervisor Evaluation – the supervisor completes a performance evaluation on the student and how well tasks and goals were accomplished
2. Work report – the student completes an assessment on the nuts and bolts of their internship, including things like transportation needed, housing etc.

Work Abroad Assignments During the International Internship/Co-Op

The Work Abroad Program also created nine assignments for students to complete while they are abroad. The assignments start the third week of their internship/co-op abroad. This "extra homework" is designed to build from the knowledge learned from the pre-departure cross cultural orientation. They are cross cultural and career development assignments for students to complete while abroad. These often assist advisors in catching any issues with students, employers, miscommunication, etc. The assignments were also designed to expand upon the students' understanding of their cross cultural development while abroad. In addition, students do find notes from this work to be quite helpful in future interviews! These assignments may be viewed at: http://www.workabroad.gatech.edu/While_Abroad_Assignments.php

Intercultural Development Inventory

One of the primary means that Georgia Tech is using for assessing student growth in the outcomes of intercultural assimilation and global disciplinary practice is the Intercultural Development Inventory (IDI). The IDI was developed by Bennett and Hammer as an instrument designed to measure intercultural sensitivity and has been found to be a reliable and reasonably valid measurement tool. The IDI is based on Bennett's Developmental Model of Intercultural Sensitivity (DMIS). The DMIS postulates six stages of increasing sensitivity to cultural difference.

The IDI has been administered as a pre-test measure to incoming Georgia Tech first-year students through a first-year seminar course (in 2005), and through FASET (freshmen orientation) sessions (in 2006-07). IP participants who did not complete the IDI at orientation took the survey at an IP orientation session. To date, a total of 3,781 students completed the instrument. In Academic Year 2009, the Office of Assessment has concentrated on collected post-test data from graduating students. To date, we have administered an online-version of the IDI to 191 students, 12 of whom are students graduating with the IP designator. This sample is too small to permit any deep analysis, however we expect a larger cohort of IP students to complete their programs of study in the next academic year. We also hope to continue funding of this study to follow the last of the pre-test cohort through to graduation.

In addition, starting in 2010 the Institute will send students an exit survey, commencement survey, and later an alumni survey to all IP graduates.

Since the International Plan started in 2005, it is too early to analyze the data as the program started out quite small. In 2010 and beyond, the program will be able to come forth with data on the outcomes of work abroad assignments. Moving forward, all of these statistics and reports will be run annually and analyzed, providing feedback for continual program improvement.

CONCLUSION

Georgia Tech's International Plan and Work Abroad Program were some of the most innovative global programs at a technical university in 2005 and the years following. Admittedly, it is possible to replicate this type of program … but it is a bold step for a "U. S. top ten" public technical institute. The Division of Professional Practice's Work Abroad Program is one of the few successful and campus-pervasive Work Abroad Programs in the United States (especially one that is not for credit).

The program is just now beginning to take on "critical mass" status, and consequently, more evaluative research and assessment can begin. It is the authors' opinions that such objectively acquired data will bear out what is currently anecdotally inferred.

In the very near future, it is critical to determine how to capitalize on Georgia Tech's global footprint. Working with companies globally touches many key aspects of a university: research, alumni, and development. These are all vital parts to the institute. Being able to grow slowly and create a successful global network can greatly strengthen the school, its corporate partners, and alumni. Working globally in 30 plus countries makes it a daunting challenge, but Georgia Tech has made a commitment to carry out this plan.

REFERENCES

Clough, G. W. (2002). *Defining the technological research university of the 21st Century: The strategic plan of Georgia Tech*. Atlanta, GA: Georgia Institute of Technology, Institute Communications & Public Affairs.

Clough, G. W. (2005). *Strengthening the global competence and research experiences of undergraduate students: A quality enhancement plan*. Atlanta, GA: Georgia Institute of Technology, Institute Communications & Public Affairs.

Institute of International Education. (2009). *Open doors web site report on international education exchange, table 22*. Retrieved February 1, 2010, from http://opendoors.iienetwork.org/?p=150836

KEY TERMS AND DEFINITIONS

Co-Op: A form of work integrated learning consisting of alternating terms of school and work, with the work being directly related to the student's chosen field of study.

Experiential Learning: A method of education requiring practical application of theory in the workplace, requiring assessment methods to determine the level knowledge acquisition.

Global: Involving more than two distinct cultures or nations, as opposed to international which requires only two different cultures.

International: Involving two different cultures or nations.

Internship: A work-integrated learning experience. It could be full-time for one term, or a part-time experience. The work is directly related to the student's chosen field of study.

Work Abroad: A work-integrated learning experience that takes place in a nation outside of the student's university country. Can be a co-op or internship.

APPENDIX

Figure 1. Work Abroad Pre-Assessment Page 1

GEORGIA INSTITUTE OF TECHNOLOGY
International Internship Survey (Pre-Internship)
Help others who will "follow in your footsteps"

Student Information

Name: ______________ GTID: ________ Degree Program: ______________ Date: ________

General Knowledge, Abilities and Skills

For each of the following skills, abilities, and attributes: First, please rate how important each of these areas is to you right now. Next, please rate how well prepared you think you are right now in each of these areas.

	Extremely Important	*Very Important*	*Important*	*Somewhat Important*	*Not Important*	*Very Well Prepared*	*Well Prepared*	*Prepared*	*Somewhat Prepared*	*Not Prepared*
1. The ability to:										
a. Communicate orally, informally, and in prepared presentations	O	O	O	O	O	O	O	O	O	O
b. Communicate in writing (e.g., business letters, technical reports)	O	O	O	O	O	O	O	O	O	O
c. Use computing technology in communications	O	O	O	O	O	O	O	O	O	O
d. Use computing technology in discipline-specific analysis and design	O	O	O	O	O	O	O	O	O	O
e. Exercise leadership skills	O	O	O	O	O	O	O	O	O	O
f. Function on multi-disciplinary or cross-functional teams	O	O	O	O	O	O	O	O	O	O
g. Effectively resolve interpersonal conflict within a group or team	O	O	O	O	O	O	O	O	O	O
h. Effectively function in your host country's culture and society	O	O	O	O	O	O	O	O	O	O
i. Design and conduct experiments	O	O	O	O	O	O	O	O	O	O
j. Analyze and interpret data	O	O	O	O	O	O	O	O	O	O
k. Think critically and logically	O	O	O	O	O	O	O	O	O	O
l. Carry out projects independently	O	O	O	O	O	O	O	O	O	O
m. Identify, formulate and solve problems within your discipline	O	O	O	O	O	O	O	O	O	O
n. Design a system, component, or process to meet desired needs and quality	O	O	O	O	O	O	O	O	O	O
o. Synthesize and integrate knowledge across disciplines	O	O	O	O	O	O	O	O	O	O
p. Use techniques, skills and tools necessary for practice in your discipline	O	O	O	O	O	O	O	O	O	O
q. Engage in lifelong learning	O	O	O	O	O	O	O	O	O	O
r. Practice your discipline in different social or cultural settings	O	O	O	O	O	O	O	O	O	O
s. Communicate in your host country's language in a social setting (conversational fluency)	O	O	O	O	O	O	O	O	O	O
t. Communicate in your host country's language in a professional setting (professional / technical fluency)	O	O	O	O	O	O	O	O	O	O
u. Professionally collaborate with persons in your host country's workplace environment	O	O	O	O	O	O	O	O	O	O
v. Work effectively and efficiently in a cross-cultural environment	O	O	O	O	O	O	O	O	O	O
w. Approach problems from different perspectives	O	O	O	O	O	O	O	O	O	O
2. An understanding of:										
a. Product development or design from a business perspective	O	O	O	O	O	O	O	O	O	O
b. Professional and ethical responsibility within your discipline	O	O	O	O	O	O	O	O	O	O
c. Environmental impact of professional practice within your discipline	O	O	O	O	O	O	O	O	O	O
d. The impact your professional practice has on society and culture	O	O	O	O	O	O	O	O	O	O
e. The role of your discipline in solving global problems	O	O	O	O	O	O	O	O	O	O
f. Your host country and their culture(s) beliefs and values within a global and comparative context	O	O	O	O	O	O	O	O	O	O

Figure 2. Work Abroad Pre-Assessment Page 2

3. Answer the following questions in regard to how accurately they describe your current skills and abilities

		Not at all True	*Hardly True*	*Moderately True*	*Exactly True*
a.	I am prepared to enter the workforce after graduation....................	O	O	O	O
b.	I am confident that I will be able to successfully interview and obtain my first job after graduation..	O	O	O	O
c.	If I am in a difficult situation, I can usually think of a solution..............	O	O	O	O
d.	I can always manage to solve difficult problems if I try hard enough....	O	O	O	O
e.	Thanks to my resourcefulness, I know how to handle unforeseen situations..	O	O	O	O
f.	I am prepared to make significant professional contributions within my discipline..	O	O	O	O
g.	I am prepared to contribute to society..............................	O	O	O	O
h.	When I am confronted with a problem, I can usually find several solutions..	O	O	O	O
i.	I can usually handle whatever comes my way............................	O	O	O	O
j.	If someone opposes me, I can find the means and ways to an acceptable solution..	O	O	O	O
k.	I can remain rational when facing difficulties because I can rely on my coping abilities..	O	O	O	O
l.	It is easy for me to stick to my aims and accomplish my goals..............	O	O	O	O
m.	I can solve most problems if I invest the necessary effort..................	O	O	O	O
n.	I am confident that I could deal efficiently with unexpected events.........	O	O	O	O
o.	Right now, I have the skills and experiences necessary to practice professionally in my discipline..	O	O	O	O

Career Plans

Answer the following questions in regard to how accurately they describe your current intentions

4.	Within the next five years...	*Not at all True*	*Hardly True*	*Moderately True*	*Exactly True*
a.	I plan to work in a position related to your field of study	O	O	O	O
b.	I plan to work in a position in a foreign country	O	O	O	O
c.	I intend to participate in a study abroad experience	O	O	O	O
d.	I will pursue/continue to pursue foreign language proficiency	O	O	O	O
e.	I will work in a position with considerable international responsibilities	O	O	O	O
f.	I plan to pursue a graduate or professional degree	O	O	O	O
g.	I will travel abroad for nonacademic or non-work related reasons	O	O	O	O

Advising and Support

Please indicate your satisfaction with each of the following offices in preparing you for your international internship. If you had no experience with a particular area, fill in the circle under "No Opinion."

Quality of information and support from:	*No Opinion*	*Extremely Satisfied*	*Very Satisfied*	*Satisfied*	*Somewhat Satisfied*	*Not Satisfied*
5. Your school/program..............................	O	O	O	O	O	O
6. GT Division of Professional Practice	O	O	O	O	O	O
7. GT Office of International Education..........................	O	O	O	O	O	O
8. School of Modern Languages	O	O	O	O	O	O
9. Your Host Employer	O	O	O	O	O	O

If not satisfied with a particular office or program, please explain:

Figure 3. Work Abroad Pre-Assessment Page 3

10. Internship Position Information

Company Name: ____________________ Job City Location: ____________________

Job Country Location: ____________

11. Have you had any prior work experience similar to your upcoming internship? O Yes O No

If yes, please describe: ____________________

12. What specific competencies do you expect to acquire or strengthen as a result of your participation in this internship?

13. Is there any additional information you would like before your internship? Are there any additional services that would be of interest to you?

Additional Comments

Thank you for your participation in this survey!

Figure 4. Work Abroad Post-Assessment Page 1

GEORGIA INSTITUTE OF TECHNOLOGY
International Internship Survey (Post-Internship)
Help others who will "follow in your footsteps"

Student Information

Name: ______________ Degree Program: ______________ Date: ________

GTID: ________

General Knowledge, Abilities and Skills

For each of the following skills, abilities, and attributes: First, please rate how important each of these areas is to you right now. Next, please rate how well prepared you think you are right now in each of these areas.

	Extremely Important	*Very Important*	*Important*	*Somewhat Important*	*Not Important*	*Very Well Prepared*	*Well Prepared*	*Prepared*	*Somewhat Prepared*	*Not Prepared*
1. The ability to:										
a. Communicate orally, informally, and in prepared presentations	O	O	O	O	O	O	O	O	O	O
b. Communicate in writing (e.g., business letters, technical reports)	O	O	O	O	O	O	O	O	O	O
c. Use computing technology in communications	O	O	O	O	O	O	O	O	O	O
d. Use computing technology in discipline-specific analysis and design	O	O	O	O	O	O	O	O	O	O
e. Exercise leadership skills	O	O	O	O	O	O	O	O	O	O
f. Function on multi-disciplinary or cross-functional teams	O	O	O	O	O	O	O	O	O	O
g. Effectively resolve interpersonal conflict within a group or team	O	O	O	O	O	O	O	O	O	O
h. Effectively function in your host country's culture and society	O	O	O	O	O	O	O	O	O	O
i. Design and conduct experiments	O	O	O	O	O	O	O	O	O	O
j. Analyze and interpret data	O	O	O	O	O	O	O	O	O	O
k. Think critically and logically	O	O	O	O	O	O	O	O	O	O
l. Carry out projects independently	O	O	O	O	O	O	O	O	O	O
m. Identify, formulate and solve problems within your discipline	O	O	O	O	O	O	O	O	O	O
n. Design a system, component, or process to meet desired needs and quality	O	O	O	O	O	O	O	O	O	O
o. Synthesize and integrate knowledge across disciplines	O	O	O	O	O	O	O	O	O	O
p. Use techniques, skills and tools necessary for practice in your discipline	O	O	O	O	O	O	O	O	O	O
q. Engage in lifelong learning	O	O	O	O	O	O	O	O	O	O
r. Practice your discipline in different social or cultural settings	O	O	O	O	O	O	O	O	O	O
s. Communicate in Your host country's language in a social setting (conversational fluency)	O	O	O	O	O	O	O	O	O	O
t. Communicate in your host country's language in a professional setting (professional / technical fluency)	O	O	O	O	O	O	O	O	O	O
u. Professionally collaborate with persons in your host country's workplace environment	O	O	O	O	O	O	O	O	O	O
v. Work effectively and efficiently in a cross-cultural environment	O	O	O	O	O	O	O	O	O	O
w. Approach problems from different perspectives	O	O	O	O	O	O	O	O	O	O
2. An understanding of:										
a. Product development or design from a business perspective	O	O	O	O	O	O	O	O	O	O
b. Professional and ethical responsibility within your discipline	O	O	O	O	O	O	O	O	O	O
c. Environmental impact of professional practice within your discipline	O	O	O	O	O	O	O	O	O	O
d. The impact your professional practice has on society and culture	O	O	O	O	O	O	O	O	O	O
e. The role of your discipline in solving global problems	O	O	O	O	O	O	O	O	O	O
f. Your host country and their culture(s) beliefs and values within a global and comparative context	O	O	O	O	O	O	O	O	O	O

	Extremely Satisfied	*Very Satisfied*	*Satisfied*	*Somewhat Satisfied*	*Not Satisfied*
3. Overall satisfaction with your internship experience	O	O	O	O	O

Figure 5. Work Abroad Post -Assessment Page 2

4. Answer the following questions in regard to how accurately they describe your current skills and abilities

		Not at all True	Hardly True	Moderately True	Exactly True
a.	I am prepared to enter the workforce after graduation......................	O	O	O	O
b.	I am confident that I will be able to successfully interview and obtain my first job after graduation..	O	O	O	O
c.	If I am in a difficult situation, I can usually think of a solution............	O	O	O	O
d.	I can always manage to solve difficult problems if I try hard enough....	O	O	O	O
e.	Thanks to my resourcefulness, I know how to handle unforeseen situations..	O	O	O	O
f.	I am prepared to make significant professional contributions within my discipline..	O	O	O	O
g.	I am prepared to contribute to society.............................	O	O	O	O
h.	When I am confronted with a problem, I can usually find several solutions..	O	O	O	O
i.	I can usually handle whatever comes my way................................	O	O	O	O
j.	If someone opposes me, I can find the means and ways to an acceptable solution..	O	O	O	O
k.	I can remain rational when facing difficulties because I can rely on my coping abilities..	O	O	O	O
l.	It is easy for me to stick to my aims and accomplish my goals.............	O	O	O	O
m.	I can solve most problems if I invest the necessary effort.................	O	O	O	O
n.	I am confident that I could deal efficiently with unexpected events.......	O	O	O	O
o.	Right now, I have the skills and experiences necessary to practice professionally in my discipline..	O	O	O	O

Career Plans

Answer the following questions in regard to how accurately they describe your current intentions

5.	Within the next five years...	Not at all True	Hardly True	Moderately True	Exactly True
a.	I will most likely work in a position related to your field of study	O	O	O	O
b.	I plan to work/ have worked in a position in a foreign country	O	O	O	O
c.	I intend to participate in a study abroad experience	O	O	O	O
d.	I will pursue/continue to pursue foreign language proficiency	O	O	O	O
e.	I plan to work in a position with considerable international responsibilities	O	O	O	O
f.	I plan to pursue a graduate or professional degree..........................	O	O	O	O
g.	I will travel abroad for nonacademic or non-work related reasons	O	O	O	O

6. Have you had any prior work experience in your internship field? O Yes O No

If Yes, how would you compare the workplace culture of your internship to that of your prior experiences?

If Yes, which internship (US or International) provided you with better opportunities? Why?

7. Please use the space below for any additional comments that could help improve the international internship program

Thank you for your participation in this survey!

Figure 6. Top 5 Countries

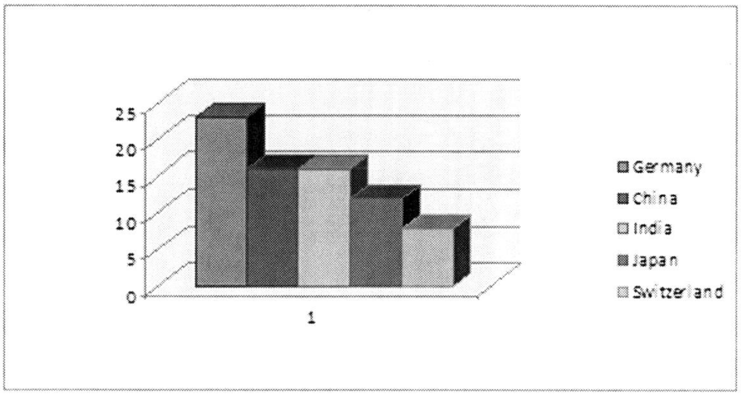

Figure 7. Top 5 Majors

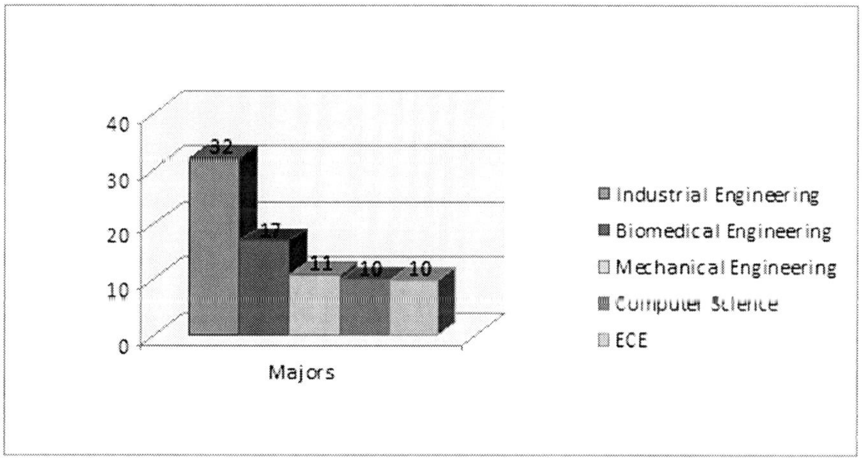

Figure 8. Male-Female Ratio

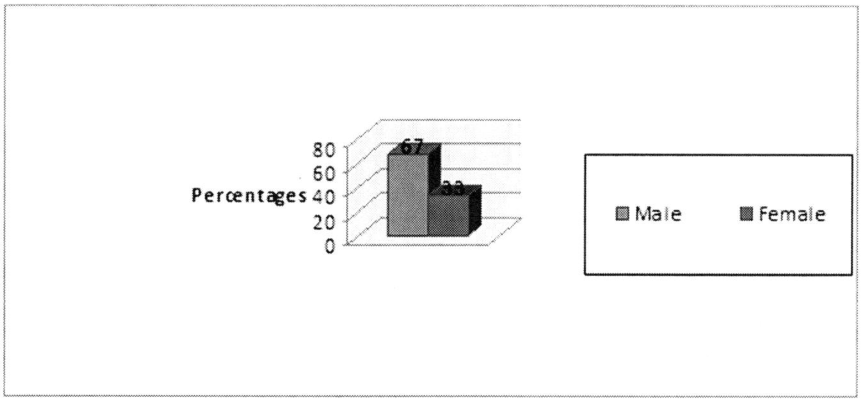

Figure 9. Ethnicities

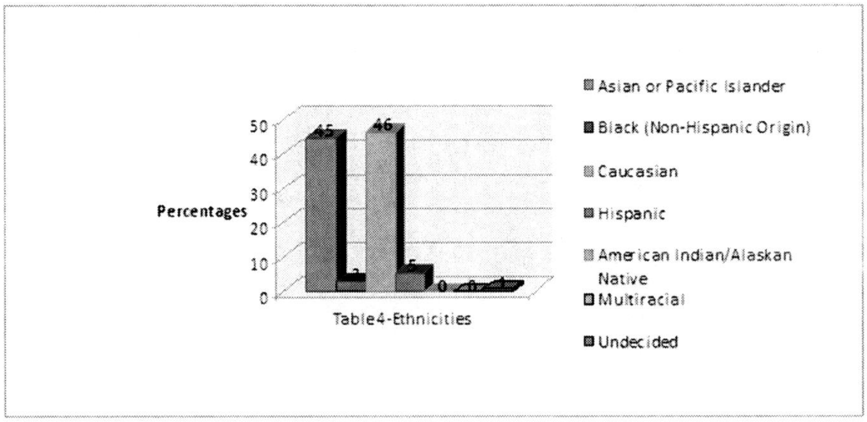

Figure 10. Countries

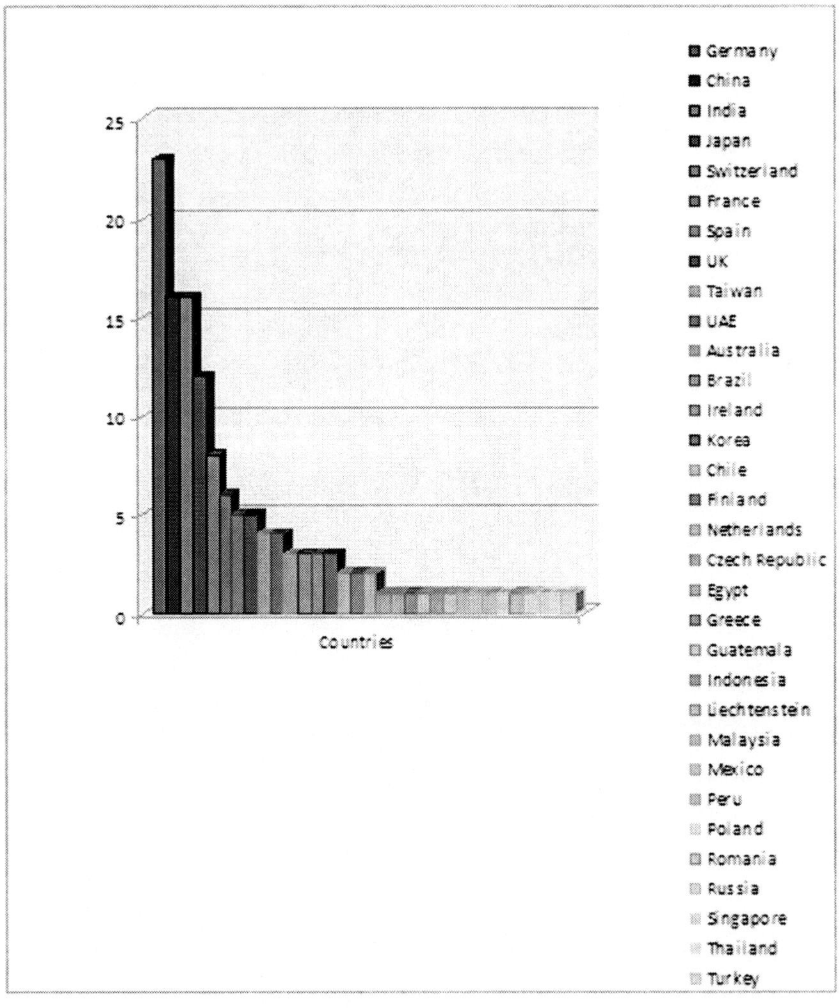

Figure 11. Majors

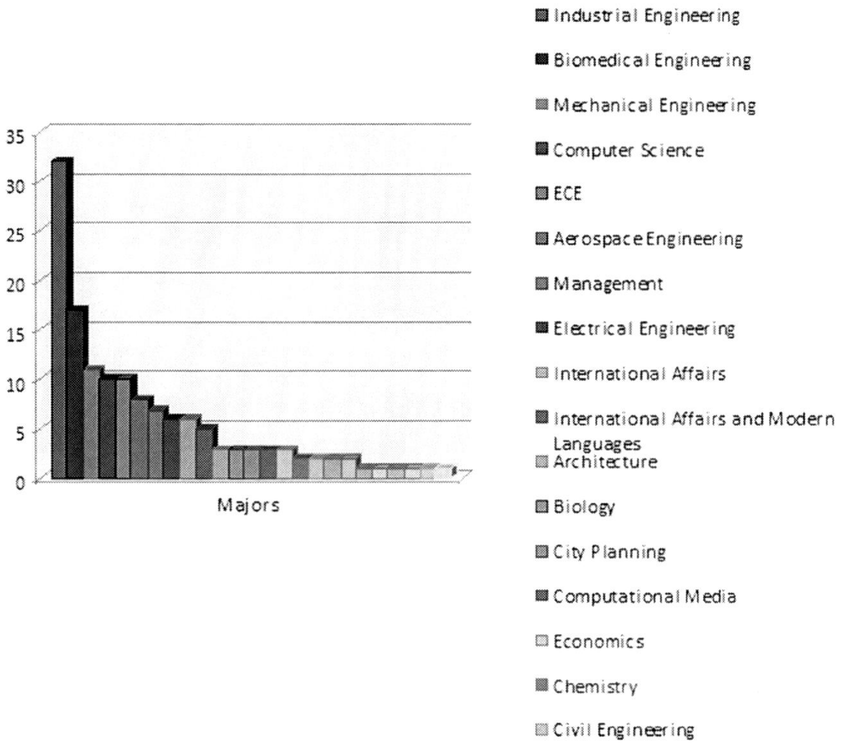

Figure 12. Types of Employers

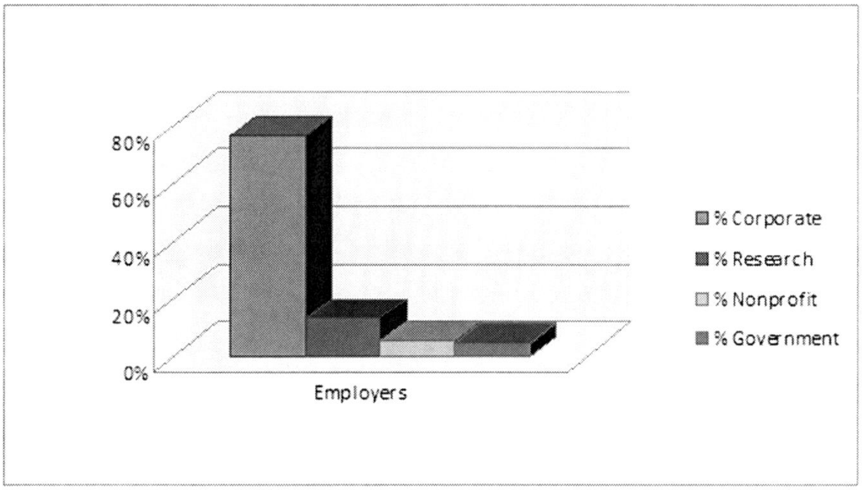

Figure 13. Year in School

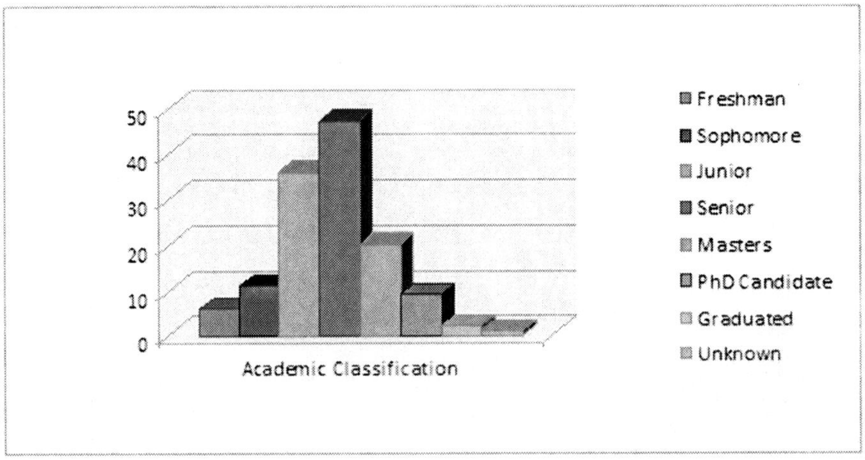

Figure 14. Level in School

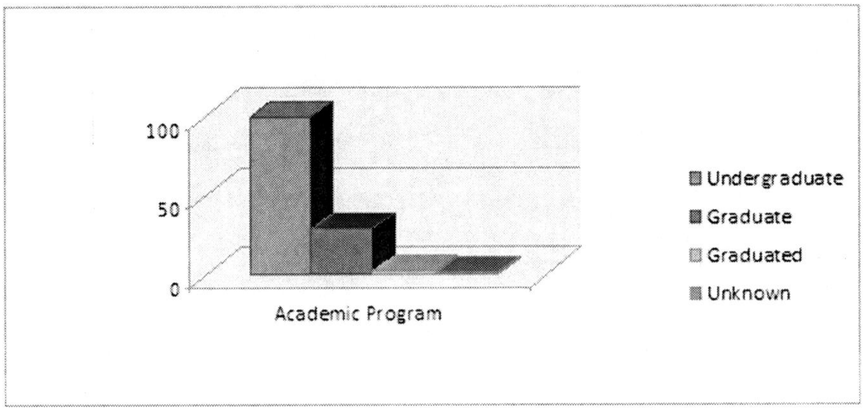

Figure 15. IP vs Non-IP

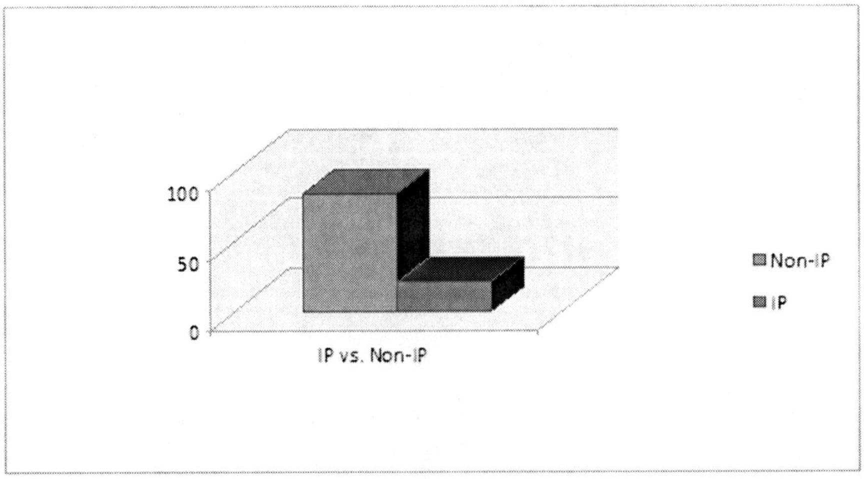

Figure 16. IP Student by Country

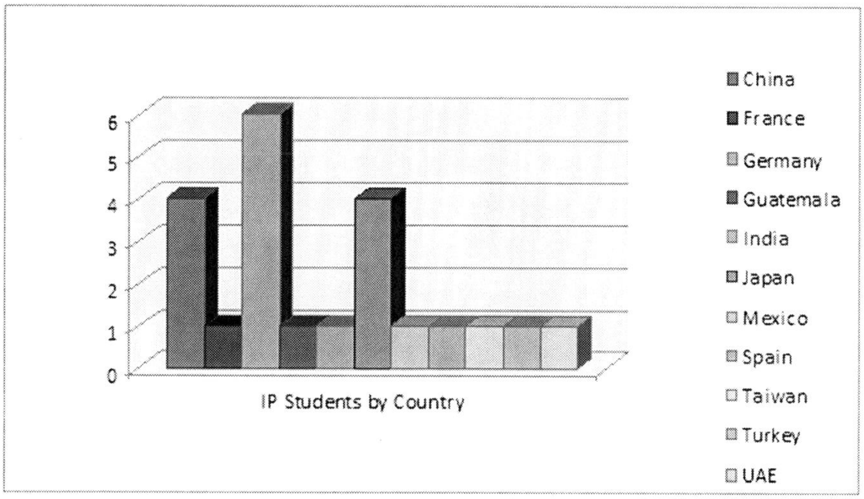

Figure 17. 2008-2009 Term Breakdown

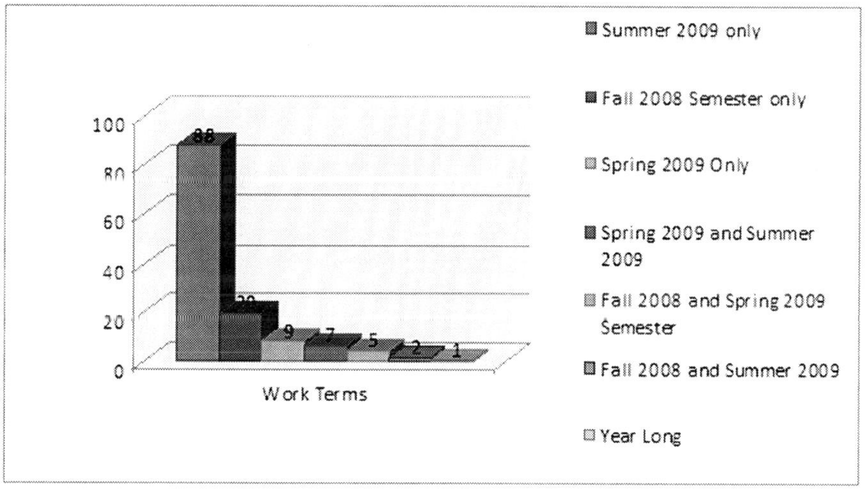

Figure 18. Students Working Abroad by Region

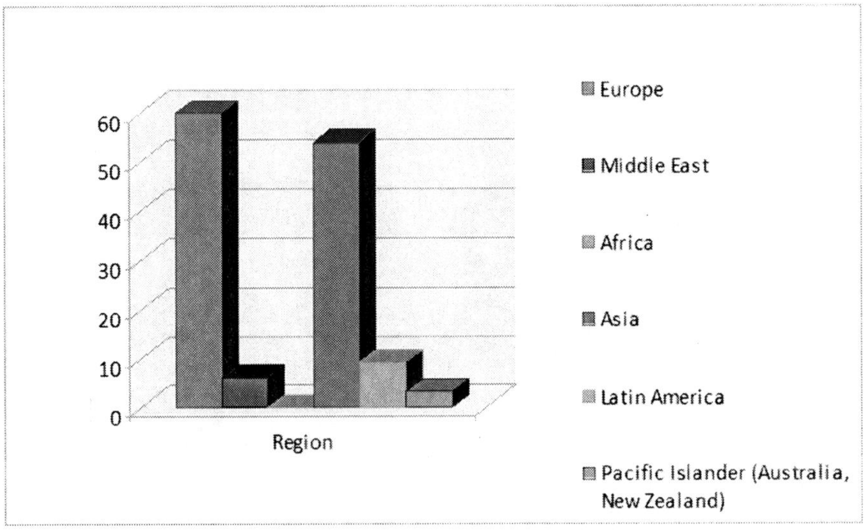

Chapter 7
Running a Successful Practice School:
Challenges and Lessons Learned

Hong-ming Ku
King Mongkut's University of Technology, Thailand

Saranya Thonglek
University of Queensland, Australia

ABSTRACT

The Chemical Engineering Practice School (ChEPS) at King Mongkut's University of Technology Thonburi (KMUTT) in Bangkok is a two-year international curriculum modeled after Massachusetts Institute of Technology's School of Chemical Engineering Practice. The aim of this Master's program is to produce professional chemical engineers with strong fundamentals, practical experience, and a good command of English. The program's uniqueness lies in its strong linkage with the industrial sector. This chapter contains a history of ChEPS and details how KMUTT operates the program. The key factors contributing to the success of the program are identified. Moreover, critical analyses gleaned from the faculty, the alumni, and the industrial sponsors are carried out to examine the current strengths of ChEPS and to identify areas for improvement. Key challenges still facing the program are also outlined. Finally, potential solutions to these challenges are recommended.

INTRODUCTION

The traditional method of learning in engineering disciplines involves classroom lectures, homework assignments, and laboratory work. Although this training is effective to a certain extent, there exists a gap in the skill set needed of students when they step into the real world. This is particularly true in Southeast Asia including Thailand, where education has traditionally been more tightly structured and teacher-directed (Ziguras, 2001). Rote learning is usually the norm, and creative thinking

DOI: 10.4018/978-1-60960-547-6.ch007

is often overlooked. *Spoon-feeding* is prevalent in classrooms even at the college level. As a result, engineering graduates in Thailand generally do not possess strong analytical and problem-solving skills. To compensate for the deficiencies in the educational system, companies are often forced to invest substantial resources on re-education and on-the-job training for starting engineers.

In addition, the English proficiency of Thai engineering students is on the average subpar. In today's global economy and with a substantial foreign investment totaling billions of dollars annually in Thailand, the importance of English cannot be overemphasized. Unfortunately, English takes a backseat in most engineering curricula. Students have very limited exposure to English, and there is little incentive for them to improve, as most programs do not have a minimum English requirement for graduation. Finally, companies often complain about the inadequate training of university graduates in communication, be it spoken or written, even in their native language. These facts are hardly surprising, given that nearly all graduate programs are taught in Thai with little emphasis on technical writing and oral presentations.

King Mongkut's University of Technology Thonburi (KMUTT) is an autonomous state institution in Bangkok with a long tradition in engineering. As early as 1996, the university recognized many shortcomings described above in its engineering programs. Shortly after, KMUTT introduced an initiative to develop a new flagship practice-based curriculum aimed at overcoming these deficiencies. The objective was to produce well-rounded engineers who possess strong technical expertise, can communicate effectively, and have good English proficiency. If proven successful, the goal was to expand the initiative to include other curricula. KMUTT chose Chemical Engineering to be the pilot program. The new curriculum called the Chemical Engineering Practice School (ChEPS) was founded in 1997 and modeled after Massachusetts Institute of Technology's (MIT) David H. Koch School of Chemical Engineering Practice. ChEPS is a two-year international master's program with one semester of compulsory industrial internship or practical training.

The practice school can be viewed as one model of Work-Integrated Learning (WIL), which can be broadly defined as educational activities that integrate theoretical learning with its application in the workplace. This learner-centric process should provide a meaningful experience of the workplace application, resulting in successful learning outcomes such as problem-solving skills, creative thinking, and other soft skills for the student. WIL models include (but are not limited to):

- Practical training via practice schools
- Internships
- Work placements
- Cooperative education
- Industry-based learning
- Community-based learning
- Student group projects

The essence of a practice school is to supplement traditional classroom learning and "instructionism" with practice-based learning (PBL) and "constructionism", which together can be succinctly described as "learning by doing." This learning takes place at industrial sites in a real work environment where students are trained to solve real-life problems. In the practice school, this practical training is an integral and compulsory part of the curriculum and earns credit hours for students to fulfill their graduation requirement.

Practice-based learning is a new pedagogy and has been found highly effective for teaching science and engineering. In higher education, well-organized PBL can tremendously benefit the three primary stakeholders, namely the university, the student, and the industry. With practice-based curricula, the university fosters strong ties with industry and is able to graduate engineers with robust skills and practical experiences. On the other hand, students understand and appreciate theories taught in classrooms better and develop strong problem-solving skills when immersed in an authentic work environment. As a result, students graduating from such programs tend to have higher self-esteem and self-confidence. Finally, industry also has a lot to gain when hiring engineers and researchers with practical experiences, because they have a relatively short learning curve and adjustment period.

It has been 14 years since the inception of ChEPS. As with any new enterprises, the road to success in creating and running the practice school at KMUTT is not without difficulties. While the ChEPS program is now tremendously popular among graduates with bachelor's degrees in chemical engineering, there is still room for improvement. There also remain challenges and hurdles which need to be overcome. Moreover, the concept of practice-based curriculum is beginning to gain traction in Thailand, both at the secondary school and in higher education levels. More educators and academics have come to espouse this new learning model. As a result, more engineering departments around the country are opening up their own practice schools, and competition for top students and for funding will certainly intensify in the years to come. To avoid complacency and resting on past laurels, steps should be taken to continually adapt and improve the program to meet the changing needs of students and industries.

This chapter contains a history of ChEPS and detailed information about how the program is operated. Moreover, critical analyses gleaned from the experiences of the faculty and interview with ChEPS's alumni and industrial sponsors are carried out to examine the current strengths and key challenges still facing the program. These difficulties are collected and will be presented as lessons learned. Each lesson examines what the program is doing right and presents future challenges. Finally, potential solutions to the problems are also suggested.

BACKGROUND

MIT is probably the pioneer in fostering learning within industrial settings, and its chemical engineering practice school is the oldest in the world. The Practice School at MIT (Mattill 1991, 2010) was established in 1916 with the goal of supplementing classroom studies with practical training in an industrial environment. The program was widely regarded as the educational "flagship" of MIT and was considered a resounding success after just a few years of operation. The Practice School at the graduate level is truly unique in the US and is found only at MIT, although in recent years, the program has expanded to international stations such as Japan (O'Connor *et al.*, 1999) as well as offering a dual-degree master's with National University of Singapore. At present, the Department of Chemical Engineering at MIT offers two parallel master's programs: a traditional thesis-based curriculum and a practice-based curriculum. Students who choose to enroll in the latter are required to do two academic semesters of coursework, followed by an additional term of industrial internship. This internship replaces the research thesis found in a conventional master's program. A Master of Science in Chemical Engineering Practice (M.S.CEP) degree is granted upon graduation. Graduates of the MIT's Practice School are some of the most sought-after engineers in the country.

Inspired by the success of MIT's learning model, educators around the world went on to create a number of practice schools, some of which are well-established now. Birla Institute of Technology & Science (BITS) in Pilani, India, operates an undergraduate practice school which encompasses all disciplines of engineering, science, and humanities (2010). In fact, BITS operates two programs of Practice School (PS), namely PS-1 and PS-2. The difference between the two programs is that the duration of internship for PS-1 is eight weeks while that of PS-2 is six months. PS-1 aims at orienting students to the professional world using national laboratories, financial institutions, R&D centers, software and healthcare organizations as practice stations. On the other hand, students in PS-2 are drawn from different disciplines, work on a much longer period on a variety of professional activities in the industrial world, and are involved in problem-solving efforts of direct interest to the host companies.

Manipal Institute of Technology, one of the constituent colleges of Manipal University (MU), in Manipal, India, has also incorporated the practice school into its engineering curricula since 2005. The internship is optional for undergraduates, but when chosen, must be at least 16 weeks long. For the graduate courses, the internship period is one full year. The percentages of students in 2009 (50.1%) and the previous year (48.0%) opting for Practice School show the popularity of the program among students (2010).

At the urging of the UK government, the Cambridge-MIT Institute (CMI) was established in 2000 at the Cambridge University to explore how academics, industrialists and educators could work together to stimulate competitiveness, productivity, and entrepreneurship in the UK. The aim of CMI serves to enhance competitiveness and innovation by improving knowledge exchange between universities and industry. Six innovative new interdisciplinary master's programs integrating science, technology, and management studies were created. Between 2000 and 2006, CMI graduated more than 350 students from these curricula. CMI issued two reports titled *Accelerating Innovation by Crossing Boundaries* and *Working in Partnership* which summarized its activities and output between 2000 and 2006. The reports can be downloaded from CMI's website (2010).

The Department of Chemical Engineering at the University of New Brunswick (UNB), Canada, offers a course called Chemical Engineering Practice School, a two-week industrial project elective which exposes students to the practical aspects of chemical engineering. Groups of students, with faculty supervisors, are assigned to engineering projects to be carried out on industrial process units. Students are required to present an oral report to plant operating and technical personnel at the end of the practice session. A written report is also required. Bendrich and Pugsley (1998) presented an approach for introducing students at UNB to real industrial problems in this industrial project course.

The Professional Engineering Placement Scholarship (PEPS) is a work-based learning program that allows final-year engineering students the University of Queensland (UQ) in Brisbane, Australia, to spend up to six months in industry while still gaining full academic credit (2010). The program is open to all engineering students taking courses in mechanical, chemical, mechatronics, mechanical/aerospace, electrical and software engineering. A similar program called PIPS (Professional Information Technology Placement Scholarship) gives final-year information technology students a chance to work in the IT industry (2010).

Through a program called Co-operative Education for Enterprise Development (CEED), Corporation Technologies Pty Ltd (2010) links students from three of Queensland's major universities, namely the University of Queensland (UQ), Queensland University of Technology (QUT), and the University of Southern Queensland (USQ), with companies and government in industry-based projects. Real-life projects are completed by students as part of their university degrees – final-year undergraduate, Masters, or Honours. Starting in 2010, the School of IT at the University of Sydney (USyd) also joined the program. CEED provides clients with a wide variety of talented students in the fields of engineering, IT, business (marketing, human resources, and finance), mathematics, and science (biotechnology, biology, and computing science) for their projects, while companies provide training and mentoring to these emerging professionals.

Victoria University in Melbourne, Australia, introduced PBL into engineering curricula in its

Schools of Architectural, Civil, and Mechanical Engineering in 2006 (Rojter, 2007). In 2010, the university implements a new PBL model for all of its Bachelor of Engineering courses (2010). The model allows students to work on small problems in year 1, community and/or industry projects in years 2 and 3, and engineering practice on industrial projects in year 4. The PBL model emphasizes active learning as well as other skills such as creativity and problem solving, interpersonal and team working, business and project management, and communication in all forms.

The Manufacturing Engineering Education Partnership (MEEP) is a unique collaboration of three major universities, namely Penn State University, University of Puerto Rico - Mayaguez, and the University of Washington. MEEP is partnered with a premier high-technology government laboratory named Sandia National Laboratories, nearly 100 corporate sponsors covering a wide spectrum of US industries, and the federal government who funds this project through the Technology Reinvestment Program (TRP). MEEP's mission is to develop and implement a new curriculum in design and manufacturing that has strong collaboration with industry. The product of this effort is called the Learning Factory, an outcome-driven program which has been successfully institutionalized at the three participating universities. Critical emphasis is also given to the development of "soft skills" such as problem solving, communication, and teamwork. Lamancusa *et al.* (1995) described the Learning Factory and presented the results from the first year of its existence.

Since the industry-based learning has been incorporated into many engineering curricular worldwide, several papers have been published on the effectiveness and outcome of such a teaching model. For example, Lamancusa *et al.* (1997) described how industry-based projects were integrated into academia with respect to the Learning Factory at Penn State University and some of the successes and failures in this effort. The target curricula consisted of mechanical engineering, industrial engineering, and electrical engineering, all at the undergraduate level. Morell *et al.* (1998) discussed assessment instruments and tools designed to evaluate the overall curricular outcomes of the Learning Factory in MEEP, including student performance such as teamwork, oral presentation, and written skills. The assessment used the ABET Engineering Criteria 2000 (EC-2000), an accreditation standard that promotes innovation and continuous improvement in engineering education, as the benchmark. ABET EC-2000 contains an outcome assessment plan that requires engineering programs to have in place a continuous process of evaluation, feedback, and continued improvement of their effectiveness. The authors concluded that the Learning Factory did comply with the ABET EC-2000. In another work, Morell *et al.* (1998) presented how MEEP designed the assessment strategy to evaluate outcomes of industry-based curricula and some of the assessment instruments and tools designed. A list of detailed assessment instruments in the form of questionnaires and evaluation sheets in areas such as oral presentation, written report, industry survey, student survey, quality of teaching, and skills knowledge was also presented. More recently, Lock *et al.* (2009) explored the attitude of undergraduate engineers towards work placements in industry and assessed the placement experience in terms of student learning outcomes and future employment aspirations by collecting both quantitative and qualitative data via survey.

It is obvious that in addition to knowledge, a well-rounded professional engineer should possess a number of additional attributes that society and employers need (Ramirez & Beauchamp, 1995). KMUTT has long recognized the importance of such skills development, which include areas such as problem-solving, self-learning, communication, English proficiency, teamwork, leadership, and social responsibilities. When KMUTT initiated the practice-based curriculum in Chemical Engineering in 1997, it was decided that the new academic program must meet the following criteria:

- *Be international.* Lectures and presentations must be conducted in English. Reports and homework assignments are written in English as well.
- *Be a graduate program*, so that it is small enough and can be efficiently managed.
- *Be practice-based*, i.e. the curriculum includes one semester (five months) of compulsory industrial internship in which practice-based learning (PBL) is emphasized.
- *Has strong linkages to the private sector*, which offers industrial sites needed for practical training.
- *Has adequate funding* to attract top-notch students.

The David H. Koch School of Chemical Engineering Practice at Massachusetts Institute of Technology (MIT) in the US offered all the desired components that KMUTT sought in its curriculum. Contacts were subsequently made at the highest level between KMUTT and MIT to import the practice-school model into Thailand. MIT was retained as an advisor, and professors from its practice school traveled to Thailand to help set up the program, assess its readiness, and teach selected courses. The first class of ChEPS consisted of 21 students, all with undergraduate degrees in chemical engineering, who were recruited from universities all over Thailand. In subsequent years, 15 - 29 students were admitted annually. Ku *et al.* (2005) gave a comprehensive review of the ChEPS program.

Figure 1 illustrates the four essential components in the practice school, namely the university, the funding agencies, the students, and the industrial linkage. The funding agencies are state, semi-private, and private organizations, which provide research grants and scholarships to academic programs. ChEPS has relied heavily on such agencies for financial support in its operations, particularly during the early years. In a traditional graduate program, the industrial component is normally missing or its role is limited. On the other hand, industrial involvements are vital to the success of a practice school to ensure that students are trained to solve real-life problems early in their studies. These industry-relevant problems are identified by industrial sponsors, and are either brought to the classroom as case studies or solved as site projects during internships.

The ChEPS curriculum consists of one academic year (two semesters and one summer) of coursework, one semester of internship at a practice site, and one semester of research, as illus-

Figure 1. The four essential components of ChEPS

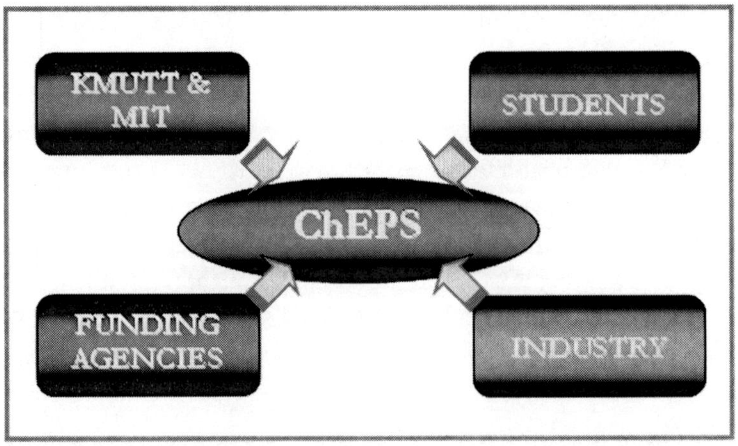

Figure 2. Timeline of ChEPS curriculum

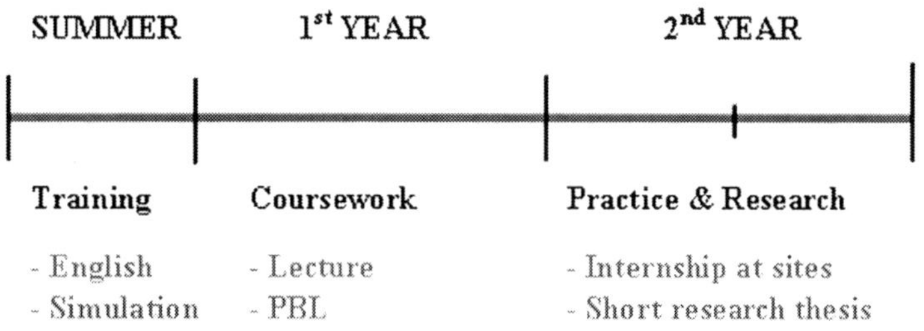

trated in Figure 2. The coursework is fixed and there are no course electives. One objective when starting ChEPS was to introduce Western-styled learning into Thai classrooms. The ChEPS curriculum emphasizes both problem-solving and intensity. The whole program can be viewed as a *learning boot camp*, where problem-based learning is emphasized and students are constantly challenged to solve problems in real plants, sometimes with limited data and many constraints.

Course lecturers consist of both local and overseas instructors, including professors from the US, Canada, Australia, and Singapore. First-year courses comprise the following core subjects:

- Advanced thermodynamics
- Applied mathematics
- Process chemistry, polymer processing, petroleum engineering, and bioprocesses
- Process control and optimization
- Process simulation and modeling
- Reaction kinetics
- Transport phenomena

In ChEPS, computers and software packages are heavily used to supplement classroom lectures. Many courses offer hands-on workshops that demonstrate applications of the theories taught. As a result, ChEPS students are well-versed in many simulation programs and programming tools such as ASPEN PLUS™, PRO/II™, MATLAB™, LINDO™, and ControlStation™.

Industrial involvements are vital to the success of ChEPS. Sponsoring companies allow ChEPS faculty and students to access their production facilities, also known as practice stations, which play a pivotal role in providing practical training for students in the second year of their study. At the same time, sponsoring companies gain valuable human resources who can work on longer-term projects, thus freeing up company engineers to focus on more urgent needs. Consequently, maintaining this win-win partnership model between the university and the private sector hinges on the successful operation and implementation of practice stations.

The duration of the practice phase is five months. Students work in teams of two or three on two projects in series, and take turns being the project leader. So each project is 10 weeks long, during which there is a proposal presentation, a progress presentation, and a final presentation. ChEPS faculty members also travel to the site and attend these presentations to provide further input. Due to time constraints, most site projects tend to be simulation-oriented, which seek to debottleneck, troubleshoot, and optimize (e.g. minimizing energy consumption) existing plants. A few projects also involve feasibility studies and design of new processes. The practice team can be likened to a consulting team who are dedicated to solving problems for the host company. Ku *et al.* (2007) discussed the operation of a chemical engineering practice station in detail.

It should be stressed that the industrial internship in a practice school is not a cooperative education program (also known as co-op). The practice model and the co-op study are different in two important aspects. One is the presence of a full-time ChEPS faculty member, called a site director, who is dedicated to a practice station. The site director lives and works with student interns in housing provided by the sponsoring company. While company engineers identify and set the scope of the projects, the site director is responsible for ensuring the academic value of the proposed work, that the project goals are attainable, and that the work is carried out as planned. Furthermore, the site director provides technical advice, prepares students for presentations, and edits students' reports.

The second difference is the commitment of the host company at every level to the practice school, beginning with basic needs such as free housing accommodations, office space, computing facilities, and Internet access. Senior management is first approached, which sets a top-down policy on sponsoring ChEPS. Plant managers, engineers, shift operators, and technicians are also consulted, since they have to interact with the students. A team of engineers is then formed to work closely with students. In a nutshell, the practice school is more systematic and more organized, and is therefore far more efficient than the traditional co-op study.

Despite the similarities, the ChEPS curriculum is not an exact duplicate of the MIT model. Certain components were added to ChEPS to enhance the MIT model and compensate for common weaknesses in Thai students. For example, ChEPS introduced a course in engineering management to enable students to better manage time, people, and projects and gain a rudimentary understanding of corporate finance. Moreover, the program is lengthened to two academic years and one summer to accommodate the intense workload and schedule of the students. Additionally, ChEPS is supplemented with the following components:

- *Presentations*. Students hone their presentation skills by giving no fewer than 30 talks by the time they graduate. Presentations are required in every phase of ChEPS, including PBL problems, research thesis, and site projects.
- *A short research thesis*. Every student must pick a thesis project to be completed in six months. The extensive networking of ChEPS inside and outside KMUTT, including overseas institutions, allows students to choose just about any chemical engineering related topic that is of interest to them. In many cases, research projects are collaborated, which offers an opportunity for students to carry out part of their research outside Thailand, e.g. in Canada, the US, and Singapore. While ChEPS' primary goal is not to produce researchers, many students find the training useful. The students learn to think critically and analytically and must devise a systematic approach to solve a research problem. In fact, many ChEPS graduates have gone on to pursue Ph.D. degrees at other institutions, both in Thailand and overseas.
- *English tuition*. ChEPS requires that all students score at least 500 on the paper-based TOEFL (called PBT) or 650 on TOEIC by the end of their second year. This is a daunting task for most students. Hence, extra English courses are provided and students' progress is monitored closely. Students are required take either the TOEFL or TOEIC at least once a year. More recently, a mentoring system has been set up in which faculty members are responsible for a small group of students, helping them improve their writing, presentations, and English in general.

ChEPS is highly competitive and admits approximately 25 top-quality students from a pool of slightly more than 100 applicants from various universities in Thailand each year. For example, about two-thirds of ChEPS applicants have an undergraduate GPA of 3.0 or above in chemical engineering. As a result, the program has an elaborate admission process to ensure that only the best are admitted. A number of criteria including undergraduate GPAs, TOEFL English test, SAT Math test, and interview scores are used to screen the applicants. Each admission criterion is given a weighting factor, with GPA having the highest impact. Viravaidya *et al.* (2007) presented findings on the relative significance of admission criteria in the ChEPS program.

The success of ChEPS as a conduit for learning has been cited by Technology and Development Program (TDP) at MIT (2010), which works to build research and academic capacity around the world through its assistance in science and technology development. The program considered the KMUTT practice-school endeavor to be its success story in Thailand. At the same time, the practice school model is gaining traction among educators in Thailand, both in higher education and in vocational schools. The practice-based learning model originating from ChEPS at KMUTT has inspired the creation of many spin-off programs, both inside and outside the university, in the past 10 years. In addition to ChEPS, KMUTT now has five additional practice-based master's programs, namely the Food Engineering Practice School (FEPS), Starch Engineering and Process Optimization Program (SEPO), Bioinformatics (BIF), Biotechnology Practice School (BIPS), and Biotechnology Business Management Program (Bioentrepreneurship). Ku *et al.* (2005) gave a comprehensive review of the graduate-level science and engineering practice schools at KMUTT.

The FEPS program (Asavasanti *et al.*, 2005) was established in 2001 to meet the demands for more food engineers, since Thailand derives a substantial amount of revenues from the exports of agricultural and food products each year and to adopt the Thai government's initiative to become the "kitchen of the world." While ChEPS students' background is exclusively chemical engineering, FEPS students have more varied backgrounds, including food science and technology, food engineering, and chemical engineering. The SEPO program was established in 2001 to develop human resources for the agro-industry, which is an important industrial sector in Thailand. SEPO students follow the standard curriculum of ChEPS or FEPS, except for six months in the second year in which they will be interned at a tapioca starch company. Their PBL projects also come from the starch industry. Site projects involve improving unit operations such as drying and hydro-cyclones and process optimization in the production of tapioca starch. The BIF program (2010) is a multi-disciplinary curriculum that combines the technical elements of computer science, biology, and biochemistry. Bioinformatics is the scientific and technical foundation of the human genome project, and promises to play a central role in life science of the coming century. BIF was established in 2003 as a joint program between the School of Bioresources and Technology and the School of Information Technology. Courses are taught jointly between the faculties of the two schools. Students in the program receive comprehensive training in genomics, algorithms for sequence analysis, database design and management, and software engineering and programming, including web-based development. Enrolled students have a background either in computer science or in biology with a few exceptions in engineering.

More recently, the School of Bioresources and Technology at KMUTT opened two practice-based master's of science curricula called Biotechnology Business Management and Biotechnology Practice School. These two curricula are multidisciplinary and innovative programs which aim to provide students with a multitude of knowledge in modern biotechnology and skills appropriate for wide career opportunities. The curricula were designed in response to rapid advances in genome science while a wide range of research projects serves the increasing demand of local biotechnology-related industries. Graduate students are trained in core biotechnology, such as molecular biotechnology, bioprocess engineering, environmental sciences, and business and legal aspects.

RUNNING THE PRACTICE SCHOOL

It is obvious that a practice-based program must provide a student-centric learning environment, and the old adage of "practice makes perfect" is an apt metaphor. Traditional lectures are still important in order to learn the fundamentals of chemical engineering, but ChEPS students learn more and learn faster when they practice what they have been taught. However, running a practice school also entails many challenges. One of them is the commitments in human and financial resources by both the university and industrial sponsors needed to make the collaboration work. For example, for a two-year curriculum such as ChEPS, which admits 20-25 students each year, 6-7 full-time equivalent faculty members and two administrative staff are needed to run the program effectively. In addition, a practice-based program requires complete dedication of its staff and special attention from the faculty. Moreover, for the program to be successful, a number of key factors must be put in place or implemented, ranging from the ability to recruit quality students, finding the necessary funding, integrating problem-based learning into classroom, to managing practice stations effectively.

Recruiting Students

The first lesson learned in running ChEPS is that recruiting top students into the program must be a priority. The old Chinese proverb "A journey of a thousand miles begins with one small step" is particularly apt when running a practice school. That "one small step" is the ability of the program to recruit quality undergraduate seniors from various chemical engineering departments around the country. It is crucial that the "raw materials" brought into a practice school, i.e. the students, be of top quality. Good students foster a culture of healthy competition within the program and serve as role models for others to emulate. With good engineering fundamentals, these top students often actualize their potentials which are brought upon by the intensity and challenges offered by the practice school. The result is the creation of well-rounded chemical engineers who are ready to serve the industry.

ChEPS accepts applications from undergraduate seniors enrolled in any university or institution in Thailand with a curriculum or department in chemical engineering. Basically, two ingredients must be put in place to attract good students. One is the availability of scholarships and financial support, which will be elaborated later. The other is the promotion of the program through "proactive recruitment". Unlike "passive recruitment" of many academic programs which wait for prospective students to apply, "proactive recruitment" targets students with "road shows" and on-campus presentations. The students will hear first-hand information from the program's staff instead of receiving the information through word of mouth and other indirect channels. In many cases, applications, already streamlined for simplicity, can be collected immediately after the presentation. At least 30% of ChEPS applications received each year are obtained this way.

Every year around November, about six months prior to the opening of a new ChEPS academic year, ChEPS will contact the chemical engineering department of a target university and request a "road show" presentation with its seniors. Generally, at least two staff or faculty members will do the road show together. The road show consists of a one-half hour slide presentation, followed by a 12-minute video presentation, and a "Question and Answer" session at the end. These presentations are often given in English to convey the international flavor of the program. In many cases, alumni of the target university who are currently enrolled in ChEPS also join the roadshow and talk to prospective students about their experiences in the program. These alumni are the program's best references.

Another added benefit of enrolling good students in the program is the attention they garner from corporate sponsors. Large chemical, petrochemical, and refinery companies see these top students as potential employees and are willing in certain years to provide full scholarships to some of the students while they study in ChEPS. In exchange, the students sign employment contracts which bind them to the sponsoring companies for a few years. This arrangement can be beneficial to all three parties. Sponsoring companies are able to sign up quality engineers earlier than their competitors, while the students are guaranteed employment upon their graduation from ChEPS. On the other hand, the scholarships provided by the sponsoring companies help alleviate the financial burden on ChEPS.

ChEPS' proactive recruitment can only be characterized as extremely successful. The program receives approximately 100 applications each year. The interest in the program remains strong even in the face of more competition from new graduate programs created in the past few years. Table 1 shows all the universities or institutions in Thailand with chemical engineering departments or curricula. Over the years, ChEPS has received applications from virtually every university displayed in Table 1.

Future Challenges in Recruitment and Solutions

In recent years, chemical engineering departments at many universities have grown more reluctant in arranging roadshow presentations for ChEPS for fear of losing their potential graduate students. Some departments are more concerned than others. Those located in the provinces with small or no graduate programs tend to welcome the ChEPS

Table 1. Universities in Thailand with chemical engineering departments

University/Institution	Abbreviation	Province
Burapha University	BU	Chonburi
Chiang Mai University	CMU	Chiang Mai
Chulalongkorn University (Chemical Engineering)	CU-ChE	Bangkok
Chulalongkorn University (Chemical Technology)	CU-ChemTech	Bangkok
Kasetsart University	KU	Bangkok
King Mongkut's Institute of Technology Ladkrabang	KMITL	Bangkok
King Mongkut's University of Technology North Bangkok	KMUTNB	Bangkok
King Mongkut's University of Technology Thonburi	KMUTT	Bangkok
Khon Kaen University	KKU	Khon Kaen
Mahidol University	MU	Bangkok
Prince of Songkla University	PSU	Songkla
Rajamangala University of Technology Thanyaburi	RMUTT	Bangkok
Silpakorn University	---	Bangkok
Srinakharinwirot Unviersity	SWU	Bangkok
Sranaree University of Technology	SUT	Nakhon Ratchasima
Thammasat University (Sirindhorn International Institute of Technology)	TU-SIIT	Bangkok
Thammasat Unviersity (Chemical Engineering)	TU-ChE	Bangkok

team. On the other end of the spectrum, a few departments will find every excuse to decline the request. In those cases, ChEPS needs to find innovative ways to circumvent the problem. One solution is to utilize its alumni to contact the seniors directly which allows ChEPS staff to correspond with the students. Departmental web boards are also a great way to advertise the program. Finally, some roadshow presentations can be arranged through personal contact with faculty members at target universities, e.g. asking their students to stay behind after class for a ChEPS presentation.

In the end, ChEPS alumni have become the program's best references as its number grows over the years. Having graduated 11 classes, ChEPS has produced a large pool of alumni base, now numbering 235 people. The majority of ChEPS alumni work for large refinery, chemical, and petrochemical companies. These alumni have become indispensable allies of ChEPS in spreading the good words about the program. In fact, many undergraduate seniors became interested in the program after hearing about ChEPS from alumni who often served as mentors or supervisors during these students' co-op study. Finally, seeing so many ChEPS alumni working in large companies by prospective students only serves to reinforce ChEPS' image and reputation as a unique program whose graduates are very much sought after by industries.

English Proficiency

ChEPS is highly competitive and admits approximately 25 students from a pool of about 100 applicants from various universities in Thailand each year. A number of admission criteria including undergraduate GPAs, paper-based TOEFL (PBT), SAT Math test, and interview scores are used to screen the applicants.

Any applicant with a PBT score below 400 (the full score is 667) is automatically eliminated from further consideration regardless of his or her GPA. It has been observed that there exists a strong correlation between the English proficiency of ChEPS students and their academic performance in the program. This is not totally surprising, since ChEPS is an international program and all lectures are delivered in English. Students with poor English skills tend to struggle academically and have problems absorbing what is taught. This is especially true with courses that are taught by English native speakers because students are not accustomed to the native accent and the speed of lecture. A fairly large number of students with excellent undergraduate academic records but poor English proficiency have turned out to be mediocre performers in ChEPS.

Recently, Buarod (2009) attempted to quantify the relationship between English proficiency of incoming students and their subsequent ChEPS GPAs. A linear regression was performed based on the PBT scores of students from Classes 4 - 10 (total sample size = 150. PBT was not administered prior to Class 4). A moderately strong positive correlation was found to exist between the English proficiency and ChEPS GPAs with a Pearson's correlation coefficient of 0.331. The correlation is significant at the 0.01 level (one-tailed). The other two admission criteria, namely undergraduate GPA and SAT-Math, were found to be better predictors of the students' final ChEPS GPAs, but not by much ($r^2 = 0.448$ and 0.339, respectively). While the study could not conclusively show that English proficiency had a large impact on the students' ChEPS GPAs because other factors such as undergraduate GPAs were not taken into account, the evidence did support the observation.

Future Challenges in English Proficiency and Solutions

As much as ChEPS tries its best to admit students with good English skills (with all things being equal), most incoming students do not possess the rudiments capable of making a coherent oral English presentation and writing an error-free technical report or thesis. The average PBT score of incoming students is approximately 440. This is far below the acceptable level when one considers that most colleges in the US require international students to score above 600 on the PBT prior to admission. Consequently, ChEPS faculty spends a disproportionately large amount of time editing students' reports, correcting their presentation slides, and coaching them on how to present. This places a huge burden on the faculty. In the past, ChEPS had provided free English tutoring to small groups of students, hired the language department at KMUTT to teach extra English courses, and even subsidized students to enroll in outside English classes. The results are mixed and these initiatives have limited success. While the overall English skills of the students generally improved, they were still below the acceptable level.

There is no quick fix to the problem. The faculty will need to show patience and encourage all students to immerse themselves in English as much as possible. For their part, the students should show more initiative in mastering the English language. With the exception of one Canadian student from the University of Waterloo, all ChEPS students have been Thai. As a result, the classroom environment is not completely conducive to learning English. Despite all coursework and lectures being in English, the students still speak Thai among themselves. One solution is for the program to enroll some foreign students, particularly those from neighboring countries such as Cambodia, Laos, Malaysia, Singapore, and Vietnam. Students from other Asian countries such as China, India, Indonesia, and the Philippines are also good prospects. It can be argued that to be truly international, ChEPS must also be able to enlist foreign students in the program, thus fostering a learning milieu in which the spoken language is English. This endeavor is no simple task and will involve careful planning and going overseas to promote the program. It will certainly take many years before ChEPS is able to establish an international reputation.

First-Year Coursework and PBL

Although the uniqueness of ChEPS lies with the practical training in the second year, it is important that the quality of teaching and the content of coursework live up to the expectations of the students in the first year. First, there must be a distinct departure from the typical undergraduate style of teaching and learning in the ChEPS curriculum. While the former tends to emphasize breadth, i.e. cramming as many courses as possible into one semester, ChEPS courses focuses on depth and intensity. Weekly lecture hours are kept short, while homework and group projects abound. Students are to make regular oral presentations, and they improve their skills by practicing a lot and receiving feedback from the faculty.

Moreover, as an international program, the students also expect the courses to be taught by qualified lecturers, both locally and from overseas. A first impression makes a lasting impression. The last thing ChEPS needs is to see its students become demoralized during the first year, even before undergoing practical training, because the coursework is dull and the teaching is subpar. Sometimes, one or two less-than-ideal courses are enough to damage the goodwill and faith of the students.

The ChEPS faculty consists of both KMUTT and outside lecturers. Nearly 50% of the courses are taught by overseas professors from Singapore, Australia, and Canada, experts outside KMUTT, and ChEPS' industrial sponsors such as those from The Siam Cement Group (SCG) and Thai Oil Public Co., Ltd. (ThaiOil). At the closing of each course, students are asked to anonymously complete an online evaluation form to assess the quality of the course and its teaching. Table 2 shows the 12 courses in the first year of ChEPS, their core contents, and the instructors.

Another important aspect of the ChEPS curriculum which helps students prepare for their subsequent internships is the incorporation of problem-based learning (PBL) into some core

Table 2. Coursework in the first year of ChEPS

Course Name	Core Contents	Instructors
LNG601: English for International Program	English grammar and writing	KMUTT
CHE610: Fundamentals of Transport Phenonmena	Transport processes of momentum, heat, and mass	Outside
CHE641: Intermediate Chemical Engineering Thermodynamics	Applied thermodynamics such as physical property models	Outside
CHE642: Chemical Reaction Engineering	Advanced reactor design and kinetics	KMUTT
CHE643: Applied Process Chemistry	Organic and organo-metallic chemistry, petroleum processing	Overseas and Industry
CHE651: Mathematical Analysis for Chemical Engineering Practice	Mathematical analyses of chemical systems via ODEs and PDEs	KMUTT
CHE654: Simulation and Modeling	MATLAB and dynamics of chemical processes, PBL-integrated	KMUTT and Industry
CHE655: Fundamentals of Chemical Engineering Practice	Plant design projects and group presentations, PBL-integrated	None
CHE656: Computer Applications for Chemical Engineering Pracice	Simulation of chemical processes with ASPEN PLUS	KMUTT
CHE657: Design Problems in Chemical Engineering Practice	Industrial projects and oral presentations, PBL-integrated	KMUTT and Industry
CHE658: Systems Engineering	Process control and process optimization	Overseas
CHE670: Engineering Management	Project management, financial analysis, marketing, and planning	Industry

courses during the first year. PBL is vital in helping students hone their problem-solving and presentation skills. In most cases, these case studies, which tend to be too narrow in scope to be site projects, are solicited from sponsoring companies. The PBL projects introduced into the classroom mimics certain aspects of the site projects. The similarities are:

- Students work in teams of three or four people.
- Typical projects involve modeling, simulation, and optimization of chemical processes or systems.
- Regular oral presentations are scheduled to keep the faculty abreast of the latest progress.

There are a number of ways for ChEPS to solicit PBL projects. One is through its site directors. Sponsoring companies such as ThaiOil, SCG subsidiaries, and PTT Chemical Public Co., Ltd. (PTT-Chem) often propose several potential site projects at a time. Some of these site proposals, if not selected as site projects, could be given to the first-year students as PBL projects (also known as design problems). All of these require careful planning, coordination, and many meetings, usually many months in advance, before the PBL projects are ready to begin.

Despite the similarities, one major difference between PBL and site projects is that, after an initial briefing by the sponsors, their involvement is kept to a minimum so as not to disrupt their routine work schedule. Instead, a ChEPS faculty member is assigned to provide technical advice on a regular basis. Students are encouraged to use only emails to contact the sponsors for further input and clarifications. On the other hand, sponsors usually ask for one progress presentation. After eight weeks, a final presentation is held at the

Table 3. Companies sponsoring ChEPS' PBL projects

PBL Sponsoring Company	Abbreviation
Alliance Refining Co., Ltd.	ARC
PTT Chemical Public Co., Ltd.	PTT-Chem
Rayong Olefins Co., Ltd.	ROC
Thai Lube Base Public Co., Ltd.	TLB
Thai Oil Public Co., Ltd.	TOP
Thai Plastic and Chemicals Public Co., Ltd.	TPC
Thai Polyethylene Co., Ltd.	TPE
The Aromatics (Thailand) Public Co., Ltd.*	ATC

*Now called PTT Aromatics and Refining Public Co., Ltd. (PTTAR)

sponsoring company, by which time it will also receive a copy of the final report.

At present, PBL is integrated into one course in each academic term. In the summer, groups of students work on plant design projects as if they were employed by an engineering firm. In the first semester, PBL is incorporated into a modeling/ simulation course which teaches MATLAB and dynamics of chemical processes. In the second semester, PBL is part of an optimization course which is a crucial element in process engineering. In all cases, oral presentations and written reports are compulsory. The students are also critiqued on their presentation skills.

The use of PBL is a great way for ChEPS to network with industries. In fact, not all case studies in PBL come from companies that serve as practice stations. In many cases, ChEPS is able to solicit projects from other companies through its alumni who work there. In addition, some companies are interested in sponsoring PBL because of the possibility of their becoming future practice stations. To these companies, sponsoring PBL is an ideal way to test-drive ChEPS and its practice model before more substantial resources are committed. Table 3 shows past contributors to ChEPS' PBL projects and Figure 3 shows the number of PBL problems sponsored by each company.

Future Challenges in Coursework and Solutions

ChEPS students are by and large very happy with the coursework. This was not always the case, especially in the early years of the program. A lot of efforts have gone into finding the appropriate people to teach the courses and making certain that all the courses follow the same or similar format. Grading must be kept fair and objective. One feature of ChEPS classes is that they tend not to follow a fixed timetable. In fact, many classes are taught on weekends and many courses are crammed into just a few weeks in order to accommodate the busy schedule of overseas lecturers and instructors from industry. This sometimes poses a bit of a challenge to the ChEPS administrative staff in reserving classrooms and confirming all classes. Furthermore, students tend to suffer from more stress during these busy block courses. However, the benefits of having a diverse and qualified teaching body in the program far outweigh the minor inconveniences. In fact, block courses can help prepare ChEPS graduates for the real world, since the ability of professional engineers to absorb a large amount of materials in a short time is an important skill. Most subsequent courses in professional development are run in this manner.

On the other hand, implementing PBL in a course is extremely time-consuming and has yielded mixed results. An ideal PBL project should be sponsored by a company, is well-defined, and is supported by fairly complete plant data. In reality, it is difficult to solicit enough PBL projects from the industry. It is even more difficult to find projects that meet all of the three stated criteria. Even when such a problem exists, the efforts and time needed to implement PBL can be enormous. Students must work closely with their KMUTT advisors and the sponsors, since the problem can be open-ended with no single correct answer. One constant challenge is finding well-defined design problems that are supported by reliable plant data. When the input information in a design problem is

Figure 3. ChEPS PBL projects sponsored by companies

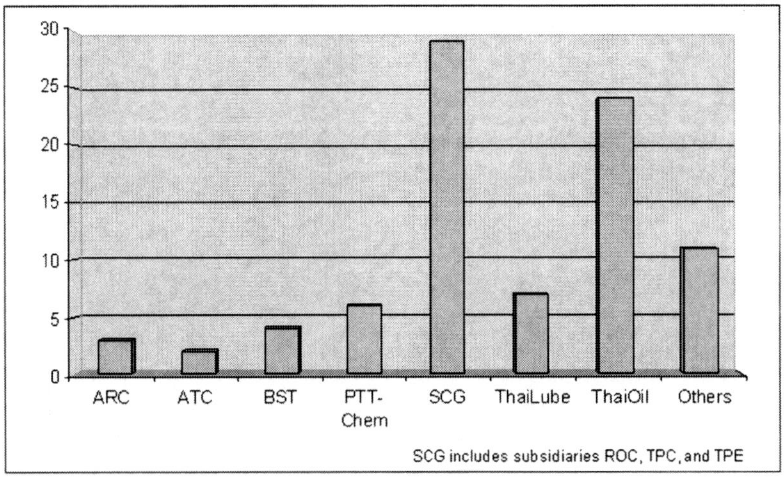

incomplete, assumptions need to be made which render the solutions less meaningful. Another challenge is the commitments of the sponsoring engineers of PBL, who are usually tied up with their routine work and have little time to interact with the students. Finally, because the nature of the design problems is very varied and spans the full spectrum of the chemical engineering fields, finding the faculty with the appropriate expertise to advise the students is not easy.

Whenever ChEPS runs into a dearth of design problems, the workaround is to substitute one with a problem from the literature. While this is less than ideal because of the problem's academic nature, the students at least have an opportunity to work as a team on a problem which is well-defined. In conclusion, ChEPS students generally find the integration of PBL projects into their core courses useful but are sometimes frustrated by some of the obstacles, which are actually important for their learning.

Financial Support and Scholarships

There are many graduate chemical engineering programs in Thailand, some of which have even longer history than ChEPS. The competition for quality students is therefore fierce, and has recently intensified as more programs are offered. Scholarships and financial support in the form of monthly stipends are needed in order to attract good students, as those with outstanding GPAs do not expect to pay for their graduate studies. As a result, ChEPS needs to secure financial resources to afford these scholarships.

ChEPS offers scholarships and monthly stipends to qualified candidates. The financial support comes from numerous sources, including funding agencies, sponsoring companies, donations, and loan payback from alumni. The funding agencies are:

- Energy Policy and Planning Office (EPPO), formerly National Energy Policy Office (NEPO)
- National Science and Technology Development Agency (NSTDA)
- Petroleum Institute of Thailand (PTIT)
- The Suksapattana Foundation

The mission of EPPO under the Ministry of Energy is to foster energy conservation and promote public awareness regarding energy savings. Since ChEPS engineers often help the industry to save money by minimizing energy consumption, EPPO has a direct interest in funding the program. NSTDA is a public organization whose mission is to advance and sustain the economic development of Thailand through research, technology development, and the promotion of collaboration between the public and the private sectors. PTIT is an independent non-profit organization supported by the government, academic and private sectors. PTIT's members comprise of petroleum and petrochemical companies, and the institute's main mission is to assist with the development of the two industries in Thailand in areas of human resource development, information service, policy and regulatory issues. Finally, the Suksapattana Foundation is a non-profit organization dedicated to the development of education and innovation in learning.

Companies who have contributed resources, both in-kind and cash, to ChEPS include:

- BST Elastomers Co., Ltd.
- ExxonMobil Exploration and Production Khorat Inc.
- ExxonMobil Limited
- PTT Aromatics and Refining Public Co., Ltd.
- PTT Chemical Public Co., Ltd.
- PTT Group and PTT Research and Technology Institute
- Thai Lube Base Public Co., Ltd.
- Thai Oil Public Co., Ltd.
- The Siam Cement Group and its subsidiaries

More recently, the Thai government channeled a substantial amount of money to support all practice-based programs in the country via NSTDA, which initiated the University-Industry Research Collaboration Program (U-IRC). The objective is to encourage more collaborative research between the universities and industries, which will improve corporate Research and Development and help enhance their competitiveness. KMUTT was the largest recipient of this funding.

ChEPS in turn offers three types of scholarships to students: full scholarship, half scholarship, and soft-loan. Under no circumstances are students required to pay while they study in ChEPS. Those who are not qualified for full scholarships are offered no-interest soft loans, which are to be paid back each month amounting to 10% of their salaries once they start working. Full scholarships are unconditional, while half-scholarship recipients must pay back the other half given out as loans. Together, full and half scholarship recipients account for about one-half of the students. The soft-loan with 0% interest has been one of the attractive selling points of ChEPS, enabling needy students to be admitted into the program without having to take a private loan.

Future Challenges in Funding and Solutions

It is not realistic to expect the funding agencies and the government to finance ChEPS indefinitely. For example, the U-IRC program is only good for two years or one class of students, and its future funding appears uncertain. Sooner or later, all practice schools must demonstrate their ability to sustain themselves financially. For future financial resources, ChEPS will need to rely on subsidies from KMUTT, financial contributions from its alumni, and more grants from industries. For that to happen requires a larger alumni base and the willingness of industrial sponsors to invest in higher education. At present, ChEPS continues to receive payback from its alumni who were on soft-loan and half-scholarships. While this payback represents a relatively small percentage of ChEPS' total annual operating budget, the amount is expected to grow substantially as more students graduate each year and join the workforce. It is projected that in a few more years, the alumni fund will grow sufficiently large to cover 50% of the budget.

One problem facing the program, especially in the early years, is the soft-loan payback from the alumni. Because the soft-loan agreements between ChEPS and students are not contracts and therefore are not legally binding, many students were unscrupulous in their payments. The payments were often intermittent or late. Fortunately, ChEPS maintains a database that tracks employment records and personal information, such as home addresses, email addresses, and mobile numbers, of its alumni. So the program has a good idea of the whereabouts and job functions of its alumni. The database is useful in sending out friendly reminders about overdue loan payments. Moreover, many ChEPS alumni end up working for the program's industrial sponsors who have good working relationships with ChEPS. Usually, a more aggressive reminder by email or letter is enough to convince most alumni to fulfill their obligations. A future ChEPS alumni association is desirable and would prove useful as a networking forum among the alumni, the program, and the industrial sponsors.

ChEPS has in the past asked and continues to ask all industrial sponsors to contribute more financially to the program, particularly in the form of scholarships and service fees for site projects. Having been shown the positive impact of practice school on industries, these companies are generally willing to shoulder some of the financial burden with some scholarships, housing accommodations, student allowances, and other in-kind contributions. However, while the sponsoring companies agree that the practice school benefits them, they do not think they should bear the financial burden alone. First is the issue of how to accurately quantify the savings to the companies after site projects have been completed. Second is the prevailing thinking that the cost of educating graduate students should partially lie with the government, the university, or even the students themselves. Thailand does not have a tradition of big endowments on higher education by the private sector. Furthermore, industrial sponsors of ChEPS tend to be big public companies which are constrained by their boards of directors, shareholders, and highly regulated policies. Consequently, securing substantial endowments from these companies has always been a challenge.

ChEPS will continue to convince companies to see the importance of funding higher education and the values of the site projects. It is now customary for ChEPS interns to calculate and attach a cost saving or profit increase to every site project. The purpose is to demonstrate to the sponsoring companies about the values of research and its potential benefits to the companies' bottom lines, which hopefully will encourage them to invest more financially in ChEPS. As more petrochemical and refinery companies embark on developing their own in-house Research and Development, it becomes obvious that the required human resources, know-how, and knowledge must come from universities. It is hoped that these companies will eventually develop a tradition of supporting universities throughout the country with more endowments.

In the end, for long-term sustainability a practice school should be run more akin to a business joint venture. As in any business model, there are stakeholders. In the practice school, the three stakeholders, namely the university, the student, and the industry, should contribute and share the cost of running the practice school since they are the primary beneficiaries. To a certain extent, this business model is working well at KMUTT, at least in the short term. However, more needs to be done to integrate all practice-based curricula at KMUTT into one practice school with shared resources and to convince all stakeholders to sustain the school over the long term.

Operating Practice Stations

For the practice school to be successful, a sponsoring company must be completely committed to the partnership. This means that the company's staff at all levels must be brought in early in the decision-making process, beginning with the managing director, plant managers, engineers, all the way down to technicians and shift operators. The company's management sets the policy and allocates the necessary resources to run the practice station. But it is the engineers and operators with whom the students have to work daily. In other words, everyone in the company must be a firm believer of the practice school model and is totally committed to making it work.

Initially, the partnership between ChEPS and the companies are officially cemented through the signing of a memorandum of understanding (MOU) or a contract. The agreement clearly spells out responsibilities of each party, expectations from each side, details about financial obligations, and issues regarding safety and ownership of intellectual property. For example, it is a standard practice for companies to offer in-kind contributions, such as office facilities, computers, and administrative support, and pay for housing of the practice team. In some instances, the companies pay monthly stipends to the students. In the past, ChEPS has been able to charge some companies consulting fees on the per-project basis as well. In any case, each practice station is somewhat different. There is no fixed formula on cost sharing, which is dealt with on a case-by-case basis.

What distinguishes a practice school from a regular co-operative program is the presence of a full-time site director at a practice station. The site director is a faculty member with a PhD degree who runs a practice station. In the first few years of ChEPS, the site directors received their training through mentoring and working closely with their counterparts at MIT's practice stations in the US for six months. In subsequent years, new site directors were trained in a similar fashion by spending a few months at ChEPS practice stations. At present, there are two full-time site directors in ChEPS who supervise 20-25 students at various practice stations.

The site director lives and works with student interns in housing provided by the sponsoring company. While company engineers identify and set the scope of the projects, the site director is responsible for ensuring the academic value of the proposed work, that the project goals are attainable, and that the work is carried out as planned. Furthermore, the site director provides technical advice, prepares students for presentations, and edits students' reports.

The site director has the following specific responsibilities:

- *Identify projects*: The site director must work closely with the industrial sponsors at the outset to identify potential site problems. Here, the site director's main job is to screen the proposed projects and to ensure that the ones chosen contain academic values. For example, asking students to survey existing tray types because the company wants to design a new distillation column may be of interest to the sponsors, but the project offers little technical value from which the students can learn.
- *Prepare a working timetable*: The site director is responsible for forming student groups and planning a work schedule that includes all the presentations.
- *Provide technical advice*: While the students rely on company sponsors for their technical expertise and advice, the site director should be able to advise students on the more general technical issues. The frequent meetings allow company engineers and the site director to exchange ideas and learn from each other.
- *Review and edit reports*: The site director is expected to spend a great deal of time reading and editing students' reports. Since all written documents are in English which is not ChEPS students' native language, the site director must also edit grammar and diction in all reports. Moreover, the site director ensures that all oral presentations go smoothly by reviewing students'

presentation slides and listening to their rehearsals.

Because of their close proximity, site directors have more access and more opportunities to see the sponsors. Although not their main job function, it is customary for site directors to elicit research topics from the sponsors as ChEPS theses to be carried out in the second year. These individual research problems, which tend to be too big to be site projects, may be advised by any faculty in ChEPS including site directors. This joint research adds another dimension to the collaboration between the practice program and the company.

A few weeks before practical training is to begin, the sponsoring company will form an advisory committee, typically consisting of division heads, engineers, and plant operators. The site director will then meet with this committee to identify projects and produce a detailed work plan. The duration of the practice phase is five months (one semester). Students work in teams of two or three on two projects in series (two phases), and take turns being the project leader. As a result, each project is ten weeks long, during which there is one proposal, one progress, and one final presentation. Figure 4 shows the working dynamics among the ChEPS faculty, the sponsoring companies, and the practicing students.

The operating procedure of site projects is outlined below:

- *Problem Assignment:* The problem statement will provide students with background, motivations, and a list of objectives for the project. In addition, names of company employees who might be of help to the students are given. Also included in this document are possible solution methods and issues to consider which will help the students get started.
- *Project Planning and Organization:* Since time is of essence in solving a site project, students must place special emphasis on planning and organizing the project. The first logical step is to investigate thoroughly the background of the given problem. Specific objectives of the project are then defined, followed by the scope of work and a systemic solution approach. It is interesting to note that information feedback is a vital part of this process, i.e. the initial plan of actions must be continually reviewed in light of the most current results.
- *Sponsors Meeting:* A meeting with station sponsors will take place one or two days after the problems have been assigned. This meeting is crucial and requires good preparation to ensure the students get a strong

Figure 4. Working relations in a practice school

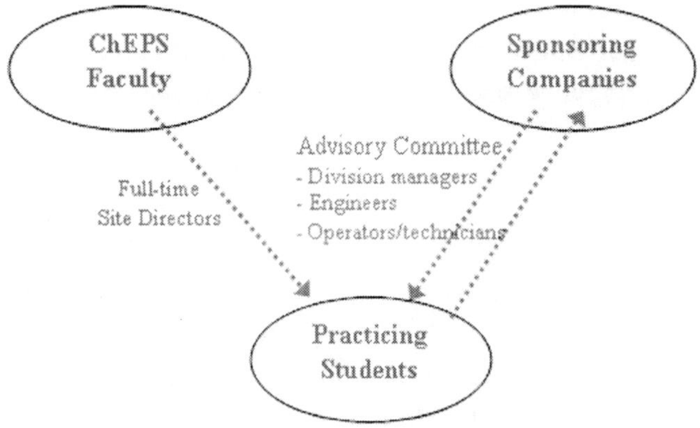

start on their projects. The group leader is expected to run that meeting. Progress meetings may be scheduled if such need arises, e.g. when students run into serious obstacles or need to air major complaints.

- *Investigative Memorandum (IM):* All student groups are required to write a brief description of the problems and their solution methodology during the week after the problems have been assigned. Safety is an important part of the IM, and special attention is given to sections on operational procedures and hazards of materials. The site director will review the first draft of the IM. Company sponsors will then comment on the document. An IM contains the following sections:
  1. Background
  2. Objective(s)
  3. Method of Approach
  4. Safety
  5. Hazards of Materials
  6. References
  7. Work Schedule
- *Proposal Conference:* Shortly after the IM has been submitted, students will schedule a proposal presentation with the sponsors, in which objectives and a detailed solution methodology will be presented. This is an important forum for sponsors to ask tough questions and critique the proposed work. By this time, the students should have familiarized themselves with the appropriate theories, previous experimental work if any, and other relevant technical information.
- *Progress Report:* Each project will have a mid-phase oral presentation in which the students update the company with the general status of their work, including what is planned.
- *Final Report:* The final oral presentation and report complete the engineering work. The final report should clearly state the conclusions and recommendations. The body and appendices of the report should also include sufficient technical materials and data for other engineers to duplicate or verify the students' findings.

The following is a list of past and present practice stations in the ChEPS program. Except for ThaiOil and ExxonMobil which are located in the Sri-racha Province (about 150 kilometers southeast of Bangkok), the remaining companies are all located in the Mathaphut Industrial Estate in the Rayong Province (about 180 kilometers southeast of Bangkok).

Future Challenges in Operating Practice Stations and Solutions

The cooperation and support that ChEPS receives from the sponsoring companies at all levels have been exceptional. In nearly all cases, ChEPS site directors and students have excellent working relationships with the engineers and operators. ChEPS is not aware of any major conflicts between the two parties that could have jeopardized the site projects. However, the operators and even engineers may occasionally feel slighted when the final presentations given by ChEPS students were well attended by the top management. In those cases, a joint presentation given by both the ChEPS students and the company's working team should be considered.

At the same time, site directors play a vital role in the success of ChEPS. They are the "eyes and ears" of the program who interact directly with the practicing students and the sponsors. Not only do they need to possess the necessary technical skills, preferably in process engineering, simulation, and optimization, they must also have good people skills and a good command of the English language. On the other hand, not all ChEPS site directors are well-versed in process engineering and simulation. Moreover, convincing the faculty to become site directors is not easy, as

Table 4. Past and Present Practice Stations of ChEPS

Company Served as a Practice Station	Service Period
Aromatics (Thailand) Public Co., Ltd. (The)	2002-2005, 2007
BST ELastomers Co., Ltd.	2001-2002
Exxon Mobil Limited	2007
PTT Chemical Public Co., Ltd.	2008-Present
Rayong OlefinsCo., Ltd.	2002-Present
Thai Oil Public Co., Ltd.	1998-Present
Thai Polyethylene Co., Ltd.	1998-1999, 2001, 2007-Present
Siam Mitsui PTA Co., Ltd.	2003, 2007-Present

most of them prefer to focus on their research on campus and are reluctant to re-locate to provinces. Monetary compensation and the chance to network with industries are the key incentives. Also, it was found that younger and new faculty staff are more likely to agree to move to the site because of the need for industrial exposure and their propensity to try new things. One reason ChEPS has been unable to expand its class size beyond 30 students each year despite strong interest in the program is a lack of qualified site directors.

ChEPS and industrial sponsors do not wish to see a separation between site and campus activities, which arises naturally because of the long physical separation between practice stations and the KMUTT campus. Sponsors value the input and participations of people from campus with site activities. In some site projects, the site director does not have all the expertise to advise the students. Instead, the faculty on campus may be able to offer new perspectives or better insights into solving a problem. Every time there is a presentation at a practice station, ChEPS does its best to send at least one faculty member from KMUTT to attend. Sometimes first-year students are asked to accompany this faculty member so they can see first-hand the activities at the sites. Because the presentation forums occur fairly frequently at practice stations, finding someone to travel to the sites each time can be difficult. Moreover, the expense in terms of time and money in transporting ChEPS faculty to the remote practice sites to attend the presentations is significant. Since ChEPS often has at least two site directors working concurrently at two different locations that are in close proximity, one site director is sometimes asked to cover the presentations of another site director. That helps reduce the number of trips between the campus and the practice stations.

Another issue that ChEPS must contend with is the fact that because all of ChEPS' practical stations are large refinery and petrochemical complexes, only a small section of the plants can be studied at a time. Moreover, companies are reluctant to divulge any confidential data or information, particularly those pertaining to costs, prices, or savings. Thus, it is not easy to put the value of a site project in monetary terms, although ChEPS strongly encourages the students and the industrial sponsors to include economic evaluation as part of each site project. The aim is to quantify and put a dollar figure on the success of each project.

ChEPS has complete confidence that site projects are valuable to the sponsoring companies. Some companies such as ThaiOil have openly stated that ChEPS students have over the years helped the company save millions of dollars in costs and energy consumptions. However, there is little ChEPS can do if a sponsoring company is not willing to share more precise figures or reveal how much an impact a project has on its bottom line. At the same time, it must be admitted that not all site projects are of high-impact or high-value in nature. Even in cases where the savings can be quantified, they are simply potential savings unless the recommendations are implemented. Consequently, as much as ChEPS tries to glean financial information from its sponsors with questionnaire and personal interview, it is extremely difficult for the program to assess the impact of site projects in monetary terms.

Thesis Projects and Overseas Collaboration

The ChEPS curriculum requires that all students undertake a short research thesis (6 credit hours as opposed to the regular 12 credit hours) for six months in the second year. This component is missing in the MIT's practice school. ChEPS feels that the inclusion of a thesis in its curriculum adds another dimension towards improving students' critical and analytical thinking. This training is particularly vital to any ChEPS students who wish to pursue doctorate degrees. Research topics in ChEPS span just about any field in chemical engineering, such as process simulation and optimization, fuel cells, catalysis, environmental engineering and clean technology, biomolecular engineering, biotechnology, waste engineering, energy, polymers, and membranes. These thesis projects are mainly supervised by faculty within the chemical engineering department, although a few are collaborated with or supervised by people outside the department or the university.

When ChEPS was first envisioned, there was little emphasis on collaborative research with overseas institutions. As the year progressed, it turned out that the opportunity to conduct research overseas was the main incentive in attracting students to enroll in the program. One of ChEPS main competitors in Thailand has overseas academic partners and allows some of their students to carry out their research overseas. This has been a strong selling point of that program. When recruiting students, ChEPS is often asked about the opportunities for overseas research.

In recent years, about one quarter of ChEPS students carry out their research during the second year outside Thailand. The students typically spend 4 - 5 months conducting research under the supervision of ChEPS overseas partners. Since all of the faculty members in ChEPS have doctorate degrees from overseas institutions, mostly from North America, the program has an extensive network of overseas connections. Moreover, academicians tend to be open to collaborative research. As a result, it is not difficult to find overseas academic collaborators who are willing to take ChEPS students. Table 5 shows past and present overseas research partners of the ChEPS program.

Future Challenges in Overseas Collaboration and Solutions

Every ChEPS research thesis is required to have a KMUTT advisor. However, because overseas research topics usually originate from ChEPS' counterpart, it is often difficult to find an advisor at KMUTT or in ChEPS who is specialized in the same field. This could sometimes be problematic when the students have no expert to advise them prior to defending their research proposals and going overseas. Furthermore, once the students go abroad, it is difficult to keep track of their progress which could jeopardize the chance of their completing the thesis in 5 months. One solution is to require that all overseas students submit bi-monthly progress reports of their research to their KMUTT advisors.

Another potential problem with overseas collaboration is to ensure that the research projects are well-defined and that their scopes are sufficiently narrow to be completed in 4 - 5 months. The standard procedure is for ChEPS to first screen the projects scrutinizing their scopes and methodologies, find co-advisors within KMUTT

Table 5. Past and present overseas research partners of ChEPS

Overseas Research Partners	Location
Carnegie-Mellon University	Pennsylvania, USA
Clemson University	South Carolina, USA
EnerDel, Inc.	Indiana, USA
James Cook University	Townsville, Australia
National University of Singapore	Singapore
University of South Florida	Florida, USA
University of Toronto	Toronto, Canada
University of Waterloo	Waterloo, Canada

who have the appropriate expertise in the proposed fields, and match the projects with ChEPS students who have interest in the fields. On the other hand, an individual research project often represents the first time a student is to conduct research independently. The student must be capable of understanding technical papers and assimilating new ideas with little supervision. The time constraint and the alien environment in overseas collaboration pose unprecedented challenges to the student, not to mention the language barrier. However, when this arrangement works well, the student benefits tremendously from the overseas experience, including a chance to interact with other international graduate students, work closely with experts in the research field, and improve her English proficiency.

At present, the overseas research collaboration seems too haphazard. The reason is that there is no official agreement between KMUTT and the overseas partners, either at the university level, the faculty level, or even the departmental level. Instead, all collaborations have been based on personal contact and connections. Consequently, it is difficult for ChEPS to match the needs of the students with those of the overseas institutions. For example, in certain years there could be a surplus of overseas projects, while in other years, many people vie for only a few available projects. If more students continue to show a strong interest in overseas research, the collaboration should be made more formal via MOU agreements between KMUTT and overseas institutions.

Program Output and Evaluations

The success of ChEPS can be judged on many levels. One is the academic records of the admitted students. Table 6 shows the distribution of universities and the average GPA of admitted students from Class 1 to Class 14. The table shows that ChEPS students are well represented by every university in Thailand that has a chemical engineering department, and that the caliber of the students applying to and admitted into ChEPS has been exceptional. Academically, admitted students are typically ranked in the top 15% of their respective classes with an average GPA of 3.20.

An objective assessment of site projects after their completion is also crucial and is an integral part of the practice school used to judge the success of the program. This documentation helps students identify their areas of weakness. At the same time, sponsors also use this evaluation to document the impact of the students' work and convey the findings to the senior management, who will decide if follow-up studies or implementations of the proposed ideas are warranted. Moreover, some financial supporters of ChEPS requires that the program submit the findings, often in the form of short abstracts, to demonstrate how their funding benefits the chemical and pet-

Table 6. Academic profile of admitted ChEPS Students, Classes 1 - 14

University	No. of Students	Avg. GPA
Burapha University	5	3.05
Chiangmai University	11	3.33
Chulalongkorn University (Chemical Technology)	16	3.19
Chulalongkorn University (Chemical Engineering)	24	3.14
Kasetsart University	42	3.15
King Mongkut's Institute of Technology Ladkrabang	39	3.20
King Mongkut's University of Technology North Bangkok	19	3.16
King Mongkut's Universityof Technology Thonburi	98	3.25
Khon Kaen University	1	3.17
Mahidol University	13	3.13
Prince of Songkla University	4	3.37
Silpakorn University	1	2.90
Srinakharinwirot University	1	3.45
Suranaree University of Technology	1	3.09
Thammasart University	19	3.19
University of Waterloo	1	Excellent Stand.
Total	295	3.20

rochemical industries, e.g. in reducing energy consumptions or boosting the industries' competitiveness.

The assessment of site projects is divided into two parts. The first is the evaluation of the students' performance by the site director alone. Site directors then work closely with the sponsors in the second part to evaluate the overall success of the projects, including the amount of money the company can save if recommendations in the study are implemented. Further calculations may be needed here, since not every project will contain cost-saving analyses. It should be noted that ChEPS is not privy to this information; as such data are considered confidential by the company. Typically, the students are judged in the following four broad categories:

- Problem-solving approach
- Accuracy and completeness of work
- Presentation skills
- Command of the English language including final reports

There are five numerical ratings in each category with 5 being the best. Some site directors prefer to conduct a student evaluation right after the presentation, in which each student presenter is critiqued by her peers. Figure 5 shows the performance evaluation form used by ChEPS for site projects.

Another indicator that can be used to gauge the success of the ChEPS program is the track record of employment of its alumni. Figure 6 shows the employment profile, categorized by industrial sectors, of ChEPS alumni based on 235 students from 11 classes. About half of the graduates currently work for large chemical, petrochemical, and refinery companies. A sizable number also work for small-to-medium enterprises (SMEs), e.g. those in the sugar and food industries. Finally, about one-tenth of ChEPS

Figure 5. ChEPS performance evaluation form for site projects

KING MONGKUT'S UNIVERSITY OF TECHNOLOGY, THONBURI

Chemical Engineering Practice School (ChEPS)

Performance Evaluation Sheet

Project: ______________________________

Student Names: ______________________ Phase: ____ Semester: ____

Overall Evaluation:

<u>Major Strengths</u>

<u>Areas for Improvement</u>

	Needs improvement	Outstanding		Needs improvement	Outstanding
	1 2 3	4 5		1 2 3	4 5
Engineering Skills			Organization skills		
Quality of technical work			Communication:		
Creativity			Oral		
Leadership			Technical writing		
Dependability			Personal		
Motivation			Engineering judgment		
Sense of responsibility			Ability to make decisions		
Response to technical guidance			Ability to work under pressure		

Name and Signature of Evaluator: ______________________ Date: __________

alumni has become affiliated with universities or pursued PhD degrees.

As a financial supporter of ChEPS, NSTDA in 2002 appointed Chulalongkorn University Intellectual Property Institute (CUIPI) to conduct an independent audit of ChEPS and evaluate the outcomes of its collaboration with industry. At the time, ChEPS had been running for five years, and there was considerable interest from all stakeholders including KMUTT to know if the program was on the right track. CUIPI set up a working team, who visited KMUTT, interviewed ChEPS students and alumni, talked to industrial sponsors, and reviewed important documents in the program. CUIPI scrutinized four aspects of ChEPS, namely financials, external administration, internal administration, and knowledge management. Each aspect was further divided into sub-categories, which were rated against benchmark numbers. The study concluded that in most cases ChEPS was able to meet or exceed the target numbers. The only weakness was in the knowledge management where the students failed to benefit sufficiently from the short research in the curriculum. The study put the blame on the time constraint and the limited scope of research, which resulted in few conference papers and publications.

Finally, ChEPS also conducts periodic surveys with site companies and employers of ChEPS graduates, although these surveys tend to be

Figure 6. ChEPS graduates employment profile, Classes 1 - 11

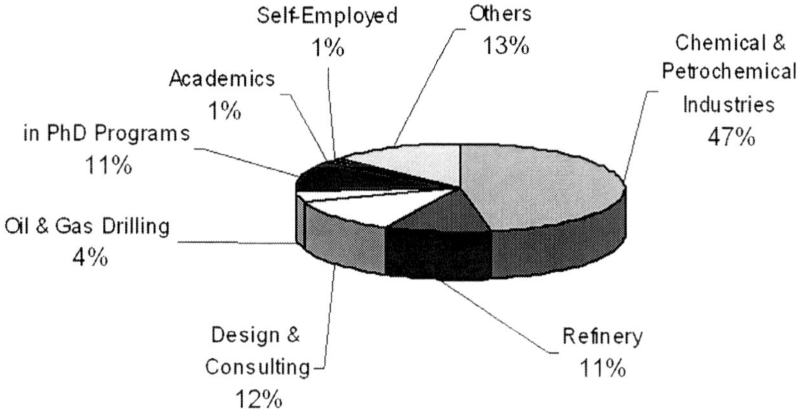

informal. The feedback so far has been positive, and preliminary data show that companies are generally happy with the performance of ChEPS graduates. The strengths of the ChEPS graduates often cited are good English proficiency, good presentation skills, self-confidence, and a short learning curve. The fact that most sponsoring companies continue to make their sites available is a testimony to the benefits and values of the practice school.

Future Challenges in Program Evaluations and Solutions

It is obvious that based on several criteria, such as the academic credentials of incoming students, their subsequent employment record, and the feedback from the employers, the ChEPS program can be regarded as successful. When a large company looks to hire a chemical engineer with a master's degree, ChEPS graduates tend to be strong candidates. In addition, if ChEPS alumni were treated as customers, their level of satisfaction would be ranked very high. To date, very few alumni regret enrolling in ChEPS. The majority only have good things to say about the program.

On the other hand, much of ChEPS output has not been quantified or has not been done so for a long time, thus making it hard to evaluate the program objectively. The core issues are the impact of site projects on the sponsoring companies' bottom line and the extent at which the program meets the students' expectations in terms of skills and knowledge acquisition. It has been almost 10 years since ChEPS was last evaluated. Another systematic and comprehensive study should be carried out in the near future to assess the academic rigor of ChEPS and determine the impact of the program on the chemical, petrochemical, and refinery industries in Thailand.

After more than a decade in operation, the ChEPS curriculum remains largely the same. Work is currently underway to revamp the curriculum to reflect the changing needs. For example, applied thermodynamics will be emphasized in the Intermediate Thermodynamics course. Process Control and Process Optimization in the Systems Engineering course will be split into two smaller courses, since the two subjects are quite different and involve two different instructors. Other soft-skill developments such as mind-mapping and Lego Mindstorms workshops will also be introduced into ChEPS as part of the students' orientation. Awareness of self, others, and the environment, e.g. social and ethical responsibilities, should also be an important part of the students' engineering education. Community services such as visiting orphanages and building houses for the homeless and needy will be included in the students' schedule. Finally, to strengthen the focus and bring peace of mind to the students before being immersed in the intensity of the curriculum, an extended meditation trip to a temple in the province will be organized. Some of these initiatives were implemented in the past, but the activities were too few and far between and were never implemented as parts of students' routine.

FUTURE RESEARCH DIRECTIONS

The success of the practice-based learning model originating from ChEPS at KMUTT has created and inspired many spin-off programs both inside and outside the university in the past 10 years. In addition to ChEPS, practice-based curricula in food engineering, bioinformatics, biotechnology, and biotechnology business management were subsequently opened at KMUTT. Moreover, a number of engineering departments outside KMUTT in Thailand have also introduced industrial projects into their curricula, although none of them were as organized and on the same scale as ChEPS and those at KMUTT.

Starting in 2005, ChEPS also expanded its program to include a PhD component, called PhD-ChEPS. The objective is to link academically outstanding students with companies who are developing in-house R&D. Qualified PhD candidates would enroll in the Department of Chemical Engineering at KMUTT, but are sponsored financially by host companies and work on research topics originating from them. In some cases, the PhD students are also co-supported by The Royal Golden Jubilee Ph.D. Program (RGJPhD) provided by the Thailand Research Fund (TRF). TRF is a juristic body that is technically a part of the government system but operates outside the government administrative bureaucracy for the sake of efficiency. The mission of TRF is to foster basic research and technology development at the graduate level and address the severe shortage of highly qualified human resources in science and technology in Thailand. Some funding in the the RGJPhD Program explicitly requires that the PhD projects be collaborated with industry. So far, PhD-ChEPS has started small and admits only 1 or 2 students each year because of the rather stringent constraints. The SCG and the PTT Group are the two corporate sponsors.

In the private sector, the SCG has been ChEPS' industrial sponsor since the beginning and saw first-hand how the practice-based model could help learners. Consequently, the company partnered with the Department of Chemical Engineering at KMUTT and created C-ChEPS (Constructionism-ChEPS) in 2000 and C-Pulp in 2001 for its petrochemical and pulp and paper subsidiaries, respectively. In 2007, another collaborative program called PI-ChEPS (Productivity-Improvement ChEPS) involving KMUTT, PTIT, and PTT-Chem was launched. The aim of these three programs is to train and upgrade process operators and technicians at SCG and PTT-Chem, who have years of working experience but only vocational degrees, in the fundamentals of chemical engineering. The hallmark of C-ChEPS, C-Pulp, and PI-ChEPS is the emphasis on project-based learning and the use of "facilitators" (equivalent to site directors in a practice school) to help teach, guide, and advise the trainees. In most cases, these PBL projects come from real-life problems that these trainees encounter on a daily basis. The duration of these special programs is about eight months. This new learning model has been so successful that recently a number of petrochemical and chemical companies in Mathaphut Industrial Estate in the Rayong Province in Thailand have jointly set up V-ChEPS (Vocational-ChEPS), a vocational training institute. The institute grants vocational degrees in several engineering fields and produces qualified vocational personnel to serve the local industries.

As the practice school model gains more traction among educators in Thailand, more academic programs, both new and existing ones, will certainly adopt this new learning approach. KMUTT has considered establishing an undergraduate chemical engineering internship program called U-ChEP (Undergraduate ChEPS) based on the ChEPS model. However, the scale of U-ChEPS will be several times bigger than that of ChEPS, and a number of issues such as manpower must be addressed before U-ChEPS can be successfully launched. Finally, the Thai government has begun to recognize the crucial role of the practice model in higher learning. As a result, the government is now making the practice model one of its initiatives and part of its national policy to develop human resources in Thailand.

Despite the interest and the number of practice-based curricula that exist, the amount of research and their effectiveness on learning and impact on industry, especially at the graduate-school level, is relatively small. One key problem is how to quantify program outcomes and evaluations such as students' satisfaction, the extent of soft-skills acquired, and financial impact on host companies. Comparative studies among engineering practice-based programs are also largely missing. While all practice-based programs share many commonalities, local academic and industrial cultures may entail different kinds of challenges. For example, Asian students tend to be more reserved and passive in classroom and are less likely to question those in authority such as industrial mentors. Another example is how differently public and medium-sized companies operate. Industrial sponsors of some programs such as FEPS are mostly small-to-medium-sized enterprises, and administratively they operate differently from large public companies. Small companies are less rigidly structured, and key decisions regarding practice school sponsorships are often made by the owners or only a handful of people.

A more quantitative study on how to evaluate the practice-based programs is currently underway at the University of Queensland (UQ), Australia. This is a joint PhD research between the two chemical engineering departments at UQ and KMUTT. The objective is to evaluate the two practice-based programs (ChEPS and PEPS) at KMUTT and UQ and conduct a comparative study between the two. The evaluations will be based on the perspectives of the stakeholders, namely the alumni and the sponsoring companies. Possible evaluation methods are in the form of questionnaire surveys and focus group interviews as well as analyses of cost savings in site projects. For example, the alumni would be asked to indicate whether the curricula met their expectations, enhanced their self-confidence, and motivated them to become lifelong learners. On the other hand, sponsoring companies will be asked to grade the alumni who work for them and whether these graduates of the programs possess the skills and attributes needed to succeed as chemical engineers. Other indirect benefits from the university-industry collaboration will also be explored. Finally, the research should help identify areas for improvement as well as organizational and administrative issues that may presently hinder the operation of the programs. As stakeholders, the alumni and the sponsoring companies are also in the best position to offer assistance and solutions for the challenges still facing the programs.

CONCLUSION

MIT in the US has successfully operated its master's degree chemical engineering practice school for nearly 100 years. The uniqueness of the practice model lies in the industrial internship, which trains students to be effective problem solvers. In 1997, KMUTT brought the practice-school model into Thailand as part of a reform initiative to improve the quality of its engineering students and to better serve the needs of industry. Thus, the Chemical Engineering Practice School or ChEPS was born. ChEPS incorporated the essential elements of the MIT model but was expanded to reflect the needs of Thai students. Problem-based learning and work intensity, together with industrial involvement, are the keys that drive the students to excel and contribute to the success of the program.

ChEPS is now in its 14th year since inception, and has since established a reputation as a premier chemical engineering program in the country. Each year, nearly 100 applicants compete for 25 openings. Admitted students are generally ranked in the top 15% of their graduating classes. Surveys have shown that companies sponsoring ChEPS believe the practice-school model to be a win-win partnership between the university and the private sector. The feedback from companies who employ ChEPS alumni is also generally positive.

The success story of ChEPS prompted KMUTT to extend the practice model to other departments, resulting in the creation of more practice schools in food engineering, bioinformatics, starch engineering, and biotechnology. KMUTT has also been able to parlay the practice school model into partnerships with industry, giving rise to many in-house training programs such as C-ChEPS and PI-ChEPS, that upgrade shift operators and technicians in the companies.

Despite the initial success of the practice schools at KMUTT, many challenges remain. The most pressing issue is the long-term sustainability of these programs and how to convince the private sector to contribute more financially. In the end, all parties, namely KMUTT, the government, the private sector, and even the students themselves, must play an active role in ensuring the continued success of the practice school. After all, they are the stakeholders who will reap the benefits that ensue in this partnership.

REFERENCES

Asavasanti, S., Bhumitratana, S., & Chayawattana, W. (2005). *FEPS: A successful story of constructionism learning approach in Thailand.* Paper presented at ICLORD2005: 19th International Conference, Bangkok, Thailand.

Bendrich, G., & Pugsley, T. S. (1998). The practical side of engineering at the University of New Brunswick. *Chemical Engineering Education, 32*(3), 208–213.

Bioinformatic. (2010). *About Bioinformatic.* Retrieved February 6, 2010, from http://www.bioinformatics.kmutt.ac.th/about.php

Birla Institute of Technology. (2010). In *Wikipedia, the free encyclopedia.* Retrieved February 13, 2010, from http://en.wikipedia.org/w/index.php ?title=Birla_Institute_of_Technology &oldid=342663877

Buarod, P. (2009). *An application of statistical models to the significance of admission criteria for the chemical engineering practice school (ChEPS).* Unpublished master's thesis, King Mongkut's University of Technology Thonburi, Bangkok.

CEED Program Qld. (2010). *Company-based student training.* Retrieved February 13, 2010, from http://www.corptech.com.au/ home/

Chem, E. (2010). *Practice school-history.* Retrieved February 13, 2010, from http://web.mit.edu/cheme/ academics/practice/ history.html

Ku, H. M., Sriwatanapongse, W., & Thonglek, S. (2007). Operating a chemical engineering practice school. In Z. J. Pudlowski (Ed.), *Monash engineering education series: 10th UICEE Annual Conference on Engineering Education* (pp. 251-254). Melbourne, Australia: Monash University.

Ku, H. M., Thonglek, S., & Bhumiratana, S. (2005). A graduate-level chemical engineering practice school model in Thailand. In Pudlowski, Z. J. (Ed.), *Monash engineering education series: 4th Asia-Pacific Forum on Engineering and Technology Education* (pp. 43–46). Melbourne, Australia: Monash University.

Ku, H. M., Thonglek, S., Thipayarat, A., & Cheevadhanarak, S. (2005). *A graduate-level science and engineering practice school model in Thailand*. Paper presented at the 2005 Regional Conference on Engineering Education, Jahor, Malaysia.

Lamancusa, J. S., George, R. T., & Soyster, A. L. (1997). *Industry-based projects in academia-what works and what doesn't*. Paper presented at 1997 ASEE Annual Conference, Milwaukee, WI.

Lamancusa, J. S., Jorgensen, J. E., Zayas-Castro, J. L., & Ratner, J. R. (1995). *The learning factory-a new approach to integrating design and manufacturing into the engineering curriculum*. Paper presented at 1995 ASEE Annual Conference, Anaheim, CA.

Lock, G., Bullock, K., Gould, V., & Hejmadi, M. (2009). Exploring the industrial placement experience for mechanical engineering undergraduates. *English Education*, *4*(1), 42–51.

Manipal Institute of Technology. (2010). In *Wikipedia, the free encyclopedia*. Retrieved February 4, 2010, from http://en.wikipedia.org/w/ index.php?title=Manipal_Institute _of_ Technology&oldid =341735957

Mattill, J. (1991). *The flagship: The M.I.T. School of Chemical Engineering Practice, 1916 - 1991*. Cambridge, MA: MIT Press.

MIT|TPD (2010). *Technology and development program*. Retrieved February 3, 2010, from http://web.mit.edu/mit-tdp /www/thailand.html

Morell, L., Zayas-Castro, J., & Velz-Arocho, J. (1998). *Some assessment tools for evaluating curricular innovations outcomes*. Paper presented at 1998 ASEE Annual Conference, Seattle, Washington.

Morell, L., Zayas-Castro, J., Velz-Arocho, J., & Torres, M. (1998). *Curricular innovations outcomes assessment and ABET 2000*. Paper presented at International Conference on Engineering Education, Rio de Janiero, Brazil.

O'Connor, A. J., Kandas, A. W., Natori, Y., & Hatton, T. A. (1999). Internationalizing practical ChE education: The M.I.T. practice school in Japan. *Chemical Engineering Education*, *33*(2), 162–165.

PIPS. (2010). *The Professional Engineering Placement Scholarship at The University of Queensland*. Retrieved February 13, 2010, from http://www.uq.edu.au/ peps/pips

Ramirez, L. M., & Beauchamp, G. (1995). *Integration of skills development across the engineering curriculum*. Paper presented at 1995 ASEE Annual Conference, Anaheim, CA.

Rojter, J. (2007). *PBL as means to better engineering education?* Paper presented at International Conference on Engineering Education and Research, Melbourne, Australia.

The Cambridge-MIT Institute. (2010). *Home page information*. Retrieved February 10, 2010, from http://www.cmi.cam.ac.uk

Victoria University. (2010). *Problem-based learning*. Retrieved February 16, 2010, from http://www.vu.edu.au/ higher-ed-and-tafe/higher-education /health-engineering-and-science/problem-based-learning

Viravaidya, W., Panyakul, W. R., & Thonglek, S. (2007). Relative significance of admission criteria for the chemical engineering practice school (ChEPS). In Z. J. Pudlowski (Ed.), *Monash engineering education series: 10th UICEE Annual Conference on Engineering Education* (pp. 147-150). Melbourne, Australia: Monash University.

Ziguras, C. (2001). Educational technology in transnational higher education in South East Asia: The cultural politics of flexible learning. *Journal of Educational Technology & Society*, *4*(4), 8–18.

ADDITIONAL READING

Benjamin, C., & Keenan, C. (2006). Implications of introducing problem-based learning in a traditionally taught course. *English Education*, *1*(1), 2–7.

Boud, D., & Feletti, G. (1998). *The challenge of problem based learning* (2nd ed.). London: Routledge.

Dettman, M. A. (2004, June). *Development and assessment of a new project / practice based civil engineering curriculum*. Paper presented at Proceedings of the 2004 American Society for Engineering Education Annual Conference & Exposition, Salt Lake City, UT.

Doty, R. T., & Williams, S. M. (2002, June). *A practice-based senior design experience*. Paper presented at Proceedings of the 2002 American Society for Engineering Education Annual Conference & Exposition, Montreal, Quebec.

Fenwick, J. T. (1998). Boldly solving the world: a critical analysis of problem-based learning in professional education. *Studies in Education*. (1), 53-56.

Harper, J. I. (1954). *Chemical engineering in practice*. New York: Reinhold Publishing Corp.

Jorgensen, J. E., Lamancusa, J. S., & Zayas-Castro, J. L. (1995, October). *The learning factory - curriculum integration of design and manufacturing*. Paper presented at Proceedings of the World Conference on Engineering Education. Minneapolis-Saint Paul, MN.

Lopez, D. A. (1997, June). *Project based instruction in manufacturing: a new approach*. Paper presented at Proceedings of 1997 ASEE Annual Conference, Seattle, Washington.

Maddocks, A. P. (2007). EASIMAP: a coherent approach to the assessment of learning outcomes on engineering degree programmes. *English Education*, *2*(2), 26–32.

Mason, W. H., & Grossman, B. (1996, June). *Virginia Tech's new practice-oriented aerospace engineering master's degree*. Paper presented at Proceedings of the 1996 ASEE Annual Conference, Washington, DC.

Mathew, M., & Madhavan V. (2008). Academia, industry and government: who must lead? *CURIE 1(2)*.

Mills, J. E., & Treagust, D. F. (2003). Engineering education - is problem-based or project-based learning the answer. *Australasian Journal of Engineering Education*. 2-17.

Newberry, B., & Farison, J. (2002, June). *An interdisciplinary practice-based design course for first-semester juniors*. Paper presented at Proceedings of the 2002 American Society for Engineering Education Annual Conference & Exposition, Montreal, Quebec.

Olds, B. M., & Miller R. L. (1998). An assessment matrix for evaluating engineering programs. *Journal of Engineering Education*. 173-178.

Ramirez, L. M., Colucci, J. A., Carter, L. F., Padron, F., Colon, G., & Estevez, L. A. (1996, June). *An interdisciplinary undergraduate manufacturing engineering option for chemical engineering*. Paper presented at Proceedings of 1996 ASEE Annual Conference, Washington, DC.

Ramirez, L. M., Zayas, J. L., Lamancusa, J., & Jorgensen, J. (1997, April). *Making a partnership work: outcomes assessment of a multi-task, multi-institutional project*. Paper presented at Proceedings of the NSF TRP/MET Annual Conference, Washington, DC.

Smith, R. P., Barton, R. R., Nowack, C. A., & Zayas-Castro, J. L. (1996, June). *Concurrent engineering: a partnership approach*. Paper presented at Proceedings of 1996 ASEE Annual Conference, Washington, DC.

Soyster, A. L. (1997, April). *Experiences in organizing and managing engineering education coalitions*. Paper presented at Proceedings of the NSF TRP/MET Annual Conference, Washington, DC.

Walter, D., Mu, X., & Berry, C. (2008, October). *Work in progress – a study of how real-world engineering experience can affect women's academic career*. Paper presented at the 38th ASEE/IEEE Frontiers in Education Conference, Saratoga Springs, NY.

Woods, D., Felder, R., Rugarcia, A., & Stice, J. (2000). The future of engineering education III. developing critical skills. *Chemical Engineering Education*, *34*(2), 108–117.

Woods, D. R. (1994). *Problem-based learning: how to gain the most from PBL*. Waterdown, Ontario: Donald Woods Publishers.

KEY TERMS AND DEFINITIONS

Proactive Recruitment: Recruiting new target students, usually seniors, by staging roadshows and on-campus presentations at the departmental level.

Practice-Based learning (PBL): A teaching model that emphasizes solving real-life problems through collaboration with industry.

Practice School: A curriculum, usually in science and engineering, which incorporates problem-based learning and involves industrial projects.

Practice Station: A sponsoring company, usually a plant in the field of chemical engineering, where students are interned and receive practical training.

Site Director: A faculty member from the practice school who is permanently stationed at a practice station and work closely with industrial sponsors and students on site projects.

Soft Skills: Additional attributes that engineers should possess, such as teamwork, presentation, written, and awareness skills.

Work-Integrated Learning: A student-centric learning process that integrates theoretical learning with its application in the workplace while emphasizing work readiness, creative thinking, life-long learning, and other desirable soft skills.

Chapter 8
Work-Integrated Learning in Postgraduate Design Research:
Regional Collaboration between the Chinese Mainland and Hong Kong

Kin Wai Michael Siu
The Hong Kong Polytechnic University, China

ABSTRACT

Instead of only staying in the university to carry out research, postgraduate research students nowadays are expected to gain knowledge and experience through work-integrated learning. The advantages of this kind of learning include better support and facilities for research and more comprehensive and in-depth experience in the research area. The learning also provides an opportunity for students to gain other research experience and explore other research interests. However, sometimes such kind of learning opportunity is not available or not the best available locally, therefore work-integrated learning is necessary or better to be carried out in remote regions. Taking regional collaboration of work-integrated learning for postgraduate design research students between the Chinese mainland and Hong Kong as a case study, this chapter discusses the advantages, merits, issues, and problems of regional collaboration. The chapter then identifies possibilities for improvement and directions for further investigation.

INTRODUCTION

Conventionally, postgraduate research students stay in the university only to carry out their research and then finalise their theses. In recent years, this learning practice and requirement has changed. Applied research in collaboration with industry has become more recognised in postgraduate research studies, in particular in subjects such as design, engineering, and technology (Etzkowitz, 1999; Leung, 2004; Smith, 1999; Reynolds, 1997). Students are also expected and required to have a

DOI: 10.4018/978-1-60960-547-6.ch008

wider exposure to the areas related to their research topics (Bourner, Katz & Watson, 2000; Siu, 2009a, 2010). All these changes in research direction and new educational goals imply that more research students are expected and required to carry out their research outside the university and have a tighter connection with industry (Dadashzadeh, Saber & Saber, 2002; Geenhuizen, 2009; Hyland, 1998; Leung, 2004; Poppins & Singh, 2005; Siu, 2009b).

There are two common modes and arrangements for research learning outside the university. First, a research student may stay in a particular place outside the university (that is, including another university where the student is not studying) to carry out research work for a long period of time. Most of the time, this kind of learning is directly related to the students' research topics. Sometimes the attachment is carried out for nearly the whole period of study. The major reason for this arrangement most of the time is due to the particular environment, facility, supervision, or data available outside the home university (Assiter, 1995; Hyland, 1998; Siu, 2009b). For example, a software engineering research student may stay in a software engineering company for most of his or her study period to develop a new computer language as the expensive hardware and confidential data are only available in this particular company. A science research student may stay in a geographical science lab at the North Pole to collect specimens and carry out analysis as particular expert advice is only available in that lab and specific data are only available at that particular site. Alternatively, a student may stay in a particular place in order to gain research experience which is expected to benefit the student's study (Hodgson, 1993; Poppins & Singh, 2005; Siu, 2009b). The places for this kind of work-integrated learning experience may be companies related to research students' research topics, or research labs outside the university. This kind of experience may not be a specific requirement or element of the students' research topics, but may be useful for enriching the experience of the students by widening their vision, perspective, and experience in the research topics, which will then benefit their future career. In general, such learning activities are carried out for a relatively short period of time. For example, a design research student interested in studying the urban redevelopment of a city can benefit from working in a planning and design company for several weeks or months to understand the professional practice and concerns of planners and designers. An architectural engineering research student may need to stay in a deprived rural community for a short period of time to understand the relationship between nature, the built environment, and human daily life.

The latter mode has become more popular and has gained more attention from educators because the aim of postgraduate research today is about not only the production of a thesis at the final stage, but also about the process of gaining both knowledge and wider experience to enable research students to become more well-rounded experts in their discipline. It also implies that a wider learning experience for research students is expected. However, constraints and difficulties are always encountered in the arrangement of this mode of learning. One of the key constraints and difficulties is the provision of this kind of learning experience fitting the needs and preferences of students, in particular many of the students' research directions and topics are quite specific and focused, therefore places for providing suitable work-integrated learning activities are sometimes not available in the local industry or community. For example, since the late 1980s, the manufacturing industry has been phased out in Hong Kong. Most of the design firms and manufacturing facilities have moved their main offices to the Chinese mainland. Design research students often find it difficult to gain work-integrated experience (that is, practice based experience) in their studies. At the beginning, this change in the economic situation had less influence on research students compared to the undergraduate students.

This is because research students at that time had less connection with the industry. However, the situation started to change in the mid 1990s as research students were increasingly expected to have a connection with the sectors outside the university since applied research is highly valued. Moreover, instead of remaining at the university, more research graduates are expected to work in industry and such work-integrated learning experience allows them to explore a wider scope of interests and become aware of different possibilities. As such, this causes enormous pressure and difficulties for research students to find suitable places to gain this kind of work-integrated learning experience.

In the following sections, a case study in Hong Kong is presented. The case study shows how a design school works with research centres, design labs, both non-profit and profit making organisations and the industry to provide a range of work-integrated experience for postgraduate design research students. The following identifies and discusses the advantages, merits, issues, and problems of the collaboration, and then makes recommendations for possible solutions as well as future research directions for the topic.

BACKGROUND

Since early 2003, the School of Design of The Hong Kong Polytechnic University has provided wider choices of work-integrated learning experience for design research students, in particular those research topics that are about industrial and product design, and product engineering and technology. However, most of the time the provision of places in Hong Kong for research students to have this kind of learning experience is not sufficient. One of the reasons is that the local manufacturing industry is in decline. This is creating a ripple effect to some of the industrial and product design companies mainly working for the manufacturing industry as their business is not in a good state, and some may not even survive. In fact, starting from the late 1990s, medium- and large-size design companies in Hong Kong have faced the pressure of manpower reduction. As pointed out by the managing director of a medium-size product design company when interviewed, even their employees cannot find sufficient projects on which to work. In addition, many medium- and large-size design companies have reduced their company size in order to cut expenses and be more flexible in business operation. Some are only run by two to three key designers for daily administration operation, and then most of the design work is offered out in freelance mode. Thus, it is difficult for the companies to provide places for research students to work as part of their learning.

On the other hand, there are a few design companies that are still capable and willing to provide places for students to stay for a short period of time to gain working as well as other kinds of research experience. However, the companies mainly prefer undergraduate design students who can practice in the companies as junior designers or trainees rather than design research students. Thus, the expectations of design research students to gain more research experience is extremely difficult to address in Hong Kong.

Moreover, Hong Kong is a very small city and the business nature of industrial and design companies are quite similar. Thus, the variety of choice for the research students for gaining work-integrated learning experience is very limited and narrow. For example, in the last ten years most of the surviving industrial and product design companies in Hong Kong produce products that are simple in form and small in manufacturing volume, such as; low-cost premium products and daily gadgets. There are few other kinds of design companies. As design research students expect to carry out research topics and gain experience more related to new and specific design directions this adds to the difficulty in matching research students with suitable companies for work-integrated learning.

To overcome the situation, the opportunity for work-integrated learning experience outside Hong Kong has been explored. Among neighbouring regions and cities, the Chinese mainland is conveniently located and has high potential to meet this need. In fact, since the early 1980s industry in Hong Kong began to establish a tighter business relationship with the Chinese mainland. Gaining the benefit of the low labour cost, rich resource and generous support of the central government, rapidly increasing numbers of design and manufacturing companies have established their branches, and later headquarters, on the Chinese mainland. This situation attracted the Hong Kong Polytechnic University to consider the importance of exploring places for design research students to have work-integrated learning there.

On the Chinese mainland, the available places for work-integrated learning are not restricted to those offered by industry. Instead, research labs and centres in, or related to universities have potential for design research students to gain short-term work experience. Different from conventional labs in the universities, many of the university research labs and centres on the Chinese mainland nowadays have tight connections and business relationships with outside sectors, for example, design projects working with the industry, and consultancy research projects tendered by or working with the government, related departments and organisations, professional bodies, social groups and community organisations. The nature and scale of research and design work available in these labs and centres is wide-ranging and varied. Some of the projects as well as work experience are only available in some big cities on the Chinese mainland due to the particular economic, political and cultural settings. As stated in an interview with a design manager, this kind of opportunity is impossible to get in a small city such as Hong Kong. Thus, all these situations allow Hong Kong design research students to have more flexible choices to gain work experience direct or indirectly related to their research topic.

CASE STUDY OF WORK-INTEGRATED LEARNING EXPERIENCE ON THE CHINESE MAINLAND

An in-depth case study was conducted between 2008 and 2010 to evaluate the successes and limitations of regional collaboration in providing work-related learning experience for design research students from which are proposed future directions and solutions for improvement. The case study also aimed at providing hints and inspirations for extending a similar kind of collaboration to other design and engineering disciplines.

In the following segment, four examples of work-integrated learning carried out at different kinds of working places with varied natures of working and learning contents are presented to illustrate the different possible arrangements and to assist discussion. Each of the examples described below included a design research student from Hong Kong who was attached to a particular working place and participated in one or some particular projects. The durations of four work-integrated learning activities were varied depending on the arrangements and needs of the collaborating partners (that is, the companies and research centres on the Chinese mainland) and the availability of the research students. In general, the main supervision of the work-integrated learning was given by the supervisor on the Chinese mainland while the Hong Kong supervisor provided coordination and supplementary support. As indicated above, it was not necessary for this kind of learning experience to be tightly bonded with the students' research topics and theses. The supervisor in Hong Kong gave a high degree of exploratory freedom to the students. Nevertheless, the Hong Kong supervisor still needed to visit the working places in order to evaluate the working environment and performance of the students.

- The first student worked in the research and design lab of a leading university in Beijing. Besides field study, the main office and working places were inside the university. The duration of the project was about three months. The student was required to participate in several industrial projects focusing on universal design. He needed to work with a team of engineers and design research students to conduct field research and handle practical design work, such as; 3-D rendering and graphic presentations. The student was also responsible for the English documents and presentations of the project.
- The second student worked in a design centre jointly managed by the organiser of an important international event, a university and several private companies in Beijing. The design centre had been formed only for the preparation and organisation of the important event and it was closed after all projects related to the event had been completed. The student did not work in a fixed office or lab. Instead, she needed to carry out the work at different places, including the office of the event organiser and several labs in the university. She worked with several teams of designers from different disciplines together under the supervision of several management personnel from the organiser and the senior designers from the university. She had the chance to participate in several design projects about the facilities in public space and the event venues. She was responsible for some specific research and design work, including archive research, data analysis, design development, and the preparation of presentations and documentation of the final report. She also needed to visit the private companies for meetings to obtain technical support. She stayed in Beijing for about five months, and returned to Hong Kong once for her research report to comply with the university study requirement for an oral presentation in confirmation of her study. Due to the short duration of her attachment, instead of following the projects from the very beginning to the end stage, the student could only participate in limited stages of the projects. Before the event started, the student was required to return to Hong Kong to re-start her own research study.
- The third student worked on an environmental and industrial design project in Tianjin. The project was included as part of the city government's coastal area development project which was managed by a local property development company. The student needed to work with team members from different disciplines, including external team members from several small companies and a university which provided technical support for the project. The focus of the student's involvement in the project was the final touches to the design for a large size sculpture/construction located at a key roundabout in the city. The project also included the surrounding landscape and public facilities, in which the student did not need to be involved. This student was responsible for the 3-D concept development with which other team members were not familiar. The culmination of the project was a medium term exhibition conducted in the urban renewal office and exhibition complex of the city. The student joined the project mid-way through and stayed there for about three months.
- The fourth student worked for a lighting company in Guangzhou as a research designer for about two months. After about one week obtaining mutual understanding

of the business direction of the company and the ability of the student, the student was assigned to design a new set of road lights. During his work-integrated learning period, half of the time he was required to return to Hong Kong to collect data which were only available there and from where contacts for expert opinions were easily accessed. The remaining time he needed to stay in the factory in Guangzhou to execute the project. While the supervisor on the mainland provided the student with opinions on lighting business and engineering matters, the Hong Kong supervisor provided advice related to a new design approach and international perspective in design. The Hong Kong supervisor in this project participated actively since he was the expert in some design areas related to the project. It was also one of the reasons the company agreed to offer the work-integrated learning opportunity to the research student.

A qualitative evaluation approach was adopted for the in-depth case study. Semi-structured and un-structured interviews were conducted with the students, Hong Kong supervisor, mainland supervisors, and persons working with the students in the workplaces. These kinds of interviews were selected because the characteristics of semi-structured and unstructured interviews have advantages for in-depth understanding of the topic. These interviews were open-ended, assumed a conversational manner, and followed a certain set of questions (Yin, 1994). In Berger's (1998) words, these interviews allowed the respondents plenty of room to speculate, offer opinions, and so on. The interviews with the students attempted to explore their personal opinions of their work experience on the mainland where most of the time it was a new place for them and a new challenge for their research and design areas. The interviews were expected to reveal the value of such kind of work experience. The interviews with the mainland supervisors and the persons working with the students focused on the performance and attitude of the students in the work place. Interviews with the Hong Kong supervisor focused on students' performance and the views of the students during the work-integrated learning period. The Hong Kong supervisor was also requested to comment on the students' research performance after their work on the mainland.

Besides interviews, observations by the Hong Kong supervisor in the workplaces were sometimes carried out in order to verify the findings collected by interviews. Comparative analysis of the natures and settings of the workplaces were conducted, and the findings could form a good database for reference, in particular with regard to particular students' research topics, their previous experience, and the expectations of the work-integrated learning.

ADVANTAGES AND MERITS

According to the in-depth case study, regional collaboration in providing work-integrated learning indeed benefits design research students. It is a fact that individual students could get different benefits and experience from their working places. What a research student gained depended on the nature of the work, project and commitment opportunity, facility, arrangement and requirement of the job opportunity, mainland supervisor(s) and working team, and the student's own experience, expectation, contribution and performance, etc. The key advantages and merits of such kind of regional collaboration are summarised as follows.

The students' experience can strengthen their research scope and develop new research interests in a particular research environment. On the whole this kind of environment is not available in the university or in a particular city (Hyland, 1998; Varga, 2009). This situation is more obvious since research students' research topics and interests are quite specific. Besides, the students can have a variety of choice to carry work-integrated learning. On the whole that variety is not available in a particular place, such as Hong Kong, which is a small city with very limited types of working opportunity for research students (Siu, 2009b, 2010).

The students can also have a better exposure to some new areas which may not directly relate to their research topics. This benefit meets the current education goals in that research students are increasingly expected to be rounded in different aspects and have the ability to explore new directions (Boud & Lee, 2009; Ehrenberg & Kuh, 2009; Siu, 2009b).

The students can gain more non-local experience, insight and opinion from working places and supervision out of the university. Nowadays this kind of exposure becomes more important to research students since many of them will go on to other places to have further study and work after graduation (Ehrenberg & Kuh, 2009; Haselgrove, 1994; see also Segal, 2003).

The students can also widen their experience in research areas in different environmental settings. The environment implies not only physical environment, but also social, cultural economic and political environments. As stated by many researchers nowadays, having research experience in different environmental settings can widen the insight of research students (Bartlett & Mercer, 2001; Boud & Lee, 2009; Ketteridge, Marshall & Fry, 2002; Leung & Siu, 2005; Siu, 2005, 2010). In addition, the students can gain experience working with persons of different backgrounds, for example, different educational, social, religious, and political backgrounds (Noe, 2008; Siu, 2010).

The students can establish good research networks outside the university and with other persons. This is a particular advantage for research students as most of them need to establish a wider and more senior level of contact for their future study or work (Boud & Lee, 2009; Siu, 2009b). There have also been some successful cases where research students can continue their postdoctoral research in the work-integrated learning places (Siu, 2010).

Regional collaboration in work-related learning is not a one-way benefit to the students. Instead, the students can bring new insights and experience to the workplaces. Sometimes a new member from another place and with no internal conflicts can easily bring positive dynamics to the work place (Jackson & Schuler, 2000). Also, some of the design academies and new design companies quite welcome non-local students to have work-integrated learning as well as other types of work attachment.

The research personnel from two places (for example, Hong Kong and the Chinese mainland) can also help establish better and mutually respectful ways to share information (Smith, 1999). As emphasised by the interviewed supervisor and employer from a private company, most of the time research data of a university (and of a research student) are quite limited for access by outside parties. Establishing a co-work and co-research manner through students' work-integrated arrangement and collaboration can provide a breakthrough in the research protection barrier.

Such a resource can create better sharing and generate mutual benefit through regional collaboration between the university and the outside parties (Siu, 2009a; see also Etzkowitz, 1999; Noe, 2008; Smith, 1999). Sometimes the resource available in the university and the outside parties may not be of the same nature. As in one of the case studies discussed above, participation in the design work for a large international event was not easy for Hong Kong students to obtain. The regional collaboration allowed the student to benefit, not in the physical form and matters, but opportunity. On the other hand, the Hong Kong student's ability and experience could bring practical benefit to the project development on the Chinese mainland. As one of the interviewed mainland supervisors pointed out, the student brought new design directions to the team though she only worked there for a short period of time.

The supervisors and also universities can establish more research networks with the persons in other places. Sometimes two-way work-integrated learning programmes can be established between two universities. For example, design research students can work in another university under an exchange agreement.

Most of the time, students' work contribution in a non-local university or company can be taken as a kind of barometer for a larger scale and more in-depth collaboration between the two parties in two places (Siu, 2009b, 2010). This situation is particularly significant when the parties in two regions clearly do not know each other (Siu, 2010). A small scale of participation can provide better mutual understanding between two parties (Smith, 1999; Varga, 2009).

It cannot be denied that competition in resource and research output exists among local universities, and even between the universities and industry. Regional collaboration most of the time has relatively less concern about the matters of local competition. This situation allows a university to have a more secure feeling thus making it more willing to have higher commitment and work with external parties (Geenhuizen, 2009; Varga, 2009). In addition, regional collaboration most of the time is considered as a kind of external resource for a university or a company to have better competition with local parties (Heshmati, Sohn & Kim, 2007; Siu, 2009b).

IMPORTANT ISSUES AND PROBLEMS

The case study identified some issues and problems in the planning and implementation of regional collaboration for design research students to undertake work-integrated learning in another place. One of the key issues and problems is related to time management matters. The duration of work-integrated learning in general is short. The students find it difficult to have sufficient time to meet their planned goals. For example, a research student pointed out that he could only understand the structure and working procedure of his workplace after two to three weeks. After that, student only had a very short period of time to participate in the actual work with the persons, and as a result students have the minimum chance to learn something from the workplace.

The short duration of the placement also causes some of the workplaces (and their supervisors) to be unwilling to give important or new jobs and tasks to the students. As commented by a student, the supervisor on the Chinese mainland tended to give him some easy jobs to let him *survive* happily in the company. Most of the time, this situation is particularly popular in the mid-level supervision in that the supervisors may not know the objective of students' learning. If the mid-level supervisors are unwilling to take up the supervision role assigned by the senior management, the situation is quite unpleasant. Under pressure from the senior management of the company to take care of this kind of short-term *visitor*, the mid-level supervisors prefer to keep their supervised working department/section unchanged and in turn just give easy jobs to the students. This situation means that the students are unable to gain real experience in the position and unable to develop their research interests.

While the students are attached to the workplace in a short-term mode, their involvement sometimes cannot match with the timetable of the projects. This situation may cause the students to miss some important stages of the project or students learning experience may be less comprehensive than anticipated.

On the other hand, due to the study requirement, some students are required to return to Hong Kong to report on their study during the work-integrated learning period. This situation makes it difficult for the students to undertake a longer term of work on the Chinese mainland. For example, research students may be at the stage where it is necessary to have verbal reporting of their study to the university. These administrative and study requirements sometimes affect the work flow and experience of the students in other places. Although this situation can be eliminated by better learning coordination and arrangement, it also causes some limitations to the arrangement of particular working places and projects for the students.

These students may face difficulties in time management as current postgraduate research is demanding and intensive and they are required to complete their study within a relatively shorter period of time compared to ten years ago. In general, students are expected to submit a report for the work-integrated learning. If the reporting exercise is complicated and demanding, it also increases the workload of research students and in turn diverts their concentration on their own research study. This situation will de-motivate the students to participate in work-integrated learning. This is also why some of the research students refuse to take up the opportunity for work-integrated learning.

In addition, some students need to face practical problems in working in a remote place. This situation is particularly significant at the beginning when the students start to work in a new place. For example, although Hong Kong students in general can manage Cantonese and English, and some are able to speak Putonghua, the students who work on the Chinese mainland still faced communication problems with persons there, including those who did not know Putonghua well. Some students also face difficulties in adapting to life in a new place; such as finding accommodation if the working company does not arrange this for the students. This situation is particularly significant in that most of the work-integrated learning is carried out by individual students. Thus, they need to face all problems alone.

Differences in the social, cultural, political and economic environment can be an advantage and merit of work-integrated learning. On the other hand, these differences can also be problematic issues for the research students when working in another place outside their home town (David & Bathmaker, 2010; Siu, 2010).

Moreover, sometimes cultural and social differences may cause difficulties in planning the work-integrated activities (and programmes) for the students. For example, the dates of some holidays on the Chinese mainland are different from those in Hong Kong. For example, the long non-working holidays on the mainland in general are not on the same dates in Hong Kong. University holidays on the mainland are also long and with very little administrative support for work-integrated learning. All these create difficulty for the planning of work-integrated learning for students in Hong Kong.

Another major issue and difficulty in work-integrated learning coordination is matching the talents and abilities of a particular student with the actual needs of a particular company. Generally such matching is not easy. Although the number of design research students is still small in Hong Kong, finding a suitable workplace in which the students can gain experience is one of the most difficult parts in programme/study management. As discussed above, research students' areas of interest for work-integrated learning are quite specific. This situation makes the matching more difficult.

Difficulty in matching suitable jobs with research students is also caused by the special study nature and mode of research study. One of the major reasons is that the progress of research study is bounded by quite a lot of administration procedures and assessment requirements. For example, design research students in general take about three years to finish their studies and complete their theses. Thus, some periods of time are not suitable for the students to have work-integrated learning, such as the commencement of the project and the final stage of writing the thesis. It is easy to see that the available time for a student to take up work outside Hong Kong is somewhat limited.

Many design companies and research and design centres generally welcome Hong Kong students to have short-term placements. In addition, due to the relationship with the supervisors in Hong Kong, on the whole the companies and centres do not say no to the request of placement. However, whether the companies and centres have a good work plan or sufficient resource to support the students to work there is another issue. Sometimes, the students indicated that they only wasted time on the Chinese mainland and that the companies and centres only put them aside to stay there, but not work there.

Although internal conflicts and competition between research students with the persons in the workplace is infrequent, sometimes it still exists. This situation is more obvious if the workplace provides some special arrangement as well as some privilege for the students who most of the time are considered as visitors or guests.

Intellectual property (IP) is complicated and sometimes can be a troublesome issue to handle in work-integrated learning coordination. Different from undergraduate students' work attachment that in general does not generate critical and significant output, research students' work has a high chance to be put into real implementation and then IP may cause disagreements (sometimes taken to the legal level) between the university and the working place. This situation is more complicated if a student is engaged in some other research projects in the university with other researchers. Regional collaboration also means the chance to solve such disagreements become more difficult due to the different interpretation of IP matters.

Networking and arranging students to work with different partners in the same region sometimes cause difficulty and embarrassment. For example, a research student may work in a design firm that may be the business competitor of another design firm that the university collaborates with, and sometimes it would cause problems with future collaboration. This situation is particularly obvious in the Chinese community. In addition, the territory of the design industry in China is still relatively small and very competitive. For example, there are quite a small number of big-scale lighting design companies on the Chinese mainland. Having students attached to a company means that it is not so good to have a student attached to another one, in particular if the student is involved in research and development work in the company.

There is a general misunderstanding that work-integrated learning is expense-free. In fact, sometimes the university as well as the research students also need to take up responsibility for some expense. This situation is particularly significant if the working places are non-profit making companies or organisations. Some research projects require the university to bear a certain percentage of the project funding. For example, some design research projects for the deprived communities supported by the United Nations do not support the research expense. The funding only goes to the practical implementation of the research outputs in the needed places. In addition, this situation is more obvious if the work-integrated learning is conducted at a place with a higher living standard. Sometimes students are required to take care of the inter-regional travel expenses and daily expenses during the working period. As mentioned by one of the interviewed students, this kind of practical consideration sometimes de-motivates research students to take up work-integrated learning outside their hometown.

SOLUTIONS AND RECOMMENDATIONS

The important issues and problems identified above can be considered from three aspects: practical matters between two regions, students themselves, and arrangement and other matters between the university and the party providing the work opportunity. These three aspects look independent, but most of the time they are co-related to each other.

Among all, resource for collaboration is one of the critical matters which affect the effectiveness of work-integrated learning. A good plan with detailed agreement between the collaboration partners in two regions assure a good start and eliminate disagreements during the work-integrated learning process of the students. Take the regional collaboration between the university in Hong Kong and the design companies or research centres on the Chinese mainland as an example, most of the time the inter-regional travel expense is funded by the Hong Kong side or the students, while the other financial support related to the work, such as the consumables and facility cost, is funded by the mainland side. The students' salary or other financial support most of the time depends on the nature of the companies or research centres. Some university research centres and non-profit making organisations on the mainland require the students to take care of their daily expense. In addition, collaboration can be understood as mutual support. According to the case study, the different natures and amounts of resource can be well-arranged for two regions with different living standards. For example, to overcome the difficulty of the students from a region with lower living standard, work-integrated learning can be implemented in an exchange mode. That is, when a mainland student's living place and daily expense is taken care of by the Hong Kong side, a Hong Kong student's living place and daily expense will be taken care of in the same way by the mainland side. Since 2005, this kind of arrangement has been run quite effectively among some design research labs on the Chinese mainland and in Hong Kong. Nevertheless, no matter what kind of arrangement, good and detailed discussion and detailed memorandum or agreements are all necessary.

Regional collaboration more or less implies a need for consideration of the routine timetables of both places. As stated above, a good plan between the parties in two regions can eliminate or at least minimise the problems in time conflict. It can make the work-integrated learning of the students more effective under such a short period of attachment.

Aside from the planning and coordination between the parties of two regions, personal time management of students is also important. To have good time management and not to have any conflict with the research study, it is necessary for the students to have a thorough plan and preparation for the work-integrated learning. Although it is a fact that some students find it necessary or have the opportunity to get work experience from other places outside the university after they start their research studies, planning as early as possible with their supervisors is the first rule of thumb. Ad hoc action in work-integrated learning sometimes causes disaster to the research projects.

More or less influence to the research study will occur if a research student needs to work outside the university for a period of time, in particular if a place is outside the city/region. As identified above, regular and official administrative procedures and requirements can affect the plan of the work-integrated learning. Thus, prior special consideration and arrangement is necessary. Most of the time, allowing flexibility in some non-critical administrative procedures and requirements is helpful. According to the case studies, such flexibility is possible and acceptable since research students are less bound by a rigid timetable. If there is really a need to fulfil some requirements, the most effective way is to fulfil it first. One of the reasons is that it can prevent complaints of unfairness from other research students. Obviously, it implies that the students expecting to go out to gain work-integrated experience need to be better prepared as mentioned above.

Due to the short period of attachment to a working place, the students as well as their supervisors should undergo psychological preparation in that the students should not expect or be expected to have a very significant contribution to the work force and also achieve outstanding breakthrough and fruitful output. On the contrary, the students should put more attention on the "experience" and "exploration". As discussed above, widening the insight of the students for later benefit in further study and work is the key goal of this kind of work-integrated learning.

As work-integrated learning (that is, the second mode mentioned above) for research students is short in nature and aims to provide students with other related experience, allowing students to enjoy and treasure the learning is recommended. As pointed out by the interviewed students, a long and detailed report of the working contents is not expected. This is because the additional workload would de-motivate the students to participate in this kind of learning.

Instead of a long report, it would be better to request the students to produce a short report of self-reflection about their experience or insight for their current and future study and work. If a long and detailed work report is necessary, the reflection part at the end of the report can help the research students. More or less research study is a very self-initiated and self-motivated learning, and thus too much pressure on routine work would only cause negative results.

In the same way, the workplaces as well as the coordinators and supervisors there, need to understand the objectives of the students attached to their workplaces. On the one hand, the students should not be seen to have redundant roles in the company. On the other hand, the students cannot be seen as a critical working force to fulfil a critical task. This type of senior management's appropriate and feasible view of the students also provides a more positive impression to the persons working with the students who will not be treated as a threat to the original working team.

Due to the nature of design, as well as other technology and engineering subjects, preparation such as technical equipment for the students to have work-integrated learning is important. Obviously, in general the working places do not need to have substantial change since the students going there do so for the particular working environment and available facilities. However, sometimes the students may need to use special equipment for certain projects, and as such the workplace must prepare in advance.

Good preparation for work-integrated learning at another place/region implies not only the preparation of technical and work-related mattes, but also psychological preparation and the preparation to fit/adapt to the new working environment and mode of life are all essential. As discussed above, cultural and social differences are key factors affecting the successful outcome of the work conducted by the students. Language preparation, cultural understanding of daily patterns, and even physical fitness preparation for new climatic conditions should help the students to work more effectively in the new workplace.

Preparation is not only the responsibility of the students. The supervisors, parties providing working opportunities and the students are better able to work together to produce a practical and acceptable plan. The plan can help a student consider the suitability of the work and adapt to the new workplace if the student takes up the opportunity. In fact, quite a lot of the time, such as in the Hong Kong situation, supervisors would give an "order" to the students to take up some external research jobs which are considered by the supervisors as good and rare "opportunities" for students. However, whether the students would think in the same way is another issue. In fact, some research students complained that much of the research work at remote places was undertaken due to the will of the supervisors as they wanted to establish good research networks with the persons of the workplaces. As such, once this kind of thinking is rooted in the students' mind, the successful outcome of the work-integrated learning will be in question.

Good planning also implies a good match with a student's talent and also interest in the need and requirement of the job as well as the potential opportunity in the working place. In particular, for arranging the students to work in a remote place/region, more details of the workplace and responsibilities are essential. According to the findings of the case study, sometimes the supervisors at the workplaces have not adequately prepared or planned for the students. The coordinators or supervisors in the workplaces may simply reply that matters such as working place and facilities would be ready when the students arrive. This kind of situation is particularly risky since flexible arrangements in general are not easy to fulfil as promised within a short period of time. One of the consequences is that the students stay in the workplace with nothing to do and in turn waste their time.

To ensure the smooth running of the learning process, prior visits to the working places for meetings and fine tuning of the project objectives, learning/working contents, and arrangement and setting of facilities are essential. This situation is particularly important if it is the first collaboration between the university and a new regional partner. As an interviewed supervisor mentioned, he had heard of a case from another discipline where a student found out after he arrived there, that the equipment at the factory had not been installed. The worst thing was that the student was also told that his accommodation was not ready.

Besides prior visits, visits by a supervisor during the students' working periods are beneficial, if they can be arranged. Such visits have many advantages, including having a better quality assurance of students' work-integrated learning, developing better communication with the regional partners, and also providing important physical support and psychological comfort to the students. The last point is especially important if the students need to stay in a remote or deprived area and to work individually by themselves.

A friendly and supportive working environment for the students can come from the host company as mentioned above. It can also come from the students' side. The students must prepare well and take practical action to *understand* and *respect* the local culture and practice. According to the case studies, most of the conflicts occurring in work-integrated learning came from misunderstanding and irrelevant practice. For example in one of the cases mentioned above, a research student wished to work late in the research centre since he expected to use the limited time to learn and achieve more. However, this situation caused disharmony with the original work force there. The situation did not change until the student realised the situation and altered his methods.

Resources and credit most of the time are two major incentives that make the collaboration successful. However, conversely they are also the two major causes that can make the collaboration unsuccessful. As a Chinese saying goes, "Make it clear and to be mean first, and to be generous at the end". As the persons working physically in a remote place, the students should be also informed about the resources (for example, funding for the students to carry out the work) and credit (for example, IP matters), if possible. Referring to quite a lot of students' feedback, most of the time they were not informed about this kind of arrangement. It is common practice that supervisors sometimes would just tell the students to go ahead to work and promise all arrangements should be all right. However, if any problems and disagreements occur during the working period, the students most of the time do not know how to handle them, and in turn the immediate sufferers will be the students. And such kind of disagreements will also cause un-resolvable problems between not only the supervisor and the coordinator at the working place, but also at the higher level between the university and the party providing the work opportunity.

Last but not least, ethics and integrity are serious matters in collaborating with other parties and should be handled carefully. As mentioned above, the university should have a good overall plan and long-term direction in collaboration with the external parties, who may be competitors, in order to avoid a negative image to the university as well as the entire field.

FUTURE DIRECTIONS

Regional collaboration in providing design research students with work-integrated learning experience has a lot of advantages and merits. In particular design is a discipline where it is necessary to provide students the opportunity to experience and explore, and the chance to work in other places is beneficial for the students. Different from the traditional thesis-centric attachment to some special working places and research labs, the discussions above describe another mode of work-integrated learning which aims to provide other experiences for the research students and allows them to explore wider research interests through work experience.

According to the experience of collaboration between the Chinese mainland and Hong Kong, regional collaboration in planning and implementing work-integrated learning experience is indeed an art. It is not only about the matters of practical resources and manpower, it is also more about the consideration of different cultural, social, economic, political, religious and other factors. As identified above, regional collaboration in providing work-integrated learning experience for design research students has quite a lot of important issues and problems that need to be addressed. Sometimes, those matters are not easy to handle, in particular because the study requirement and working environment are in a constant state of change. In recent years, research students have received increasing pressure to finish their study on time while the working environment on the Chinese mainland is also constantly changing to be more technology-oriented and a great deal of team work is expected. The scale and duration of the projects handled by the design and manufacturing companies on the Chinese mainland are also becoming larger and larger. Not counting some other personal interests and conflicts, all these make the planning, arrangement and practical implementation of work-integrated learning more complex and difficult.

We cannot deny that the pros and cons of regional collaboration are like the two faces of a coin. Most of the time, the advantages and merits also cause some issues and problems. Thus, to have a better quality and development of work-integrated learning for research students, it is important to have further studies on the topic. These should focus on planning, implementation and evaluation. The levels should cover the regional government and education policy, university and non-university sectors, and individual supervisors and students' needs and preferences. For example, further investigation can focus on how regional and education sectors can work together to enhance and support regional collaboration, how the overall resource and research atmosphere can be strengthened to support regional collaboration, what the objectives and goals of universities and non-university sectors are, and the detailed needs and preferences of students and supervisors. Obviously, more comprehensive and in-depth studies and reviews of the topic are necessary. In short, as in the discussion in the last section, only through good planning, open and continuous communication between the parties of two regions, mutual consideration and support, and high flexibility in handling the issues and matters that may occur can enable the collaboration to be more successful and eventually provide real benefits to the students, universities and collaboration partners.

ACKNOWLEDGMENT

I would like to thank The Hong Kong Polytechnic University and the K. C. Wong Education Foundation for the support of the study discussed above. The Asian Scholarship Foundation provided me support for the visits to other Asian universities to collect references and opinions for the improvement of the management of regional collaboration. The researchers and design research students of the Public Design Lab provided me with a lot of support and assistance for the study. Tsinghua

University, the Central Academy of Fine Arts and the Tianjin Academy of Fine Arts in China also provided important support to the students and participated in part of the study. Although some of the private companies, event organisers and research centres do not want their names to appear in this paper, I must acknowledge the research and work-integrated learning partners on the Chinese mainland. They have provided a lot of valuable opportunities to the design research students in Hong Kong.

REFERENCES

Assiter, A. (Ed.). (1995). *Transferable skills in higher education*. London, UK: Kogan Page.

Bartlett, A., & Mercer, G. (Eds.). (2001). *Postgraduate research supervision: Transforming (r) elations*. New York: Peter Lang.

Berger, A. A. (1998). *Media research techniques* (2nd ed.). Thousand Oaks, CA: Sage.

Boud, D., & Lee, A. (Eds.). (2009). *Changing practices of doctoral education*. London, UK: Routledge.

Bourner, T., Katz, T., & Watson, D. (Eds.). (2000). *New directions in professional higher education*. Buckingham, UK: SRHE and Open University Press.

Dadashzadeh, M., Saber, A., & Saber, S. (Eds.). (2002). *Information technology education in the new millennium*. Hershey, PA: IRM Press.

David, M. E., & Bathmaker, A. M. (Eds.). (2010). *Improving learning by widening participation in higher education*. Abingdon, UK: Routledge.

Ehrenberg, R. G., & Kuh, C. V. (Eds.). (2009). *Doctoral education and the faculty of the future*. Ithaca, NY: Cornell University Press.

Etzkowitz, H. (1999). Bridging the gap: The evolution of industry-university links in the United States. In Branscomb, L. M., Kodama, F., & Florida, R. (Eds.), *Industrializing knowledge: University-industry linkages in Japan and the United States* (pp. 203–234). Cambridge, MA: MIT Press.

Haselgrove, S. (Ed.). (1994). *The student experience*. Buckingham, UK: Society for Research into Higher Education, Open University Press.

Heshmati, A., Sohn, Y. B., & Kim, Y. R. (Eds.). (2007). *Commercialization and transfer of technology: Major country case studies*. New York: Nova Science Publishers.

Hodgson, I. (1993). Learning contracts for the industrial experience of hospitality management students. In Stephenson, J., & Laycock, M. (Eds.), *Using learning contracts in higher education* (pp. 107–114). London, UK: Kogan Page.

Hyland, T. (1998). Work-based experience and higher professional learning in British universities. In James, J. J. F. (Ed.), *University teaching: International perspectives* (pp. 305–324). New York: Garland Publishers.

Jackson, S. E., & Schuler, R. S. (2000). *Managing human resources: A partnership perspective*. Cincinnati, OH: South-Western College Publishing.

Ketteridge, S., Marshall, S., & Fry, H. (Eds.). (2002). *The effective academic: A handbook for enhanced academic practice*. London, UK: Kogan Page.

Leung, T. P., & Siu, K. W. M. (2005). Design for new urban needs: Balanced consideration in design. In H. I. Lei & D. C. Dai (Eds.), *Selected papers from the international symposium on design in urban spaces* (Part I, pp. 42-48). Macao: Macao Polytechnic Institute.

Noe, R. A. (2008). *Human resource management: Gaining a competitive advantage* (6th ed.). Boston, MA: McGraw-Hill/Irwin.

Poppins, P., & Singh, M. (2005). Work integrated learning in information technology education. In van Weert, T. J., & Tatnall, A. (Eds.), *Information and communication technologies and real-life learning: New education for the knowledge society* (pp. 223–230). New York: Springer. doi:10.1007/0-387-25997-X_25

Reynolds, A. J. (1997). Interdisciplinary programs in community and applied research. In O'Donnell, C. R., & Ferrari, J. R. (Eds.), *Education in community psychology: Models for graduate and undergraduate programs* (pp. 65–82). New York: Haworth Press.

Segal, N. (2003). *International jobs: Where they are, how to get them* (6th ed.). New York: Basic Books.

Siu, K. W. M. (2005). Learning opportunity for product designers and engineers. In Acland, T., Samuel, J., Saunders, D., & Storan, J. (Eds.), *Access, retention and employability: Transforming higher education* (pp. 108–114). London, UK: Forum for the Advancement of Continuing Education, University of East London.

Siu, K. W. M. (2009a). Review on the development of design education in Hong Kong: The need to nurture the problem finding capability of design students. *Educational Research Journal*, *23*(2), 179–202.

Siu, K. W. M. (2009b). New trends of research in postgraduate design education in China. *US-China Education Review*, *6*(9), 16–21.

Siu, K. W. M. (2010). Meeting the new needs: Design research education in China. *Research in Higher Education Journal*, *6*(March), 157-173.

Smith, S. W. (1999). The industrial perspective on university-industry relationships in Japan and the United States. In Branscomb, L. M., Kodama, F., & Florida, R. (Eds.), *Industrializing knowledge: University-industry linkages in Japan and the United States* (pp. 579–588). Cambridge, MA: MIT Press.

van Geenhuizen, M. S. (Ed.). (2009). *Value-added partnering and innovation in a changing world*. West Lafayett, IN: Purdue University Press.

Varga, A. (Ed.). (2009). *Universities, knowledge transfer and regional development: Geography, entrepreneurship and policy*. Cheltenham, UK: Edward Elgar.

Yin, R. K. (1994). *Case study research: Design and methods* (2nd ed.). Thousand Oaks, CA: Sage.

ADDITIONAL READING

Assiter, A. (Ed.). (1995). *Transferable skills in higher education*. London, UK: Kogan Page.

Boud, D., & Lee, A. (Eds.). (2009). *Changing practices of doctoral education*. London, UK: Routledge.

Branscomb, L. M., Kodama, F., & Florida, R. (Eds.). (1999). *Industrializing knowledge: University-industry linkages in Japan and the United States*. Cambridge, MA: MIT Press.

Castleton, G., Gerber, R., & Pillay, H. (Eds.). (2006). *Improving workplace learning: Emerging international perspectives*. New York: Nova Science Publishers.

Chan, M. Y. (2003). *An evaluation of an industrial placement scheme on engineering students: Benefits and issues*. Hong Kong: The Hong Kong Polytechnic University.

Chen, J. F. (2006). *Theories of cooperative education and the practice in China: Research on work-integrated learning* (Chinese ed.). Shanghai: Shanghai Jiao Tong University Press.

Clifford, J., & Thorpe, S. (2007). *Workplace learning & development: Delivering competitive advantage for your organization*. London, UK: Kogan Page.

Coll, R. K., & Eames, C. (Eds.). (2004). *International handbook for cooperative education: An international perspective of the theory, research, and practice of work-integrated learning*. Boston, MA: World Association for Cooperative Education.

David, M. E., & Bathmaker, A. M. (Eds.). (2010). *Improving learning by widening participation in higher education*. Abingdon, UK: Routledge.

Duignan, J. (2005). *Undergraduate work placement and academic performance: An investigation into the relationship between learning transfer and the architecture of the internship in a business undergraduate programme*. Wetherby, UK: Document Supply Centre, British Library.

Ehrenberg, R. G., & Kuh, C. V. (Eds.). (2009). *Doctoral education and the faculty of the future*. Ithaca, NY: Cornell University Press.

Evans, K. (2006). *Improving workplace learning*. London, UK: Routledge.

Gibson, D. V., Heitor, M. V., & Ibarra-Yunez, A. (2007). *Connecting people, ideas, and resources across communities*. West Lafayette, IN: Purdue University Press.

Goldfarb, R. (2001). *Careers by design* (3rd ed.). New York: Allworth Press.

Gowing, R., McGregor, H., & Taylor, E. (1997). *Making your work placement effective: A student guide to enriching workplace learning*. Melbourne: RMIT Publishers.

Hans-Joachim Gögl, H.-J., & Schedler, C. (Eds.). (2009). *Knowledge loves company: Successful models of cooperation between universities and companies in Europe*. Basingstoke, UK: Palgrave Macmillan.

Hutcheson, P. (Ed.). (1996). *Directory of college cooperative education programs*. Phoenix, AZ: Oryx Press.

Hyland, T. (1998). Work-based experience and higher professional learning in British universities. In James, J. J. F. (Ed.), *University teaching: International perspectives* (pp. 305–324). New York: Garland Publishers.

Kalay, P., & Chen, D. (1995). Sheshet: Decision support system for guidance and placement of enrolling high school students. In B-Z. Barta, T, M. Telem & Y. Gev (Eds.), *Information technology in educational* management (pp.147-152). London, UK: Chapman & Hall.

Kneller, R. (2007). *Bridging islands: Venture companies and the future of Japanese and American industry*. Oxford, UK: Oxford University Press.

Leung, T. P. (Ed.). (2004). *Hong Kong: Better by design*. Hong Kong: School of Design, The Hong Kong Polytechnic University.

Linn, P. L., Howard, A., & Miller, E. (Eds.). (2004). *Handbook for research in cooperative education and internships*. Mahwah, NJ: Lawrence Erlbaum Associates.

Mankin, D. A. (2004). *Business without boundaries: An action framework for collaboration across time, distance, organization, and culture*. San Francisco, CA: Jossey-Bass.

Oh, S. A. (2001). *From ivory tower to knowledge factory? The impact of university-industry links on universities in China and England*. Wetherby, UK: Document Supply Centre, British Library.

Rainbird, H., Fuller, A., & Munro, A. (2004). *Workplace learning in context*. London, UK: Routledge.

Reeve, R. S. (2001). *Employers' guide to work-integrated learning*. Boston, MA: World Association for Cooperative Education.

Rowden, R. W. (2007). *Workplace learning: Principles and practice*. Malabar, FL: Krieger Publishers.

Royer, D., & Gilles, R. (Eds.). (2003). *Directed self-placement: Principles and practices*. Cresskill, NJ: Hampton Press.

Siu, K. W. M. (2005). Learning opportunity for product designers and engineers. In Acland, T., Samuel, J., Saunders, D., & Storan, J. (Eds.), *Access, retention and employability: Transforming higher education* (pp. 108–114). London, UK: Forum for the Advancement of Continuing Education, University of East London.

Siu, K. W. M. (2009). Review on the development of design education in Hong Kong: The need to nurture the problem finding capability of design students. *Educational Research Journal, 23*(2), 179–202.

Siu, K. W. M. (2009). New trends of research in postgraduate design education in China. *US-China Education Review, 6*(9), 16–21.

Siu, K. W. M. (2010). Meeting the new needs: Design research education in China. *Research in Higher Education Journal, 6*(March issue), 157-173.

Stenström, M.-L., & Tynjälä, P. (2009). *Towards integration of work and learning: Strategies for connectivity and transformation*. Dordrecht, The Netherlands: Springer. doi:10.1007/978-1-4020-8962-6

Stephenson, J. (2009). *Capable workplace learning*. Rotterdam: Sense Publishers.

Stern, E., & Sommerlad, E. (1999). *Workplace learning, culture and performance*. London, UK: Institute of Personnel and Development.

The Hong Kong Polytechnic University. (2009). *Learning through off-campus work-integrated education: Handbook for employers*. Hong Kong: The Hong Kong Polytechnic University.

The Hong Kong Polytechnic University. (2009). *Reaching for the stars: Knowledge transfer at PolyU*. Hong Kong: The Hong Kong Polytechnic University.

Thompson, N. (2006). *Promoting workplace learning*. Bristol, UK: Policy.

Ujang, B. (2000). *A study of collaboration between university industrial design departments and corporate firms*. Unpublished thesis. Columbus, OH: Ohio State University.

van Geenhuizen, M. S. (Ed.). (2009). *Value-added partnering and innovation in a changing world*. West Lafayett, IN: Purdue University Press.

Weert, T. J. van., & Tatnall, A. (Eds.) (2005). *Information and communication technologies and real-life learning: New education for the knowledge society*. New York: Springer.

Wilson, J. W. (2002). *Creating and initiating a cooperative education program*. Boston, MA: The World Association for Cooperative Education.

KEY TERMS AND DEFINITIONS

Applied Design Research: Is research that is applied, accessing and using some part of the design research communities' accumulated theories (including the areas of design, social sciences, sciences, engineering), knowledge, methods, and techniques, for a specific, often state, commercial, or client driven purpose. Applied design research is often opposed to pure research in debates about design ideals, programs, and projects. In general, applied design research can be considered as a kind of research which deals with practical problems and is generally empirical.

Basic Design Research: Basic (also known as fundamental or pure) design research is theoretical and/or experimental work undertaken primarily to acquire new design and design-related knowledge of the underlying foundation of phenomena and observable facts, without any particular application or use in view. More design researchers have indentified that basic design research is driven by a researcher's curiosity or interest in a design question. The main motivation is to expand man's knowledge, not to create, invent and/or design something, e.g. a chair, a poster, a home interior environment. Some design researchers also point out that there is no obvious commercial value to the discoveries that result from basic design research.

Experimental Design Research: Or development is systematic work, drawing on existing design other related knowledge gained from research and/or practical experience, which is directed to producing new systems, environments, materials, products, devices, graphics, images and messages, to installing new design processes, systems and services, or to improving substantially those already produced or installed.

Industrial Design: In general is considered as a combination of applied art and applied science. Aesthetics, ergonomics and usability of products (i.e. most of the time referring to mass-produced products) are key design considerations. Industrial designers are expected to create and execute design solutions towards problems of form, usability, user ergonomics, engineering, marketing, brand development and sales (see also http://en.wikipedia.org/wiki/Industrial_design). Some professional industrial design organisations and institutes (e.g. Chartered Society of Designers, International Council of Societies of Industrial Design, Carnegie Mellon's School of Design) also define that industrial design is the professional service of creating and developing concepts and specifications that optimize the function, value and appearance of products and systems for the mutual benefit of both user and manufacturer.

Postgraduate Research: Represents a formal area of study that is recognized by a university or institute of higher learning (or, sub-degree education). It is also commonly referred to as graduate research in the United States, and most of the North America regions. The notion of "postgraduate" carries the implication that the student undertaking such research has already completed a formal degree, diploma of certificate study (e.g. BA, BSc, BEd, BEng, BSocSci, BPhil) at an accredited university or tertiary institution. The resulting qualifications arising from postgraduate research vary from traditional Doctorates (PhD, DPh) and Master's degrees through to professional doctorates or higher doctorates (e.g. DDes, DSc, DEng, DSocSci). The structure of postgraduate research programs can vary significantly from one place to another.

Product Design: Is concerned with the efficient and effective generation and development of ideas through a process that leads to new products. In general, product designers identify issues and problems, conceptualize and evaluate ideas, making them tangible through products in a more systematic approach. The key role of product designers is to integrate and/or combine art, science and technology to create tangible three-dimensional products. Product design is sometimes confused with industrial design, and up-to-date there is no absolute definition to distinguish the difference between product design and industrial design.

Regional Collaboration: In the educational field, the fundamental definition of regional collaboration means the co-work in research, teaching and learning between two or among some places not located in the same region, i.e. cross-boundary work. The ways of collaboration are flexible and varied that there is no single model for collaboration, no universal approach that works in all situations. However, the principles and tools developed in the recent years can help educators and researchers work across boundaries.

Work-Integrated Learning: In design means the educational activities that integrate theoretical learning in design with its application in the workplace. These educational activities aim to provide a meaningful experience of the workplace application that is intentional, organised and recognised by the institution, in order to secure learning outcomes for the design student that are both transferable and applied. Some design educators also further identify that work-integrated learning in design focuses on design theory and practice in relevant authentic, work-based contexts. It is differentiated from laboratory and field practicals. It develops specific competence that forms part of the overall design curriculum. In addition, a good quality work-integrated learning can help the design student socialise into the workplace environment.

Chapter 9
Integrating Work and Learning in a Postgraduate Maintenance Management Program

Patrick Keleher
Central Queensland University, Australia

Arun Patil
Central Queensland University, Australia

Gopinath Chattopadhyay
Central Queensland University, Australia

ABSTRACT

Central Queensland (CQ) University conducts a suite of postgraduate programs in maintenance management. There is an emphasis upon Work Integrated Learning, and the programs are delivered in a flexible mode by academics and lecturer-practitioners so that learners are provided with an authentic learning experience. The learners are mature aged, experienced practitioners who are either graduate engineers or trades qualified, working in the maintenance management area of their organisation. Study materials have been designed and developed through the collaboration and consultation with industry, university, and practitioner stakeholders to ensure the student's learning and the assessment of that learning is incorporated into the tasks and responsibilities of learners in their workplace. A blended learning delivery model includes the opportunity for students to attend a two day residential school. Ongoing improvement of the academic program, consultancy, and research opportunities arises from engagement with stakeholders through a number of mechanisms including a conference to showcase innovative practices of physical asset and maintenance management.

INTRODUCTION

Globally universities are under pressure to change as their traditional client base has drastically altered and this is a consequence of the influences of massification (Winckler, 2009; Guri-Rosenblit, Sebkova & Teichler, 2007; Alexander, 2000) and post-massification (Teicher, 2003; Reiko, 2001; Gumport, Nozzi, Hamah, & Zemsky, 1997) of higher education. The student demography now includes students who in the past would not usually have attended university. This includes

DOI: 10.4018/978-1-60960-547-6.ch009

more mature students, not only seeking technical knowledge, but coupled with their life and professional experiences seek a learning environment with an emphasis upon life-long learning. In order to be nationally and internationally competitive universities have become more concerned with knowledge production, innovation and relevance of their activities to the 'real world' environment.

Kerr (1995) describes the modern university as a 'multiversity' which comprises a complex, conglomerate of many communities engaging for related purposes. Coupled with the process of massification, an explosion of knowledge and the ability of the public to access it more readily through technology have led to the increasing sophistication of the public. Consequently, universities are no longer viewed as bastions of rarefied knowledge, only accessible to a privileged few. Further Treby and Shah (2005, p.16) contend 'skills development is enhanced by the inclusion of teaching staff with industry experience, the addition of visiting speakers, other good contact and input from relevant professionals and the opportunity to undertake work-based learning'. Indeed, Wright (1990, as cited in Bradney, 1992) emphasise 'the view that industry and higher education should work more closely together has come to be an all-but-unshakeable part of the dominant, conventional wisdom'. CQUniversity (CQU) recognised these aspects as a significant opportunity to challenge the traditional university model and to 'remake' such a model through collaboration, exploration and engagement with the 'real world' environment of industry and their educational needs. From their inception the suite of postgraduate Maintenance Management Programs have operated as a nexus between CQUniversity and industry. This nexus is established as a result of demonstrated respect, recognition of experience and the sharing of knowledge from differing perspectives between its members. The resulting learning environment represents well the foundations of Mode 2 knowledge development whereby aspects such as contextual learning, trans-disciplinary approaches and multiplicities of perspective are central and supported (Fook, Ryan & Hawkins, 2000, p213). Mode 2 is typified as the interactive production of knowledge within the context of application, while Mode 1 knowledge production coincides with the traditional '*ivory tower*' image of university knowledge production. Table 1 highlights the comparative differences between Mode 1 and 2 knowledge production.

This chapter focuses upon the CQUniversity-industry nexus and the interaction between the disciplines of education and engineering practice. Taking into consideration the relationship between working and learning and the importance of workplaces as sites of learning and knowledge production (Symes & McIntyre, 2000; Boud & Solomon, 2001) and employees as worker/learners (Chappell et al., 2003; Tennant, 2000; Usher & Solomon, 1999). The suite of postgraduate programs comprises the Graduate Certificate, Graduate Diploma and the Masters by coursework. The focus is upon authentic learning (Gulikers, Bastiaens & Kirschner, 2004; Birenbaum, 2003; Gielen, Dochy & Dierick, 2003; Dochy, 2001;

Table 1. Attributes of Mode 1 and Mode 2 knowledge production (Source: Hessels & Van Lente 2008)

Mode 1	Mode 2
Academic context	Context of application
Disciplinary	Trans-disciplinary
Homogeneity	Heterogeneity
Autonomy	Reflexivity/social accountability
Traditional quality control (peer review)	Novel quality control

Petraglia, 1998; Birenbaum & Dochy, 1996) and its assessment, trans-disciplinary approaches in solving real-life problems and students act as change agents through their understanding of theoretical knowledge and direct application of knowledge and understandings to their workplaces.

A common criticism leveled at academics operating in a university environment, is that they are perceived to be only occupied in teaching and research, that is far removed from industry and the '*real world*' of the practitioner. Likewise, industry practitioners (students and casual lecturers), are considered to be engrossed in day-to-day operations and profit margins, perceived as being far removed from knowledge attainment processes and critical thinking and the '*real world*' of the educator.

As Keleher, Wong and Howard (2006) highlight the defining features of the programs as the university-industry collaborative initial and ongoing developmental process, the authentic assessment approach, the lecturer-practitioner teaching team and the blended learning (Allen, Seaman & Garrett, 2007; Albrecht, 2006; Valiathan, 2002) delivery mode. All serve to bridge the perceived divide between academics in '*ivory towers*' and practitioners in '*muddy trenches*'. The university-industry collaborative approach meets a strong and clearly voiced industry need for up-skilling maintenance management practitioners. Authentic assessment items are extremely practice-orientated and are closely linked to an individual student's current workplace roles and responsibilities. The lecturer-practitioner teaching team ensures that university lecturers oversee the academic credibility and relevance and currency of the study materials. The majority of lecturers delivering the courses are practitioners, employed on a part-time teaching contract. This ensures industry-based practitioners are establishing a dialogue that models practices and mentor fellow practitioners. CQUniversity's student demography in these programs consists of mature students, operating as practitioners, who are located in urban, rural, regional and remote areas. A blended delivery mode (paper-based study materials, Moodle site and residential school) accommodates study circumstances of participants for achieving expected learning outcomes. Ongoing university-industry consultation processes ensure the continuation of our *aus der praxis, fur die praxis*. (from practice for practice) approach as we embark on further program developments.

BACKGROUND

The Gladstone Advisory Committee established the Gladstone Engineering Planning Committee (GEPC) in June 1993. The GEPC involved the considered involvement and commitment from industries in the Gladstone area, the Queensland State Government and CQUniversity representatives, with the vision to establish an Engineering Centre at the Gladstone campus of the then University of Central Queensland, now CQUniversity. This original group provided the impetus for a wider representation in November 1993, with a smaller sub-committee developing a business and curriculum plan for establishing maintenance management courses (Abbott, 2004, p14).

The Gladstone campus, located 120 kilometres south of the Rockhampton campus of CQUniversity, serves a city which includes a major Queensland port and supports a dynamic and expanding level of engineering activity by national and international engineering organisations. It has an emphasis upon mineral processing (bauxite to aluminium), manufacturing (cement, industrial chemicals), electricity generation, mining and bulk handling (coal) and the recently announced significant development of a liquefied natural gas (LNG) industry in Queensland based on processing coal seam gas (CSG) from the nearby Surat and Bowen coal basins.

The university-industry-community advisory and planning committees included senior industry managers and leaders from international engineering firms. Through the advocacy of these groups the need for industry staff, including many senior managers themselves, to gain formal knowledge in the critical function of maintenance management and for them to be acknowledged for this achievement through the award of a formal qualification from a recognised university facilitated the development of these postgraduate programs. The Graduate Certificate and Graduate Diploma commenced delivery in 1995 and the Masters (by coursework) commenced in 1998. The programs have attained successful outcomes as per the Faculty's Strategic goals of establishing agendas for new spheres of knowledge and practice and also in pioneering new and innovative learning and teaching processes. Learning outcomes have also aligned well with University Strategic goals, in particular in the areas of developing networks and implementing processes to ensure effective engagement with government, industry and other stakeholders.

To ensure industry credibility, technical currency and relevance the majority of the staff involved in design, development and delivering the program are practitioners, sourced from industry as subject matter experts, currently engaged in the field of maintenance and physical asset management. The programs are coordinated and administered by academics of the university to ensure academic credibility, technical and education currency and relevance and to facilitate educational quality assurance processes. The successful management of this educational-practitioner dynamic requires stakeholders to operate as a trans-disciplinary team. Courses within the programs are delivered in a blended, flexible mode, accessed through a Moodle website, and comprising residential schools, telephone and email contacts so they are flexible for people currently working full-time in industry and are resident to a range of locations (urban, rural, regionally, remotely, nationally, internationally). Owing to the emphasis upon Work Integrated Learning (Groenewald, 2004; Beard & Wilson, 2002; Reeders, 2000, p. 205; Katula & Threnhauser, 1999) and authentic assessment, a condition of enrolment is students must have a current position as a maintenance planner or a maintenance manager. In this way they have access to a range of people, data and information within their organisation to enable them to complete assessment items directly related to the work place. In some cases, through their educational experience, the students act as change agents within their organisations precipitating changes in practice. As assessment items contain information sensitive to the competitive advantage of organisations, all information is managed with a strict code of commercial in confidence.

In Australia, while undergraduate programs are accredited by Engineers Australia postgraduate coursework programs are not. To ensure program relevancy and currency is critiqued the process has been to undertake a number of strategies. An independent consultancy report was commissioned in 2004 (Abbott, 2004) to review the curriculum and delivery and possible pathways to strengthen and extend the programs. Recommendations were critiqued and the improvements using course feedback were implemented. This process involved engagement with stakeholders; industry, lecturer-practitioners, students and academics. In 2010, stakeholders were invited to be part of a physical asset and maintenance management conference (CAMM2010, 2010) to celebrate the type of work being undertaken by stakeholders.

THE SUITE OF MAINTENANCE MANAGEMENT PROGRAMS

The CQUniversity maintenance management programs have been designed as a cohesive suite of courses that progressively build on each other to develop the essential knowledge and skills required by a maintenance manager. The programs are quite unique, as there are very few university programs in Australia (CQUniversity, Monash,

University of Western Australia and University of Wollongong) or internationally that offer postgraduate study in the management of maintenance.

The programs have in general two annual intakes which provide flexibility in timing the start of studies. This allows both students and organisations the ability to match study commitments with their workload obligations. CQUniversity conducts a three term study schedule. Each study period (Term) lasts 12 weeks. Maintenance management courses are conducted in both Term 1 and 2 and some courses are available in Term 3 to provide enhanced flexibility to students. The programs cater for professional development of maintenance practitioners, maintenance planners, supervisors, coordinators and maintenance managers. They provide a holistic business-centered view of the maintenance function and management. Organisations have the opportunity to build a robust and highly trained maintenance workforce from within. Potential maintenance leaders can also be identified within the organisation to undertake this type of training, providing reward mechanisms and motivational support for those identified as future maintenance managers, maintenance planners, supervisors and crew leaders. By completing these programs professional engineers have the opportunity to develop their skills and knowledge and to broaden their career options in the maintenance field. The programs are underpinned by the principles of business-centered maintenance that place asset maintenance as a key component and critical element of the business of an industrial organisation.

A significant advantage of the programs is that they are delivered in a workplace based flexible study mode which allows organisations to retain their workforce at the same time as educating them. Students get to use the concepts covered while they are working and all assessment items require students to consider maintenance-related issues and problems being faced within their organisation. This provides an opportunity for the organisation to be involved in identifying particular organisational areas or issues for students to investigate during their learning experience and for the student to work through practical solutions.

The learning objectives of the postgraduate maintenance management programs focus upon students gaining the following proficiencies:

- Apply the key physical asset and maintenance management concepts supported by the various management systems in their daily work.
- Identify and describe the points of integration and interfaces between the numerous management systems encountered in large industrial plants.
- Develop physical asset management strategies and monitor the effectiveness of maintenance management strategies.
- Understand and apply the business centred approach for ensuring physical assets are maintained to achieve the desired outcomes.
- Describe and explain the complex and diverse responsibilities of maintaining a large industrial physical asset.
- Develop proposals for project development that meet established viability criteria for overhaul, replacements and retrospective fittings.
- Evaluate options for project development in continual improvement in maintenance of industrial physical assets.

The Graduate Certificate in Maintenance Management program requires completion of four course units which can be completed over one to two years. It provides the foundation for the program, establishing an integrated framework of knowledge and skills essential for effective maintenance management. As outlined in Table 2, the program requires completion of four core courses.

The Graduate Diploma of Maintenance Management program requires completion of four

Table 2. The Graduate Certificate in Maintenance Management program course structure (Source: http://research.cqu.edu.au/FCWViewer/view.do?page=1887)

ENMM20010 introduction to Maintenance Engineering This course provides an overview and introduction to asset management and maintenance engineering within an industrial organisation, including: Business centred Maintenance Systems view of maintenance Models of maintenance Maintenance financial analysis Reliability centred maintenance Failure Analysis/Criticality/Root Cause Analysis Maintenance analysis techniques Maintenance and the environment	*ENMM20011 Establishing the Maintenance Strategy* This course covers the essential elements of an effective asset maintenance strategy, including: Establishment of maintenance objectives Life plans, schedules and life cycle costs Capital asset management Reliability centred maintenance Controlling asset reliability/maintainability Preventive maintenance Scheduled/unscheduled maintenance Formulating the maintenance strategy Key Performance Indicators (KPIs)
ENMM20012 Maintenance Organisation This course considers the organisational contexts associated with operationalising maintenance strategy in the modern industrial organisation, including: The complexity of industrial organisations Modelling the maintenance organisation The maintenance workload Maintenance resource structures Maintenance administrative structures Human factors in maintenance organisations Trends in maintenance organisation Total productive maintenance (TPM) Case Studies	*ENMM20013 Maintenance Systems & Documentations* This course builds on the concepts covered in the maintenance strategy and mainatenace organisation courses, to develop a comprehensive understanding of the industrial maintenance organisation. Maintenance systems Maintenance budgeting Maintenance management control Organisational effectiveness and efficiency Maintenance planning Maintenance documentation systems Computerised maintenance management systems (CMMS) Spare parts management Key Performance Indicators (KPIs)

additional course units that take another one to two years to complete. This program extends the knowledge gained in the Graduate Certificate in Maintenance Management to develop the skills required to provide leadership and continuous improvement in the industrial maintenance organisation. As per Table 3 the program requires completion of the CQUniversity Graduate Certificate in Maintenance Management plus four of the courses outlined.

For those students that wish to continue their studies and develop the skills required to conduct applied industry-based research of real-world problems, the Masters program can be completed in one year, following completion of the Graduate Diploma program. The Master of Maintenance Management focuses the previous learning and experience towards creation of opportunities to improve maintenance management practices. The Masters program requires the student to investigate and resolve a significant real-world issue affecting the maintenance organisation. As Table 4 highlights the program requires completion of the CQUniversity Graduate Diploma of Maintenance Management plus two project courses.

THE TEACHING AND LEARNING ENVIRONMENT

Student Cohort

As maintenance planners or maintenance managers the students are experienced practitioners who

Table 3. The Graduate Diploma in Maintenance Management program course structure (Source: http://research.cqu.edu.au/FCWViewer/view.do?page=1887)

ENMM20014 Maintenance Leadership This course focuses on leadership theory and practice appropriate to current and future maintenance practice. This course explores the techniques for advancing leadership as a critical business imperative, including: Management and leadership Leadership theories Skills for effective leadership Measuring and managing performance Change management Coaching and training Developing teams and groups Valuable lessons on effective leadership	*ENMM20015 Auditing Maintenance System* This course develops and understanding of audit methodology focused towards the improvement of industrial asset management and maintenance systems, including: Goals of asset management auditing An auditing methodology Auditing human factors Auditing maintenance departments Types of audits Benchmarks and performance indicators Case studies
Plus any two of the following:	
ENMM20017 Advanced Condition Monitoring This course focuses on condition monitoring as a preventive maintenance strategy. The course covers a range of condition-based maintenance techniques, including: Condition based maintenance Selection of condition monitoring techniques Parameter symptom limits Oil analysis Thermal monitoring Corrosion identification and monitoring Vibration monitoring Vibration analysis	*ENMM20018 Turnaround Management* This course covers the theoretical knowledge and technical skills required to perform turnaround events, using a systems approach to maintenance, including: Turnarounds from a business and engineering perspective Phases of a turnaround Logistics of a turnaround Initiation of a turnaround Preparation for a turnaround Execution of a turnaround Termination of a turnaround
ENMM20020 Maintenance Materials Management This course focuses on the materials supply function of asset maintenance management, including: Objectives of materials management Categorizing materials Managing inventories Warehouse management	

Table 4. The Masters of Maintenance Management program course structure (Source: http://research.cqu.edu.au/FCWViewer/view.do?page=1887)

ENMM29025 Research Project A This course requires the student to plan the investigation that they are going to undertake. This will require the student to submit a project proposal outlining the investigation they are going to undertake, the research questions, the methodology they will use and desired outcomes.	*ENMM29026 Research Project B* This course requires the student to conduct the investigation of a real-world problem in an industrial organisation, applying the methodology previously devised, gather and analyse the data, develop recommendations and a solution to the original problem identified in ENMM29025 Research Project A. The final submission for this course will be the Research Report that is the culmination of the industry based research project. This report, written in a thesis format, will provide a complete description of all stages of the project from conception through to completion. This report will be submitted for marking and grading by the Faculty of Sciences, Engineering and Health.

are in middle management positions within their organisations. Their roles and responsibilities are diverse and involve, in the majority of cases, multi-million dollar physical asset management budgets, overseeing human resource management of technical staff, the acquisition and decommissioning of equipment and ensuring statutory environmental requirements are upheld and health and safety aspects are met to meet business objectives and to provide a safe workplace.

The program's student profile reflects a cohort consisting of 'non-traditional' learners rather than a school leaver student profile which would have a pedagogical learning environment emphasis. Owing to the student demography (mature age, graduate and trades qualified), location (urban, rural, regional and remote), their responsibility for overseeing physical and human resources in their role as a maintenance manager or planner and their working conditions (12 14 hour shifts, 7 day shifts, fly-in/fly-out, drive in/drive out) their learning environment needs to be centred upon authentic learning processes and to follow an andragogical learning approach (Knowles, 1980).

An andragogical learning approach, first proposed by Kapp (1883), is characterised by Knowles (1990, as cited in Harris, 2003) as the 'art and science of helping adults learn' and further contends such an approach should incorporate learning activities that acknowledge such learners as being, self-directed; having a great diversity of life and work experience; ready to learn when experiencing a need to know or to do something; life-centric, task-centric or problem-centric and motivated from their self-esteem, their need for recognition, the seeking of a better quality of life and their own self-actualisation.

Definitions of approaches and experiences that constitute authentic learning and its assessment are varied (Gulikers, Bastiaens & Kirschner, 2004; Birenbaum, 2003; Gielen, Dochy & Dierick, 2003; Dochy, 2001; Petraglia, 1998; Birenbaum & Dochy, 1996).

In particular, an authentic learning environment provides the opportunity to learn in a real-world relevant context, so that the learning activities and their assessment match real-world needs of professionals in practice. Additionally it also allows competing solutions and a diversity of outcomes. Further aspects of authentic learning and its assessment are discussed in the following section.

Typically, as 'non-traditional' students, these adult learners are self-supporting, mature and responsible (MacKinnon-Slaney, 1994, cited in Blakely & Tomlin, 2008, p7). They are motivated to learn more readily than traditional learners driven by internal motivators such as self-esteem, recognition or a better quality of life (O'Brien & Renner, 2000, cited in Blakely & Tomlin, 2008, p7). Furthermore they may be motivated to return to study to update their skills for career advancement (Kerka, 2001, cited in Blakely & Tomlin, 2008, p7). The consistently expressed reasons students highlight for pursuing our courses, from a personal perspective, is that they are interested in learning knowledge which would support their personal goals and career trajectory including sustained employment, promotion or employment opportunities for transference to another position within or external to their current employer.

From a knowledge acquisition perspective, the attraction of our courses to maintenance planners and managers is the delivery of theoretical aspects through a practical and pragmatic approach so that what they are learning can be tried, tested and critiqued in their workplace, in context, as they acquire the knowledge and skills. This sits well with Knowles (1978) and MacKinnnon-Slaney (1994) who highlight a further unique aspect of adult learners over traditional learners is that their expectation to apply their newly acquired knowledge in the near future.

Owing to the myriad of roles and responsibilities adult learners tend to display effective time management and do so through the mechanism of their ability to juggle these roles (Richardson & King, 1998). This mirrors the types of dialogue we have with students in juggling their work life, study life and family life. Students in our courses are studying from throughout Australia and typically are working 12-14 hour shifts on a rotating 7 day shift roster. If not living in the district where their workplace is located then they either fly from their home base, leaving their family or spouse behind, to the work site and return to their home base upon completion of the roster (fly-in/fly-out) or instead of flying undertake their travel by driving. These types of arrangements, in particular the mining industry, include food and accommodation being provided for workers only at the mine site. So workers have to pack up everything to bring to work and then bring everything home with them, so they have no permanent individual 'space' or location at the worksite. As a mode of work style, the Chamber of Minerals and Energy, Western Australia (2005, p6) in undertaking an analysis of mining operations of publicly listed companies in Western Australia showed 46.6 percent of companies utilise fly-in/fly-out. They reported, assuming if generally representative of the mining sectors reliance upon fly-in/fly-out and production value, that this correlated to an estimated $12.6 billion annually of costs of mining operations in Western Australia. This does not include the extensive mining activities in Queensland and South Australia utilising fly-in/fly-out arrangements nor does it take into account any aspect of drive-in/drive-out arrangements and occurrences. It has significant impact upon the work arrangements, lifestyle and study habits of individuals and the manner in which industry functions.

Students undertaking courses in the suite of these postgraduate programs are either engineering (or technically skilled) graduates or have commenced their careers with training in a recognised trade; in particular a mechanical or electrical trade. As either graduates or trades qualified practitioners, as they proceed with their careers they find themselves constrained in their ability to contribute and innovate, and feel the need to up-skill and enhance their knowledge and capabilities by pursuing a postgraduate education that is industry-relevant. In adopting an andragogical approach (Knowles (1990), as cited in Harris, 2003) as academics we acknowledge the diversity of life and work experience of students by providing recognition of their prior learning and experiences. Recognition of Prior Learning (RPL) is called Credit Transfer at CQUniversity and enables students to apply for credit of previous formal training that is equivalent to a CQUniversity course. This may translate as enabling students to gain entry to the programs at Graduate Certificate level without having completed an undergraduate qualification (having advance diploma with substantial experience is one of the examples) or by providing them with course exemptions and reducing the number of courses they are required to undertake to complete a qualification. All students who enrol into the program must have been working in the field of maintenance engineering or maintenance management for a minimum period of time as set in the entry requirements to gain entry to the program (most students who enter the program far exceed this minimum requirement).

In determining the courses they are going to study students are discerning in choosing courses that provide them with an authentic learning experience and which enables them to acquire and use such knowledge in their workplace. The cohorts of maintenance practitioners undertaking the postgraduate programs exhibit a broad range of the types of characteristics of adult learners. Indeed, with a greater emphasis on providing authentic learning that can be immediately applied it is as Mirvis and Hall (1996, p73) so aptly expresses it workers need to *learn a living rather than earn a living*.

Lecturer-Practitioners

The *'muddy trenches'* perspective is provided by the lecturers who have extensive experience in the area of physical asset or maintenance management or its related areas such as logistics and condition monitoring and are currently practitioners within industry. They fall within the category of being experienced maintenance managers, experienced people in a particular area of physical asset or maintenance (e.g. logistics, condition monitoring) and/or operate as consultants running their own businesses or as consultants within a firm engaging with industry assessing or undertaking aspects of that industries physical asset or maintenance management. They are subject matter experts and are employed as casual lecturers for a 12 to 14 week period to conduct courses. In some cases they have been engaged to design the course and write the study material. They include engineering graduates, people with a trade backgrounds (in particular mechanical and electrical) or have a higher degree in a related physical asset or maintenance management area.

The model adopted arises from the original ideas proposed by the Gladstone Engineering Planning Committee back in the early 1990's when the programs were being developed. The idea was to draw on the diversity of experience and knowledge already available in the Gladstone industries to ensure, while inclusive of maintenance management theory and practices, the emphasis of the courses were on the practical aspects of maintenance management with current practitioners providing the instruction rather than the courses being highly theoretical courses exploring the theory of physical asset or maintenance management.

The inclusion of practitioners as part of the teaching team fits well with the *aus der praxis, fur die praxis (*from practice for practice) philosophy adopted and represents one of the number of opportunities that are available to take advantage of this approach. The approach has peers teaching peers, sharing their knowledge and experiences of solving maintenance management problems and improving maintenance management strategies.

Academics

The academic (*'ivory tower'*) perspective is necessary to certify the program's scholarly credibility. It covers the learning objectives, the contents, study material, case studies, assessments and the grading. Any well conducted training course could provision the student with the equivalent technical material. However it requires a sound educational approach, based on scholarship, to guarantee the technical and non-technical aspects presented and the outcomes delivered are not only grounded in practice, but are substantiated by the latest research of best practice; both technical and educational. The program's scholarly credibility underpins the graduate's education and development as a professional, as opposed to merely being up-skilled as a maintenance practitioner through a training scheme. The academics involved in overseeing the programs and operate as course co-ordinators. As course co-ordinators they are responsible for ensuring the course materials are relevant and current and also oversee the delivery of the course by the lecturer-practitioner and the validation process of learning outcome is consistent, transparent and meaningful. They are responsible for generating and managing the Moodle site for the course and monitor the interactions between students and lecturers, facilitation and grading.

This is another avenue for establishing and maintaining the *aus der praxis, fur die praxis (*from practice for practice) philosophy. The academics have international reputations in being involved in the research of best practices for physical asset and maintenance management and for curriculum development and instigation of learning environments to support Work Integrated Learning in the workplace. They are involved in research projects and also supervise research higher degree students undertaking masters by research, masters by coursework and a Doctor of philosophy whose topics have a focus upon physical asset or maintenance management.

Liaison Librarian

In the transitional process of enhancing and extending the knowledge and skills of the students enrolled in the programs do not only increase and update their technical knowledge and skills but also their generic skills (communication, information literacy, computer literacy). These generic skills are developed through their use and interaction with technologies like Moodle course website, accessible as part of their enrolment to the course, and the electronic resources (databases and electronic journals) accessible through the CQUniversity library.

Our institution has Liaison Librarians for each of the major discipline areas and students are provided with opportunities to enhance their information literacy skills. The importance of information literacy cannot be underestimated. While there are databases and journals to access through the university's resources network, a sometimes unacknowledged aspect of information literacy takes place in the workplace. This is evident and arises as colleagues within the workplace become valued as primary sources of information. The course content and the assessment emphasise the necessity to access information within the student's organisation to aid them in understanding and implementing the new knowledge they are acquiring through the course.

This highlights a stark contrast between total reliance on formal information sources, typically located in the library or accessed through the library, such as books and databases, and less formal knowledge and information sources provided by the experience, knowledge and skills of workplace peers.

Blended Learning

Synchronous and asynchronous instruction and communication is adopted in delivering the courses in a blended learning model (Allen, Seaman & Garrett, 2007; Albrecht, 2006; Valiathan, 2002). Blended learning is also known as 'hybrid learning' (Martyn, 2005) or 'mixed learning' (Everett & Van Dessel, 2005; Vernon, Kirby, Allen & Sargeant, 2003).

The rise of the massification of education has seen is a greater reliance upon technology and its equitable access and development of student's computer literacy skills or what is also termed Information Communication Technologies (ICT) (International Council for Open and Distance Education, 2009; Catts, 2004, p8; Hamid, 2002; Bates, 2000). The programs have made extensive use of Web 2.0 technology platforms and has recently transferred from a Blackboard platform to the Moodle platform. This approach increases the options for greater quality and quantity of human interaction in a learning environment. Furthermore understanding the principles behind the use of technologies is critical so that learners have a basis for adapting to new technologies as they emerge and they need to adopt them in their workplace. As the course is delivered in a distance education mode it is reliant upon the appropriate use of the mixture of technology and face-to-face interactions to create and encourage an interactive and engaging learning environment. The curriculum structure provides opportunities for differing modes of engagement between students, lecturer-practitioners and academics. The blended mode provides learning opportunities through paper-based study materials (including the concepts and theoretical basis of the course content, resource materials and case studies), the Moodle site has a number of facilities to allow this occur (synchronous and asynchronous forums, asynchronous information dissemination, emails). In this way students, through forums can communicate, debate and engage with other students and academic staff and the lecturer-practitioners to discuss aspects and issues related to maintenance management. The lecturer-practitioners and the academics are also able to post information through a forum reserved for that purpose.

Learning by distance education is supported by a dedicated non-compulsory two-day residential school for each course. The residential schools provide an opportunity for students to meet the lecturer-practitioner, academics and other students in an environment where they will have access to further information and resources related to maintenance and its management. The residential school covers the concepts of the course in greater detail and provide additional opportunity for discussion and syndicate work using industry-based case studies. In this way informed critique of current practices and real solutions and possible strategies are workshopped. The new knowledge acquired can be evaluated as relevant or fit for purpose by students in their own organisation and immediate application of that knowledge can be undertaken.

The decision to make residential schools non-compulsory arises from a number of facets. A student's seniority of position and geographical location are the two key facets. Owing to the seniority of their position and the role and responsibilities of students in their organisation it is not possible for them to attend residential schools as it is imperative they are on call or on site for immediate action. Whether this is combined with their geographical location or not, gives rise to interesting scenarios. Since our student cohort is drawn from urban, rural, regional, remote, national and international landscapes then it is just not possible nor advantageous for them to attend residential schools. For instance students working in the remote areas of Western Australia will take longer travel to, attend and return from the residential school then it does to attend the residential school. So the 'wasted time' in travelling to and fro and attending and not being available 'on call' for immediate actions on site is counterproductive to the successful conduction of multi-billion dollar enterprises. To ensure equitable practices, information discussed or decided at residential schools is posted on the course Moodle site. Thus the provision of a blended learning environment is paramount in delivering the course so that all students can interact by means of the technological platform of Moodle transferring their knowledge and experiences to others and learn from others. Indeed skills of transferability can be fostered by providing learners, who are practitioners, from different settings and knowledge domains, the ability to conduct and engage in the Moodle courses electronic forums and face-to-face if they are able to attend the residential school. In this way they share, compare and contrast their knowledge domain, experience and skills and challenge and critique their own and others perspectives to seek alternatives or validation (Fook, Ryan & Hawkins, 2000, p210).

A challenge arises with the small cohort of international students whom are located in remote locations of Papua New Guinea, Malaysia and Chile. In these locations there is a reduced reliability of telecommunications and the ability to gain access to our web-based Moodle site in our blended curriculum approach. In order to exercise equitable practices these types of circumstances need to be taken into consideration and appropriate strategies put into place. This involves allowing time for assessment items to be submitted by snail mail or understanding that since telecommunication capabilities are intermittent or access may only be available for specific times of the day owing to organisational telecommunications priorities over individual employee priorities then students will not interact in the same temporal periods as other students.

As academics we are aware that some students are not as proficient in their computer literacy as others. To alleviate this we design activities which enable students to develop their familiarity with the Moodle site and their proficiency in computer literacy skills. These activities are assessed both as formative and summative assessments. Figure 1 provides an example of formative appraisal. This type of exercise provides opportunities for students to develop their computer literacy without penalising them during their learning phase of operating within an e-course. This also serves as a mechanism by which they can provide personal and career details to other students and helps them to make networks. This can also be described as co-operative networking (Catts, 2004, p10) or as a generic skill it is the act of team building and is developed through student's socialisation be it electronically via the Moodle course site or in a face-to-face scenario of the residential school. Which fits well with the drive by massification to use technology as a means of offering opportunity and as highlighted by Randall et al (2005) it provides the opportunity for students to collaborate with other members of the workforce and to share and learn knowledge.

Authentic Learning and Assessment

In the arena of professional education there is greater emphasis being placed upon facilitating an increased legitimacy of workplace generated knowledge. Consequently greater attention is being given to 'authentic' learning and its assessment.

These postgraduate programs enhance a student's learning by engaging them in authentic learning and the assessment of this learning by using their own workplace as the setting for their assessment tasks for workplace based problem solving. Their real-world tasks are used to identify and demonstrate the meaningful application of the essential knowledge and skills they require to operate as proficient practitioners. To this end, the most pertinent overarching definition used to describe our approach to authentic assessment correlates well with that of Newmann and Archibald (1992, cited in Boud & Falchikov, 2005, p35), as academics we design assessment tasks 'to ensure that there is a greater correspondence between student work and that undertaken in workplaces'. In refining this working definition, the aspects enunciated by Tennant (2000), help to better define the features our programs adopt to develop student's technical knowledge, skills (technical and generic) and attitudes through completing assessment items which focus upon their individual workplace context. This includes the ability to: learn from others; analyse workplace

Figure 1. Sample experience computer literacy development without penalty (Source: ENMM20014 Maintenance Leadership Course Profile, Term 2, 2010, p4)

Details

Moodle is the learning management system used by the university. In this course this resource is used as a centralised communication tool or 'centre' for all students to meet in one place and participate in online discussions, ask questions and access resources and other information posted by the lecturer and other students. It is VERY IMPORTANT for every student to access the site and be familiar with the resources and facilities available. You are thus required to login to the site and complete the following activity:

Use the 'Discussion Forum area' in the 'Communication' block. By the end of Week 1 you will be required to post a message to the class group in the student introduction discussion area that outlines the following:

(a) who you are
(b) where you live (you may find someone to set up a study group with)
(c) life-experience
(d) where you are employed and any experience in maintenance
(e) your interests such as sport, hobbies, family etc.
(f) what you hope to get out of this course
(g) what gives you a buzz about maintenance management.

experience; choose among multiple courses of action; act without all the facts available; learn about organisational culture; understand the competing and varied interests in the shaping of one's work or professional identity (Tennant, 2000).

The context of the learning experience and its relevance to workplace practices has significantly shaped the assessment of such learning within the postgraduate programs. Students, 'clients' or 'customers' or 'learners' are experienced and qualified practitioners currently working in the area of the maintenance management and are already considered having expertise and experience as competent practitioners. Consequently these learners are more likely to value assessment tasks they perceive to be 'real', relevant and which present serious challenges in terms of their own workplace experience in order to extend and challenge them. To address these demands the assessment items in each course in the Graduate Certificate and Graduate Diploma either consists of work-based written assignments and/or work-based projects. Both forms of assessment item are practical, realistic and challenging and focus upon identifying, investigating and proposing strategies or solutions to improve maintenance management practices and processes. The courses are conducted over a twelve week period with assessment due dates typically at week 4 and week 12. The work-based course project is aimed at a more extensive exploration and solution of an issue in the workplace. The Masters Program is an open-ended problem proposed by the student that relates to a maintenance management problem on their worksite. The Program consists of two twelve week courses with an interim report presented for the first twelve week stage and a dissertation submitted at the end of the second twelve week period.

Figure 2 outlines a sample work-based written assignment for the 2010 term 2 course offering of ENMM20012 Maintenance Organisation, while Figure 3 shows a breakdown of the associated assessment criteria used to examine it. The work-based assignment constitutes 30% and the work-based project 70% of the designated assessment for that course.

Both assessment tasks map well to the range of aspects outlined in the continuum of traditional to more authentic assessment processes as per highlighted by Randall et al (2005): by providing the opportunity for students to collaborate with other members of the workforce; its real-world relevance matching real-world tasks of professionals in practice; comprises complex tasks investigated over a sustained period of time and allows competing solutions and diversity of outcomes.

Figure 4 outlines a sample work-based project assessment for the 2010 term 2 course offering of ENMM20012 Maintenance Organisation, while Figure 4 shows the associated assessment criteria used to examine it. The project outlined in Figure 4 is more prescriptive than would be expected for what Randall et al (2005) define as an 'ill-defined (exercise) requiring students to define the tasks and subtasks needed to complete the activity'. Instead, it is employed as a learning experience, to develop skills in readiness for other assessed projects within the program, the masters project and ultimately their everyday workplace tasks, so that the student has the necessary skills to critically deconstruct the poignant aspects to be identified and investigated. This exercise, in the context of the program assessment, models the behavior as a transition between traditional and authentic assessment. Figure 5 outlines the assessment criteria used to evaluate the assessment item.

The masters program is a form of authentic assessment that enables students to undertake an 'ill-defined' problem. An 'ill-defined' problem (Daniels, Carbone, Hauer, & Moore, 2007; Huba & Freed, 2000; King & Kitchener, 1994; Simon, 1973) is one that addresses complex issues, is not able to be easily described in a concise, complete manner as there will be competing factors which may suggest several approaches to solving the problem, requiring critical analysis to determine

Figure 2. Sample work-based written assignment (Source: ENMM20012 Maintenance Organisation Course Profile, Term 2, 2010, p4)

Scenario

Contractors are widely used today to carry out a range of maintenance activities in the workplace. This may range from period contracts to repair buildings or office fittings, to carrying out major equipment overhauls during a plant turnaround. Contracts dealing with construction contracts (where the equipment is being installed and started up for the first time) should not be included in this assignment.

Select three (3) maintenance contracts that have been let in your workplace in the last three (3) years.

Part 1 (8 Marks)

Interview an **employee of the Owner/client** (preferably someone in a supervisory capacity) to determine in each case, what they see as the three (3) most significant *barriers to the contractor being able to perform an efficient service.* These barriers should be within the control of the parties involved and taken from the list shown on the contractor survey sheet attached (not included here). Your findings for each contractor should be summarised on this sheet (one per contractor).

Part 2 (8 Marks)

For the same three contracts, interview an **employee of the Contractor** (preferably someone in a supervisory capacity) to determine in each case, what they see as the three (3) most significant *barriers to the contractor being able to perform at peak efficiency.* These barriers should be within the control of the parties involved and taken from the list shown on the contractor survey sheet attached. Your findings for each contractor should be summarised on this sheet (one sheet per contractor).

Part 3 (14 Marks)

For each of the three contracts audited, provide recommendations on how best to overcome the problems identified. Your answer should contain specific actions that could be undertaken by the contractor and/or client. Describe your recommendations for each contract in no more than a page per contract.

Notes

1. Your answers to this assignment should NOT contain actual contractor names for reasons of confidentiality.
2. Use copies of the same survey sheet for contractor or client interviews and tick the box at the top of the sheet to indicate who is being interviewed.
3. You must provide copies of your completed survey sheets as part of the submission to The lecturer. There should be a total of six sheets, ie. three (3) sheets for clients and three (3) for contractors.

the best approach. In this way the learner is able to develop empirical, critical-thinking and problem-solving skills as the resolution of the problem requires judgment, planning, the use of multiple strategies, and the implementation of previously learned skills. The set of possible strategies may seem overwhelmingly large to the novice, and the whole process of developing expertise may consist of winnowing away unhelpful strategies. Consequently the learner relies upon deeper learning to create meaning and to develop understanding.

Since its inception, in 1998, the masters program has graduated some 30 students. Students are supervised by an academic supervisor with an industry based supervisor nominated by the student. Besides supervision the industry based supervisor's role is to provide support at organisational level and if necessary to provide the student with an avenue to gain access to information they

Figure 3. Assessment criteria for sample work-based written assignment (Source: ENMM20012 Maintenance Organisation Course Profile, Term 2, 2010, p5)

Assessment Criteria

This is a criterion-based assessment item. Please note that it is highly recommended that you read beyond the course materials to complete assessment items. Your submission will be evaluated against the following criteria:

Demonstrate ability to:

1. (10%) accurately collect and collate data from various sources (owner/client and an employee of the contractor).

2. (20%) identify barriers to a contractor from different points of view; view of the owner/client and view of an employee of the contractor

3. (20%) critique barriers to a contractor from different points of view; view of the owner/client and view of an employee of the contractor.

4. (50%) analyse data and information to deduce conclusions in order to formulate Recommendations.

Figure 4. Sample work-based Project Part A (Source: ENMM20012 Maintenance Organisation Course Profile, Term 2, 2010, p8)

Your have been asked to carry out a major review of the maintenance structure in your organisation (for students who do not work in a suitable sized organisation select one with which you are familiar, carry out work for, etc.)

The view has been requested by the Plant Manager as there has been very little structural change in the Maintenance Department over the last 10 years, but the nature of the company and its associated operations has changed considerably. The Plant Manager is under pressure to improve the bottom line maintenance cost per unit of production without affecting the long term viability of the business.

Part A 25 Marks

Produce a report to senior management which includes the following information:

1. Organisational chart showing where the maintenance department or function fits into the overall administrative structure (see figure 4-1 in your notes for an example) *(2 marks)*
2. A detailed organizational chart of the Maintenance and Operations Departments/Sections showing the relationship senior management to 'shopfloor' (see figure 4-20 in your notes for an example) *(2 marks)*
3. Number and type of trade unions which have representation on site and of these how many represent maintenance employees. *(1 mark)*
4. An approx % split of those maintenance employees who are paid a 'Fixed Salary' versus those who are paid for 'Overtime'. *(2 marks)*
5. The average % overtime which is worked across the plant (if applicable). *(2 marks)*
6. A comment on the amount of demarcation which exists. *(3 marks)*
7. The different day or shift rosters which are worked. *(3 marks)*
8. The approx number of employees working on each shift roster. *(2 marks)*
9. An approx % split of proactive versus reactive maintenance. *(3 marks)*
10. A comment on the amount of contractor labour which is used and whether this is constant or to cover peak workloads (or both). *(3 marks)*
11. A comment on the use of condition monitoring across the Plant. *(2 marks)*

Figure 5. Assessment criteria for sample work-based project (Source: ENMM20012 Maintenance Organisation Course Profile, Term 2, 2010, p9)

Assessment Criteria

1. (10%) collect and collate appropriate data
2. (10%) clarity of expression, including correct grammar, spelling and punctuation
3. (15%) appropriately presented technical data (eg. Chart, tabular, graphical)
4. (35%) analyse data and information to deduce conclusions in order to formulate recommendations.
5. (30%) performing project design and management course

may not be able to access themselves. This is necessary also because the student may not have the authority to commit funds or to gain access to relevant data or information or the systems that enable access. The program is conducted in flexible mode and is designated as a one year program. It consists of two courses with an interim report written at the conclusion of the first course. The interim report contains a literature review, refined project definition, assessment of the problem and preliminary discussions of possible solution strategies. Depending upon the project and the student's progress some students will progress to an implementation stage and may have also trailed and critiqued possible solutions by the conclusion of the first course. The project dissertation is completed as an outcome of the second course. To maintain credibility with our industry partners and to promote currency and relevancy the dissertation is examined by two external examiners and these are practitioners currently employed in the Asset or Maintenance Management area.

An interesting aspect evident at Masters level is the demonstration of knowledge creation that leads to efficiencies (financial, human resources and physical resources), innovation and changes in practice. Large organisations have electronic repositories and intranet capabilities to establish electronic communities of practice. This mechanism provides a fluid and vigorous environment for employees to query, discuss and report their findings of work practices and improved processes adopted at their site and to share it with a global organisational network and to influence local, national and international practices. Likewise the ideas and findings reported from their Masters research undergoes peer review and critique and is challenged and debated and the strategies they outline are repeated when posted on these organisational electronic communities of practice. This fits well with the *aus der praxis, fur die praxis* (from practice for practice) philosophy we have adopted and it also demonstrates fundamental facets of the research process. In a great many ways we are observing the emergence of Work Integrated Research (WIR) through a coursework masters as a consequence of the work undertaken to complete the program has knowledge creation outcomes. There are issues as the organisation's intranet would not be accessible by peers who are not employees of that organisation and this arises because of commercial in confidence aspects and in that sense of the body of knowledge the intranet is not open source resource. However in some ways that is no different to a discipline specific journal having its published works being only accessible by eligible members. In a more traditionally research orientated context, the research findings could be published in a discipline journal or conference, through the usual peer review process, if permission was gained by the organisation for either releasing such information or by reporting the findings in a manner in which the organisation is not identified. The issue is gaining the appropriate permissions to ensure an ethical reporting of findings so that the organisation is fully cognisant of what is being reported and that it does compromise their operations in terms of commercial sensitivities. Stewart and Chen (2009) discuss developing a framework for Work Integrated Research Higher Degrees exploring and reporting on this approach in the context of three case studies that are of a Doctor of Philosophy context and not in a WIL context as is the philosophy adopted by CQUniversity. Another non-traditional mechanism to explore the emergence of WIR from a WIL contextual environment would be some professional degrees such as Professional Doctorates with the focus on creating new knowledge with a WIL philosophy as its basis. CQ University's professional Doctorate would fall into that category.

Future Directions

In the spirit of *aus der praxis, fur die praxis* (from practice for practice) stakeholders are being continuously engaged. Recent outcomes from the dialogue with industry-academics-practitioners-students has seen interest in extending the focus on maintenance management to physical asset management and also to develop a project management program. A costed proposal for the redevelopment of the maintenance management curriculum has been submitted for consideration for enhancing the flexibility, equity and applicability for the wide spectrum of industries and participants from distance demographic areas. The inaugural international conference on Asset and Maintenance Management 2010 (CAMM2010, 2010), conducted in March 2010, demonstrated the support for the postgraduate programs and showcased the research being undertaken by the university in improving maintenance management practices. Planning for the 2011 conference has commenced.

Further investigation into the notion of Work Integrated Research arising within a Work Integrated Learning context is being investigated. The mechanisms by which this occurs through coursework masters and also professional degrees like a Professional Doctorate are the focus of further research to help create an appropriate definition.

CONCLUSION

The suite of programs developed through consultation and collaboration through the nexus of industry-university-practitioner are unique. The emphasis upon authentic learning and its assessment ensures learners are gaining skills and knowledge pertinent to their workplace and they are able to incorporate into their workplace practices. Owing to the urban, rural, regional or remote scenarios for the location of students across Australia and internationally a blended learning delivery mode ensures students have synchronous and a synchronous modes of interaction that best suits their needs as an independent learner. Continuous improvement of the programs is achieved by engagement with a cross-sectional representation from the industry-university-practitioner stakeholders.

REFERENCES

Abbott, J. (2004). *Report to James Goldston, Faculty of Engineering and Physical Systems: A strategic review of the postgraduate maintenance management courses.* Unpublished consultancy report.

Albrecht, A. (2006). *Enriching students learning through blended learning.* Educause Centre for Applied Research. *Research Bulletin (Sun Chiwawitthaya Thang Thale Phuket)*, *2006*(12). Retrieved from http://net.educause.edu/ir/ library/ pdf/ERB0612.pdf.

Alexander, F. K. (2000). The changing face of accountability. *The Journal of Higher Education, 17*(4), 411–431. doi:10.2307/2649146

Allen, I. E., Seaman, J., & Garrett, R. (2007). Blending in: Extent and promise of blended learning in the United States. Retrieved July 24, 2010, from http://www.sloanconsortium.org/sites /default/files/Blending_In.pdf

Bates, A. (2000). *Managing technological change: Strategies for college and university leaders*. San Francisco, CA: Jossey-Bass.

Beard, C., & Wilson, J. P. (2002). *The power of experiential learning: A handbook for trainers and educators*. London, UK: Kogan Page.

Birenbaum, M. (1996). Assessment 2000: Towards a pluralistic approach to assessment. In Birenbaum, M., & Dochy, F. J. R. C. (Eds.), *Alternatives in assessment of achievements, learning processes and prior knowledge* (pp. 3–29). Boston, MA: Kluwer Academic Publishers.

Birenbaum, M. (2003). New insights into learning and teaching and the implications for assessment. In Segers, M., Dochy, F. J. R. C., & Cascallar, E. (Eds.), *Optimising new modes of assessment: In search of qualities and standards*. Dordrecht, The Netherlands: Kluwer Academic Publishers. doi:10.1007/0-306-48125-1_2

Blakely, P. N., & Tomlin, A. H. (Eds.). (2008). *Adult education: Issues and developments*. New York, NY: Nova Science Publishers, Inc.

Boud, D., & Falchikov, N. (2005). Redesigning assessment for learning beyond higher education. [Sydney, Australia: HERDSA.]. *Proceedings of the HERDSA Conference, 2005*, 34–41.

Boud, D., & Solomon, N. (Eds.). (2001). *Work-based learning: A new higher education?* Buckingham, UK: SRHE and the Open University Press.

Bradney, A. (1992). Ivory towers and Satantic mills: Choices for university law schools. *Studies in Higher Education, 17*(1), 5–20. doi:10.1080/03075079212331382726

CAMM2010. (2010). *Proceedings of the Conference on Asset and Maintenance Management 2010*, Gladstone, Australia. Retrieved 29 July, 2010, from http://research.cqu.edu.au/ FCWViewer/getFile.do?id=32434

Catts, R. (2004). *Lifelong learning and higher education-reflections and prospects*. Keynote address, 3rd International Lifelong Learning Conference, Yeppoon, June 2004.

Chappell, C., Solomon, N., Rhodes, C., Tennant, M., & Yates, L. (2003). *Reconstructing the lifelong learner: Pedagogy and identity in individual, organisational and social change*. London, UK: Routledge. doi:10.4324/9780203464410

Daniels, M., Carbone, A., Hauer, A., & Moore, D. (2007). *Panel-Ill-structured problem solving in engineering education*. 37th annual Frontiers in Education Conference, Global Engineering: Knowledge Without Borders, Opportunities Without Passports. Oct. 10-13.

ENMM20012. (2010). Maintenance organisation course profile, term 2. Retrieved July 15, 2010, from http://content.cqu.edu.au/ CourseProfile/deliverPDF.jsp? docId=34902&course Code=ENMM20012

Everett, J., & Van Dessel, M. (2005). Using cases in a mixed learning environment: Meeting the challenges of higher education reform. *International Journal of Case Method Research and Application, 17*(3), 341–350.

Fook, J., Ryan, M., & Hawkins, L. (2000). *Professional expertise: Practice, theory and education for working with uncertainty*. London, UK: Whiting & Birth Ltd.

Gielen, S., Dochy, F., & Dierick, S. (2003). Evaluating the consequential validity of new modes of assessment: The influence of assessment on learning, including pre-, post- and true assessment effects. In Segers, M., Dochy, F., & Cascallar, E. (Eds.), *Optimising new modes of assessment: In search of quality and standards*. Dordrecht, The Netherlands: Kluwer Academic Publishers. doi:10.1007/0-306-48125-1_3

Groenewald, T. (2004). Towards a definition for cooperative education. In Coll, R. K., & Eames, C. (Eds.), *International handbook for cooperative education: An international perspective of the theory, research and practice of work-integrated learning* (pp. 17–25). Boston, MA: World Association for Cooperative Education.

Gulikers, J., Bastiaens, T., & Kirschner, P. (2004). A five-dimension framework for authentic assessment. *Educational Technology Research and Development*, *52*(3), 67–86. doi:10.1007/BF02504676

Gumport, P., Nozzi, M., Hamah, S., & Zemsky, R. (1997). Trends in United States higher education from massification to post massification. Retrieved 27 July, 2010, from http://www.stanford.edu/group/ncpi/documents/pdfs/1-04_massification.pdf

Guri-Rosenblit, S., Sebkova, H., & Teichler, U. (2007). Massification and diversity of higher education systems: Interplay of complex dimensions. *Higher Education Policy*, *20*, 373–389. doi:10.1057/palgrave.hep.8300158

Hamid, A. (2002). E-learning is it the 'e' or the learning that matters? *The Internet and Higher Education*, *4*, 311–316. doi:10.1016/S1096-7516(01)00072-0

Harris, S. (2003). An andragogical model: Learning through life experiences. *Kappa Delta Pi Record*, 38–41.

Hessels, L., & Van Lente, H. (2008). Re-thinking new knowledge production: A literature review and a research agenda. *Research Policy*, *37*, 740–760. doi:10.1016/j.respol.2008.01.008

Huba, M. E., & Freed, J. E. (2000). *Learner-centered assessment on college campuses*. Needham Heights, MA: Allyn & Bacon.

International Council for Open and Distance Education. (2009). *Global trends in higher education, adult and distance learning*. Retrieved 27 July, 2010, from http://www.icde.org/filestore/ Resources/Reports/FINALICDE ENVIRNOMENTALSCAN05.02.pdf

Kapp, A. (1883). *Die angragogik ober bildung im mannlichen alter. Platons Erziehungslehre, als Padagogik fur die Einzelen und als Staatspafagogik*. Germany: Minden and Leipzig.

Katula, R. A., & Threnhauser, E. (1999). Experiential education in the undergraduate curriculum. *Communication Education*, *48*, 238–255. doi:10.1080/03634529909379172

Keleher, P., Wong, K., & Howard, P. (2006). The nexus: Ivory towers and muddy trenches-studying maintenance management by distance education at postgraduate level at central Queensland University, Australia. *The Journal of Cooperative Education and Internships*, *40*(2), 26–32.

Kerr, C. (1995). *The uses of the university* (3rd ed., pp. 1–34). Cambridge, MA: Harvard University Press.

King, P. M., & Kitchener, K. S. (1994). *Developing reflective judgment: Understanding and promoting intellectual growth and critical thinking in adolescents and adults*. San Francisco, CA: Jossey-Bass.

Knowles, M. (1978). *The adult learner: A neglected species* (2nd ed.). Houston, TX: Gulf Publishing Company.

Knowles, M. S. (1980). *The modern practice of adult education*. New York, NY: Association Press & Cambridge Book Publishers.

MacKinnnon-Slaney, F. (1994). The adult persistence in learning model: A road map to counseling services for adult learners. *Journal of Counseling and Development*, *73*(2), 268–275.

Martyn, M. (2005). *The hybrid online model: Good practice, a hybrid approach to online learning offers important lessons for institutions entering the online arena*. Retrieved 24 July, 2010, from http://net.educause.edu/ir/library /pdf/EQM0313.pdf

Mirvis, P. H., & Hall, D. T. (1996). New organisational forms and the new career. In Hall, D. T. (Ed.), *The career is dead: Long live the career*. San Francisco, CA: Jossey-Bass.

Petraglia, J. (1998). *Reality by design: The rhetoric and technology of authenticity in education*. Mahwah, NJ: Lawrence Erlbaum Associates Publishers.

Randall, N., Roberts, T., & Rogers, S. (2005). *Connected assessment: Getting real with authentic assessment online*. Presented at the ETUG Spring Workshop on Educational Technologies, June.

Reeders, E. (2000). Scholarly practice in work-based learning: Fitting the glass slipper. *Higher Education Research & Development*, *19*(2), 205–220. doi:10.1080/072943600445655

Reiko, Y. (2001). University reform in the post-massification era in Japan: Analysis of government education policy for the 21st century. *Higher Education Policy*, *14*(4), 277–291. doi:10.1016/S0952-8733(01)00022-8

Richardson, J. T. E., & King, E. (1998). Adult students in higher education: Burden or boon? *The Journal of Higher Education*, *69*(1), 65–88. doi:10.2307/2649182

Simon, H. (1973). The structure of ill-structured problems. *Artificial Intelligence*, *4*, 181–201. doi:10.1016/0004-3702(73)90011-8

Stewart, R. A., & Chen, L. (2009). Developing a framework for work integrated research higher degree studies in an Australian engineering context. *European Journal of Engineering Education*, *34*(2), 155–169. doi:10.1080/03043790902833325

Symes, C., & McIntyre, J. (Eds.). (2000). *Working knowledge: The new vocationalism and higher education*. Buckingham, UK: SRHE/Open University Press.

Teicher, U. (2003). The future of higher education and the future of higher education research. *Tertiary Education and Management*, *9*(3), 171–185.

Tennant, M. (2000). Learning to work, working to learn: Theories of situational education. In Symes, C., & McIntyre, J. (Eds.), *Working knowledge: The new vocationalism and higher education*. Buckingham, UK: SRHE/Open University Press.

The Chamber of Minerals and Energy. Western Australia. (2005). *Fly in/fly out: A sustainability perspective sustainable, a discussion of the triple bottom line impact of fly in/fly out operations in Western Australia's resources sector.* Retrieved July 24, 2010, from http://www.peopleforthefuture.com.au /files/files/20_FIFO_Report.pdf

Treby, E., & Shah, A. (2005). Bridging the gap between academia and practitioners: Training coastal zone managers. *Planet*, *14*, 16–17.

Usher, R., & Solomon, N. (1999). Experiential learning and the shaping of subjectivity in the Workplace. *Studies in the Education of Adults*, *31*(2), 155–163.

Valiathan, P. (2002). *Blended learning models*. Retrieved July 24, 2010, from http://www.astd.org/LC/2002 /0802_valiathan.htm

Vernon, C., Kirby, F., Allen, M., & Sargeant, J. (2003). A mixed learning technology approach for continuing medical education. *Medical Education Online, 8*(5).

Winckler, G. (2009). Universities in the 21st century. *Information Services & Use, 29*(3), 65–72.

ADDITIONAL READING

Cooper, L., Orrell, J., & Bowden, M. (2010). *Work Integrated Learning: A Guide to Effective Practice*. London: Routledge.

Felstead, A., Fuller, A., Jewson, N., & Unwin, L. (2009). *Improving Working for Learning*. London: Routledge.

Macdonald, I. (2005), Blended learning in the workplace: Why is it so good? *International Conference on Engineering Education*, July 25–29, 2005, Gliwice, Poland.

Smith, P. J. (2000). Preparedness for flexible delivery among vocational learners. *Distance Education, 21*(1), 29–48. doi:10.1080/0158791000210103

KEY TERMS AND DEFINITIONS

Authentic Learning: Authentic learning provides students with the opportunity to explore, discuss, and meaningfully construct concepts and relationships in contexts that involve real-world problems and projects that are relevant to them as a learner.

Blended Learning: This refers to a mixing of different learning environments to provide different approaches and modes of knowledge acquisition and dissemination. In particular, a blended learning approach can combine face-to-face instruction with computer-mediated instruction.

Flexible Mode: A term also used to describe courses conducted by distance education mode of delivery.

Work Integrated Research (WIR): The creation of new knowledge and practices through research findings arising from using Work Integrated Learning philosophy as its basis.

Chapter 10
A Self-Paced Flexible 'Learning While Earning' Process

P. Kaye Clark
Central Queensland University, Australia

ABSTRACT

Three of the professions directly related to the construction industries by which Central Queensland University's undergraduate Built Environment programs are accredited, form the foci of this chapter. The students enrolled in those programs are working in the relevant industries during their part-time external studies. Although learning while working 'on the building site' has been known since human beings first began constructing shelter, relevant background theories of philosophy and psychology have been introduced here and utilised to provide substantive support for a debate regarding the mixture of formal and informal opportunities for work-integrated learning to which these students are introduced. The strengths and weaknesses of flexible external studies, as well as the nature of their workplace-based learning in these programs, are discussed at length. Greater emphasis is placed on the need to capitalise on the many opportunities for reinforcement of, and reflection about principles and practices introduced in either or both their employment and/or undergraduate studies, than merely on the advantages or disadvantages of flexible external studies. These students' truly work-integrated learning experience may be considered to be a 'self-paced flexible learning while earning' process.

INTRODUCTION

History of Construction

In education, there are few more famous and frequently quoted truisms (cited in equally frequently varied interpretations) than that attributed to Confucius (541-479 BC); "Tell me and I will forget. Show me and I may remember. Involve me and I will understand". Not surprisingly, in any of its re-interpretations, it is an eminently suitable support for work-integrated learning and is entirely relevant for students who aspire to become qualified construction managers, building designers or building surveyors and certifiers.

DOI: 10.4018/978-1-60960-547-6.ch010

History and anthropology abound with theories of cave dwellers and nomadic tribes throughout the 'hunter gatherer' phase who adapted the caves or constructed rudimentary shelter. The subsequent beginnings of agricultural and animal husbandry pursuits brought with them the creation of more permanent dwellings from timber, earth and rock in the bush and on the plains for the more settled tribes and the development of more sophisticated transportable shelter for the nomads using animal hides for cladding their tents (Risebero,1979).

The feats of the builders of Stonehenge, for which the huge stones were believed to have been shaped with stone hammers before their erection, continue to amaze us (Field 2007, p13) as do the pyramids (pp14-15). The same awe is inspired by the feats of the early designers and builders who constructed huge buildings (fortresses, cathedrals, and similar) in many countries many hundreds of years ago without any of our modern equipment such as cranes.

From the plethora of historical theories of early construction, it may safely be assumed that all of those designers and builders (whether the early primitive humans building their own shelter or slaves building huge structures under continual supervision) learned how to design and build by being actively involved in the design and building processes, thus the earliest version of 'learning on the job' (Field, 2007; Risebero, 1979).

In primitive times the building of shelter had been traditionally undertaken by the tribes; the men in some cultures and the women in others. As civilisation developed and settlements grew, so the roles changed and building began to be undertaken by specialist groups.

So too, did the ways of learning those skills change and develop from the family or community groups adapting or building their own shelter to the specialist builders encompassing the whole spectrum of the ever-widening gap between the smaller buildings constructed by a few and the huge projects requiring hundreds of workers.

The roles of master and apprentice soon came into play, and again it is assumed from the wealth of historical records that the learning process for the apprentice was undertaken 'on the job'. In more recent centuries the theoretical aspects of the technical training of building apprentices, designers and surveyors had taken place outside of the worksite. For example, in the mid-twentieth century; one night a week plus perhaps one day a fortnight of classes was prescribed for building apprentices to provide the theoretical knowledge to back up their practical skills developed on the building site.

The latter part of the twentieth century saw the growth of undergraduate programs to provide university qualifications for higher levels of the previously 'trade-based' professions. This meant that persons whose skills may have been more practical than intellectual or academic were facing new challenges. This process is continuing into the early part of the twenty first century.

Structure of the Programs

At Central Queensland University, the School of Engineering and Built Environment offers the Built Environment suite of programs, which incorporates three streams of study that are relevant to the professions of;

- Construction management
- Building design, and
- Building surveying and certification.

These are offered as the following six undergraduate awards;

- Bachelor of Construction management - comprising four years equivalent full-time study load (EFTSL)
- Bachelor of Building surveying and certification (4 years EFTSL) from which there are two exit awards;

- Bachelor of Building surveying and inspection (3 years EFTSL)
- Associate Degree of Building surveying (2 years EFTSL) which is alsoavailable for direct entry, intended for students with a lower entry level score, while the exit from the 4 year program is intended for students who find that they are struggling or who do not wish to continue their studies
- Bachelor of Building Design (3 years EFTSL)
- Associate Degree of Building Design (2 years EFTSL)

Coursework comprising almost two years EFTSL, is shared between all programs encompassing the three discipline streams. A small number of relevant courses (subjects) are shared by only two different streams or programs. This leaves sufficient coursework that is exclusively specific to a single stream or program to ensure approval of the relevant industries' professional expectations of our graduates and the university's academic standards.

As well as CQ University-based academic program reviews, all Built Environment programs are accredited by their relevant industries in five year cycles, or less if industry changes occur. The Australian Institute of Building Surveyors (AIBS) accredit the three Building Surveying awards while the Australian Institute of Building (AIB) accredits the Bachelor or Construction Management program. To date the Building Designer's Association of Australia has accredited the two Building Design programs through their Queensland state organisation.

Learning / Delivery Mode

All programs offered in the Built Environment streams are delivered in external flexible mode. The first programs were offered by Central Queensland University during the 1990's through the traditional correspondence mode of study. This incorporated fully documented study guides, the content of which is still intended to take the place of the actual lectures that would comprise 'face to face' delivery. All learning materials were posted through the mail service including the assignments from students, their feedback and final results by return.

Over the past four to five years however, this mode of delivery has progressively utilised far more online delivery via e-courses, allowing a web presence for each course/subject with online availability of course resource materials, announcements, discussion forums, plus the uploading of assessment items by students followed by their feedback from markers.

Students undertaking studies without face to face contact, would benefit from developing the self awareness of metacognition that would enable them to become a "self-regulated learner". Zimmerman claims that self-regulation is "the self-directive process by which learners transform their mental abilities into academic skills" (2002, p 65).

Strengths

For those students who wish to remain in their home region while studying and working, the flexible external delivery of programs is a considerable advantage. By the same rationale it is equally advantageous for regional employers wishing to retain trainees.

Weaknesses

Initially, the lack of face to face contact increases the challenge of undertaking university level study for new students and this is especially so for mature students.

Opportunities

Changes in the industry such as the recent economic downturn that may lead to reductions in, or complete loss of employment, can be translated into the opportunity to take up or accelerate studies towards one's preferred profession, which is distinctly possible when those studies are available in flexible mode of delivery.

Relevant Category of Work-Integrated Learning

As mentioned earlier in this introduction, persons involved in building construction-related roles in the distant past are assumed to have learned their trades and professions 'on the job' (ie; at the construction site). This process, while still relevant, has long since had formal education components added, which arrangements have been stratified into the following types of work-integrated learning, which are:

- the labourers or assistants who continue to learn in the workplace with no qualification forthcoming
- the apprentices or trainees who learn practical skills at the worksite or in the office and supplement that with additional knowledge from part-time formal instruction at technical colleges or institutions
- those aspiring tradespersons who study full-time at a technical college for a certificate or diploma qualification, then obtain work to supplement that theory with practice
- those who study full-time or part-time at University for a higher qualification; which studies may or may not include appropriately monitored and assessed workplace experience.

Students in the Built Environment programs fall into the final category listed as they are required to obtain or continue work during their studies within the profession that is relevant to the qualification to which they aspire. This may be either full-time or part-time employment, which is obviously dependent upon the opportunities available to them. Their experience may be seen as a form of uncontrolled work-integrated learning (WIL) in itself, linked to, or possibly the same as 'experiential learning' (Argyris & Schön, 1976), 'work-based learning' or other theoretical types in some instances. Further detailed discussion of these definitions and justification of our classifications and overall approaches will be included later. The term 'uncontrolled' is used above to identify that the nature and/or focus of the work experience that they obtain in their self-sought employment is outside of any 'control' by Central Queensland University.

The Construction Management and Building Surveying programs include a form of reflective project-based work experience to be assessed as part of their coursework. The two industries that accredit those programs, require students to undertake supervised workplace experience that must be logged and signed off by their employer, which logbook is then submitted as one part of the assessment of a work experience course. This logged and supervised experience will form part of the requirement for post-graduate licensing or registration similar to that which is relevant to the architectural profession. The Building Design programs encourage the use of a mentor through relevant areas of their studies.

The term; "self-paced flexible learning package" was used by, and may even have been coined by Jancauskas, Atchison, Murphy and Rose (1999). While that title was proposed for the training of academic and industry supervisors for students undertaking carefully organised work placements for practicum during their programs of study, it nevertheless describes quite accurately the unsupervised work experience and ongoing flexible studies of our students who all study externally.

Although we have a recommended schedule of part-time study to guide students as to which courses to study in which order of progression, it only needs one withdrawal from or failure in a course to wreck that careful planning for them. To clarify that further, should a student be working full-time while studying, he or she may justifiably need a reduced study load of perhaps one course per term when private life responsibilities and workloads make the recommended loads unachievable. Should they be working part-time, then the recommended loadings might be feasible, or students who are temporarily unemployed may wish to take on a study load that is close to full-time. In complete contrast to the formal order of study for face to face students, our undergraduates will often enrol in courses to suit their areas of interest in an 'ad hoc' manner that may be almost entirely 'self paced'. Thus it may be seen that the 'self-paced' aspect is truly relevant and it seems appropriate to claim that the mixed forms of work-integrated learning that our students undertake is a form of 'self-paced flexible learning while earning' WIL process.

Strengths

The majority of our students have the distinct advantage of being employed within the industry to which they aspire, before they qualify academically for their profession, which means that many will immediately become eligible on graduation, for promotion or appropriate pay rises.

Weaknesses

The uncontrolled nature of 'working while learning' (learning while earning) as recommended, is concomitant with a lack of certainty regarding which aspects of work will be experienced.

Opportunities

In spite of the above acknowledged 'weakness' there is a potential 'wealth of opportunity' for students to link studies and/or assessment processes with current or past work experiences.

THEORETICAL BACKGROUND

Philosophy

In line with the truism of Confucius with which this chapter was introduced, many of the famous early philosophers began to address the value of learning from experience, although the theories of several later 'thinkers' were rather tenuous, as we can now see many centuries later.

According to Stokes (2002); of the early philosophers who were later designated as 'academics'; it was Socrates, through the evidence of the writings of his pupil Plato, who began the development of 'continuous critical reflection' by which process many initial theories could be retained or be quickly reduced to nonsense (p.21). To put that 'theory into practice'; imagine some of the more radical designs we see in buildings today, then imagine the chaos on the building site had we not begun that 'critical reflection' process with our engineers in the conceptual design stage, and had we not then continued that process throughout its various construction phases.

Aristotle (another 'academic' philosopher) favoured observation and the strict classification of data. He drew the line between menial (doing) and liberal (mental) education and Dewey claimed that his error lay in assuming the necessary separation of the two (1944). Nevertheless, Aristotle promoted;

- teleology (purpose) and telos (goal)
- relating behaviour to goals, and
- function (Stokes, 2002, p.25).

How easy it is to see in this age, a mere two millenia later; the relevance of Aristotle's approach in the above points in the planning, design and construction of our buildings. Putting this 'theory into practice' can be easily demonstrated in the following simplistic question and answer (Q & A) scenario: Question; What do we want?

Answer; Our goal is a building. Q; What sort of a building? A; We'll work that out with planning and design (behaviour) to suit our purpose. Q; How can we construct that building? A; Through specific behaviour (and function perhaps?) by 'finding' tools, materials and workers with the skills to use them.

Those early philosophical ideas are so well-established within our educative world today, that it is difficult to comprehend how very challenging they were to those in power during their own era, but one must recall that Socrates was sentenced to death for promoting his philosophies.

While the 'scholastics' mainly debated religion, Joannes Duns Scotus, in the late thirteenth and early fourteenth centuries, was one who delved further and defined three modes of knowing with evident early connections to 'experience';

- principles known by themselves (Cicero's *a priori*)
- things known immediately from experience, and
- knowledge of our own actions

with the unfortunate reverse reaction which meant that his name become the (much later) source of the opposing term 'dunce' (Stokes, p.53). To put Duns Scotas' theories 'into practice' in this context; it is self-evident that the builder would quickly learn from his experience if he wrongly positioned stones in an arch and it collapsed.

For the purist reader and researcher, this author acknowledges that Duns Scotas' theory and those equally profound and deeply considered philosophical 'connections of knowledge to experience' supported and/or extrapolated by many later philosophers were certainly not as basic and straightforward as is their use as part of this preliminary 'argument' in support of work-integrated learning presented in this chapter.

Psychology of Learning

Rationalisation of various theories of psychology gained strength throughout the middle of the twentieth century. An early textbook by Woodworth and Marquis (Psychology - A Study of Mental Life) was first published in 1922 and continually republished until its 20th edition in 1949, to then be republished in that edition until at least 1963 (as in the referenced copy, or even later editions). This book has provided a rich source of reference for this section, to be used for the presentation of relevant early theories about learning and for cross reference and discussion later in this chapter when compared to recent educational publications and to this author's own experience in work-integrated learning.

The earliest psychological theories of learning may have been derived from observations of infants and the development of their emerging awareness, purposive movements and apparent understanding of the world around them.

Scientific experiments with animals and insects later formed the basis of comparative observational studies of human capacities for learning, given comparable, similarly defined circumstances.

According to Gregory, Piaget's work on the intellectual development of children "owed much to his early studies of water snails" in which he observed the passive and active stages of their life and the growth of their shell associated with the activity. From this he formed his "model of development through active adaptation" which became "the prototype for learning" (1987, p622).

Another fascinating theory has been introduced in Woodworth and Marquis's textbook wherein they discuss the potential value of sleep after learning. In researching memory and retention of learning in human subjects through controlled testing, it has been theorized that forgetting was slow during sleep and relatively "rapid during daytime activity". However, the authors warn that this is really only relevant to "...*barely learned* material..." and that it is "...the *brand-new traces* that are disturbed by activity and favoured by sleep" (p555). This certainly may be one advantage for our students, if (as is likely) they tend to undertake studies during the evening.

Conversely there may be a risk for those students who undertake energetic physical work (eg; while earning their living) in the claim that the undertaking of extreme activity immediately before a learning period may negatively impact on one's ability to learn. Similarly, it is claimed that intense physical activity immediately after a learning period may reduce retention. While this relation of activity to retention may have been confirmed later with observations and tests with human subjects (as inferred in the cited textbook), the source of this research by Minami and Dallenbach in 1946 was controlled experiments with cockroaches; as cited in Woodworth and Marquis, who claimed their use as subjects for those tests to be an excellent use of that "abundant resource" (1963, fig 136, p556).

In the same book, the re-learning method is promoted, in which the original learning period has been followed by an interval of time in which re-reading or intentional consideration of the original learning matter has not occurred or has been avoided as far as possible. When a period of reviewing or refreshing what one learned originally, is then undertaken, the ease of re-learning is purported to be much easier and more successful than expected. "You may recall almost nothing of a previously learned lesson and still find it easy to relearn. And you may recognize more than you recall" (p551). This aspect of recall, recognition and/or relearning may be relevant to our mature students or to those students who fail to complete earlier studies then return to them later, because those students often lack confidence and can be surprised at how much they actually remember the second time around.

From the above early theories, delayed reviewing would appear to be the best approach, given that any interference directly after a period of learning (ie; within the first hour) can greatly increase forgetting. Naturally one must keep in mind that memory in itself (thus retention, recall or recognition for example), does not equate to understanding.

In order to learn we need first to be motivated, as one cannot force anyone to learn. One may insist on attendance at a lesson but that does not mean that the student's mind is actually 'attending' to what is being imparted in the lecture. Most people would be fully aware of their potential to lose concentration or to be distracted during a lecture, or even during a private conversation, thus missing an important point. Green claims that; "for learning to occur, the learner must be aroused or activated. The forces that produce activity are called motives" (1964, p60).

Motivation can occur through an interest in the subject matter, through a perceived need for the information being imparted or from a perceived urge to compete with others. The motivation of the 'competitive urge' is well known and can result in a better quality of submissions from those who are competitively inclined. The associated, but less 'driven' form of competition is that of 'pace-making', ie; the self assessment of one's own standards with regard to; 'keeping up' with others, forging ahead, or falling behind the evident standards of other class members.

For the external student who has no face to face lectures and thus no classroom contact with other students, and for whom therefore, there can be no stimulus of the competitive impulse, the difficulties inherent in a lack of that form of motivation are obvious as how can one compete when one doesn't know 'with what it is, that one is competing'? Thus it is, that for our students, lacking that face to face contact with both lecturer and classmates, the matter of motivation may become that of 'self competition' for which feedback is important, one might even say; essential.

Theoretical Background to Work-Integrated Learning (WIL)

Kolb's "Cycle of Learning" forms an appropriate basis for the development of learning from experience as relevant to the work-integrated learning experience of our students. It comprises learning modes of; Concrete Experience (feeling), Reflective Observation (watching), Abstract Conceptualization (thinking) and Active Experimentation (doing). He links these four learning modes to the learning style types of Assimilating, Converging, Accommodating and Diverging. The accommodating style is a combination of active experimentation and concrete experience which, Kolb states, is one in which people "learn primarily from 'hands on' experience" (Osland, Kolb, Rubin & Turner, 2007, pp53-56).

Argyris and Schön (1976, p99) defined three steps as their definition of experiential learning which may be seen as another form of WIL. They are:

1. **look forward:** predict consequences of behaviour
2. **look backward:** examine the governing variables of the behaviour
3. ***identify the feedback*** *that keeps the 'actor' resistant toward change.*

The ad hoc label, that is so appropriate to the Built Environment students' work-integrated learning experience mentioned previously, is just one of four types of WIL described by Abeysekera (2006). The other three are; cooperative education, work-based programs for organizations and the "workplace-based program (WPL), in which the degree is taught at the workplace" (p9), none of which accurately relate to our modes of WIL.

Bates (2003) defines field placements as negotiated work-related projects which nevertheless have an element of unpredictability in that one cannot always anticipate workplace tasks, events and challenges. While Bates is discussing the field placements of students in criminology programs, that unpredictability is entirely relevant to the ad hoc, unstructured nature of our students' experience in employment in their study-related industries while studying in our Built Environment programs. However, it is important to clarify that our students are not given 'field placements' per se, because their employment is entirely self-sought.

Wiredu (2005, p53) states that "The purposeful and predetermined nature of WIL implies that there is a strong focus on specific skills acquisition by learners towards the solution of specific identified problems...". Certainly, for students like ours, who are studying part-time, this would also be the case for their personal work-based learning situation. However, the specific skills in focus for them may be restricted to those role-tasks required by the employer rather than those that might link directly to their studies as well. This is not however, seen as a disadvantage in the long term.

Tagg (2003) discusses problem-based learning (PBL), collaborative learning, service learning and undergraduate research which all have relative values for work-integrated learning and its assessment. In discussing Schön's (1983) work about reflection in action he states; "Certainly subject matter knowledge is important. But that knowledge is trivialized if it is merely testable, but not usable" (p162). This brings us to the important matter of assessment which will be discussed in some detail in the next section of this chapter.

GENERAL DISCUSSION

Position Statement

Because the majority of the students in the Built Environment programs are working while undertaking their studies, thus potentially including aspects of their personal work experience in assessment items in various courses, this chapter will address the roles of their lecturers, the skills that the students are developing and the assessment items by which their studies are evaluated over all coursework, rather than by selective consideration of the exclusively work experience-based courses alone.

Nature of our Student Learner

There is no single 'typical student' profile for our undergraduates who (over all enrolments in our Built Environment programs) encompass a broad range of age groups, ranging from a small number of school leavers, to an equally small, but growing number who are close to sixty years old, with the bulk of students representative of most age groups in between those two extremes.

The ratio of males to females enrolled is consistently reducing as female enrolments increase, although we have yet to attain a 50/50 balance.

Student enrolments come from all Australian States and Territories, plus students located in South East Asia, PNG and the Middle East. Middle Eastern enrolments are usually Australians working abroad, who (one assumes) may be understandably taking advantage of the building boom in that area but wishing to obtain an Australian qualification for when they return home. Students based in South East Asia (often temporarily) have generally been employed in our Australian armed forces. Capital cities, regional and remote Australian locations are all well represented as the residential locations of our students.

Our students are primarily Australian citizens. However, Australia has long been considered to be a multi-cultural society, with our indigenous Australians who were originally supplemented by the first English, Scottish and Irish colonial settlers, who were followed quickly by an influx of Chinese and Europeans. Through the second half of the twentieth century, immigration has continued from all of those countries with the addition of many from the Mediterranean region. More recently, immigrants have arrived from the Middle Eastern countries, the sub-continent (India and neighbours), Africa and South East Asia, the USA and South America as well. Needless to say the cultural diversity of all of these new Australians has considerably enhanced our appreciation of new foods, provided a greater breadth of cultural understanding and a welcome utilisation of the skills they have brought with them.

It can confidently be claimed, without having actually surveyed the cultural backgrounds of our students, but from their family names alone, that we have student representation from most of those cultures and countries of origin mentioned above. In spite of those family names that are obviously of Chinese, Scottish or Italian origin, for example, we cannot know without personal contact whether or not that is a married name only, or if not, then whether their families have been in Australia for many generations or whether they are literally new Australians for whom English may be their second language.

From personal contact through phone and email (which are our primary routes of contact with students as we have no face to face contact), we are often made aware that we do have a number of students who have specific physical, neurological or psychological challenges (disabilities) which means that their progress through their studies with us is and / or has been evidence of their specific and admirable determination and capabilities to overcome those difficulties.

Uhlbrich, Schier, Lindstaedt and Görtz's theory provides a literal, building plan-related spatial concept to support work-integrated learning that comprises three rooms; the work space (task context), the learning space (competency context) and the knowledge space (domain context) to ensure separation of events (2006). Their theory might assist our Built Environment students to plan for such divisions in their mind and its focus if not in their workplace or home.

Role of the Lecturer in WIL

The first part of this discussion covers the evident differences in the lecturer's role with external flexible delivery and the utilisation of e-courses (when compared to face to face delivery), without covering the specific issues of matters linked to the 'work-related' aspects of the student's studies. Specific reference to the lecturer's role in work-integrated learning aspects will conclude this section and be considered in some detail regarding generic skills development and assessment processes.

The staff who are the lecturers for courses in the Built Environment programs are a mixture of full-time permanent academics and external professionals who have casual lecturing contracts for each relevant term in which they teach. While the courses that are specifically work experience-based are always handled by the permanent staff, more than half of other courses are handled by those casual external lecturers, most of whom work fulltime within their relevant industries. We see this 'industry connection' as a significant advantage for our students.

When studies are offered in external delivery mode, the normal role of the university lecturer is rather different to that for traditional face to face lecturing. As mentioned previously, when each stream in our Built Environment suite of programs was first offered, they were all delivered in the old correspondence mode, and the study guides and books of required extra readings prepared by the relevant lecturers were posted to the students. This preparation of study material remains the main preparatory task of our full time lecturers and is not so very different to the preparation of lectures for face to face delivery, except that our preparations for the whole term, must be done well in advance of the start of term to allow time for printing and postage. While we are moving into the provision of an online 'presence' quickly now, with the utilisation of e-courses for most courses, not all of our lecturers will be actively using those sites and many of the study materials will still be available in print, often as well as being available online.

Although this author cannot speak with any personal authority about the experience of our casual external lecturers, the lack of face to face contact for our students most certainly results in a major change in role for our internal lecturers to a significant extent. This is because many new students tend to become high dependence or high maintenance with some phoning and emailing frequently with questions, many of which are actually addressed in the study materials or recommended readings. Naturally enough this is indicative of a lack of confidence that would be alleviated in the real lecture theatre when questions can be asked by one, and with the whole class hearing the answers and seeking further clarification immediately if required.

The biggest challenge for the lecturer handling the new students in their earliest coursework, is that those classes invariably consist of large enrolments (between 100 and 300 students), so that the task for the lecturer is magnified accordingly. It is essential in such cases that sufficient staff are available to tutor suitable sized groups in online forums to replicate the smaller, tutorial groups offered in the face to face delivery of other programs.

Given the current move towards e-courses for most subjects/courses, the main difficulty ensuing from large classes, ie; 'many students equals many queries' can be alleviated with the staffing described above. For the uninitiated, the e-courses provide opportunities for the provision of study materials online, preferably following an iterative process, plus advice and important announcements that are visible for all the class, with discussion forums or similar for student participation. It is important that the online forums or discussion groups are able to be monitored (ie; utilising the tutored groupings mentioned previously) so that the worries of many students may be addressed to their satisfaction.

The difficulty here is that, if one was trying to re-create the once a week lecture theatre scenario or even that of the weekly tutorial for large classes within a web-based medium that is available online for twenty four hours a day, every day of the week, there is a potential problem that is extremely difficult to monitor. Hypothetically, if students were to be advised that their queries would only be answered once a week, one might find, on accessing those discussions or questions that had been put forward during the intervening seven days, that the volume of items and matters of concern may well be overwhelming. The real point that is relevant here, is that some e-courses with very large enrolments may need far more regular contact and maintenance from the lecturer or tutors than a once a week intervention (or than the once a week lecture would have needed too).

Even when it is possible to provide online video recordings of lectures as one would automatically do with multi-campus delivery using ISL videoconference facilities or similar, the lack of interaction for those viewing later (as for many of our students) would be a limitation.

Because all courses in our programs are directly relevant to the professions and industries that accredit them, the lecturer's role in selecting or devising assessment items and then providing appropriate feedback to student submissions is critical to ensure appropriate academic performance while demonstrating understanding of theories and reflecting upon the necessary skills to put that theory into practice in the workplace. That ability to convert theory (from their studies) into practice is considered to be an essential skill for our students. Conversely, and equally important, is the need to be able to link practice (from their work experience) to the theories to which they are introduced in their coursework. This 'theory practice' continuum is contingent to the form of work-integrated learning that our students experience.

An additional skill that many lecturers now find necessary with our flexible external mode of delivery is that of the social worker, ideally with some lecturer-student empathy, as phone contact regarding requests for extensions or for support of an application for withdrawal without penalty from a course, may sometimes bring genuine examples of personal distress to your direct attention, that would normally have been hidden when such requests used to be re-routed to you via a 'gate-keeper' such as a receptionist or through a formal written request. There is the potential for a student peer support network to be effective, although the information promulgated therein may well be inappropriate and the prevalence of online bullying and abuse that is becoming endemic at this time may have unwelcome negative impacts.

Generic Skills Development

All courses in our built Environment programs have subject-specific learning outcomes identified in their course profiles. In addition, there is a course synopsis that gives a succinct outline that, in most instances, puts those learning outcomes into their broader context. All of these are peer reviewed and have been passed through the university's course approvals processes at some length to ensure that a sufficiently high academic standard is upheld.

The professional organizations that represent the relevant accreditation industries have prescribed schedules of professional competencies and/or university benchmarks with which our coursework and assessment items must align. All of these are mapped in great detail against our learning outcomes and the broader course content for each industry accreditation process. It is from addressing these benchmarks that assessment items are developed that will ensure appropriate skills are developed and/or assessed.

As well as simply containing essential information to encourage knowledge formation that is relevant to their current or future work practices, the assessment items for most courses include activities that should assist in the development of relevant skills. Naturally the skills in independent research, assimilation of appropriate information, review and redevelopment of that information to address new scenarios as relevant to the subject in question are expected to develop understanding of the subject matter. The skills of writing up the results of those activities in an academically appropriate manner are part of the required basic skill formation for all of our courses. As will be obvious to the reader, these academic skills will ensure that our graduates are well qualified and proficient in the production of formal reports, correspondence and office documentation.

In addition to the basic but essential academic skills, many of the suggested or required activities are directly associated with the building site which is the primary focus of the current or future roles of all our undergraduate students. These activities may include, for example, the observation and compilation of comprehensive written and photographic records of the various stages of building activity in the student's local residential area or within range of their workplace.

Similarly, the focus of a number of courses provide opportunities for students to begin building up a personal library or database of easily accessed matters to better 'inform' their current or future roles. For students motivated in that way, it can become a habit forming practice in preparing schedules and/or other records of relevant standards, building materials, examples of good and bad practice, for example, within or in connection to their industry and any other items which would be especially useful if they ever set up their own businesses.

The assessment items frequently contain research and other specifically focussed activities that are intended to facilitate the development of many skills that our students will require in their professions. Building design, documentation of the plans, inspections of sites and construction, accessing and utilising appropriate codes and regulations are but a few of those skills facilitated through the students' studies and fully developed in their workplaces.

For instance, in an introductory first year course, one topic provides the basic information about what one must observe, take note of, and understand the future implications of, when undertaking an on-site analysis of a proposed building site which is directly related to work-related tasks for their intended profession. The students are required to demonstrate their comprehension of the relevance of those observations in a small but important part of one of their assessment items. For this, they are required to find a vacant site to which they have legal access, or in lieu of that to focus on the land on which their own home is located. They must then undertake a thorough site analysis of that site, the results of which will be utilised as an essential resource in the actual design of, or choice of a design for a residence for a hypothetical client. Their decisions about the orientation and location of activity-related spaces in the floor plan and elements of the building chosen, are then required to be justified using the matters observed on or adjacent to the building site they had inspected.

The above example is one in which the students' actual comprehension is clearly demonstrated (or not) by their ability to apply what they have learned to the given scenario, in other words, by putting theory into practice.

Students will be expected to develop many more skills throughout their work- integrated learning process. These will include observation, analysis, reporting, initiating appropriate actions, to name just a few.

Assessment Processes

Assessment of our coursework is primarily via assignments or reports. A number of courses also require students to complete a workbook that is often graded as pass / fail, with a satisfactory audit (pass) required to enable students to pass that course overall, irrespective of whether their marks for the graded assignments therein rate a pass, credit, distinction or high distinction. With a pass for the workbook their real grades are allowed.

At this point in time it would appear that the widely spread locations of our students and the differing time zones therein, are the factors that have made the utilization of formal examinations somewhat difficult to convene or oversee. Therefore until recently we only had one course in which a formal exam comprised part of the assessment. This number is increasing at the request of several accreditation bodies as a means to ensure that some guarantee can be made that the work submitted is actually that of the student. Obviously that is an understandable concern even in face to face education.

Another concern that has justified the continuing prevalence of assignments rather than exams, is the tendency for lecturers (everywhere) to lean towards easily marked items, the most manageable form of which is standardized testing. This author agrees with Linkon (2005) who claims that many see that "as demeaning both our teaching and our students' learning, because it reduces education to multiple-choice questions. Scholars also argue that such generic approaches do not adequately measure students' learning in higher education (p3)." Bloom (1956) and his associates found that "over 95% of the test questions that students encounter, require them to think only at the lowest level - the recall of information."

One wonders; is this appropriate for persons whose future is an important role in the building sector and on whose skills the stability and safety of our new and refurbished buildings depend? Is this appropriate when that person's knowledge and judgement must be relevant and up to date in order to adequately perform their role, especially given the current climate of change, with continuous introduction to these professions of new concepts, new challenges and new regulations? Given the ongoing nature of their work-integrated learning, it is hoped that most students would be aware through their workplace of the ongoing need to avoid legal liability and unwanted litigation through costly mistakes. However, should that not be the case, and should our teaching strategies and assessment items fail to ensure that the appropriate knowledge is assimilated, they would certainly be at risk.

There is also a very real danger, with the utilization of easily marked exams or tests, that the intensive pre-exam study typically undertaken, while it may well lead to a good grade, may nevertheless fail to ensure that the information has been fully 'digested'. The following saying attributed to Albert Jay Nock; "the mind is like the stomach. It is not how much you put into it that counts, but how much it digests" has obviously impressed this author as its amusing analogy (which is also a good mnemonic technique) springs readily to mind. Refer back to the theories of reinforcement discussed previously.

The advent of the online e-course brings with it the novelty factor and the enthusiasm of many for the convenience of online tests while for this author, associated warning signals fire off in one's mind! While weekly online tests (in lieu of the workbooks) may be a good idea for students studying fulltime on campus, there is a need for ensuring equity for all our students in whose lives their work commitments will obviously take priority when a choice is necessary.

Positive feedback has been received from a number of our students about the reduction of stress gained from having relative freedom in personal study times and long preparation times for assignments, due dates notwithstanding. This means that the imposition of a requirement for them to undertake specific tests each week via the e-course may become an unnecessary stressor, even though on-screen advice of correct answers can be withheld until all students have responded. In spite of that, the nature of such tests clearly takes us back to the situation of the perceived problems with exams that are also the problems with such tests online especially 'point-in-time' tests.

The much maligned essay, assignment and report format of assessment items that we utilise provide the best opportunity for our students to demonstrate their ability to put theory into practice in proving that they are developing the knowledge and skills that are required for their intended future roles. The fact that they are working while studying, means that they are likely to be actively integrating their work into their learning, as the focus of their assignments intends.

CONCLUSION AND FUTURE DIRECTIONS

One might say with some truth, that life for our students is not easy, that the challenges for them of working throughout their extended years of study, while managing their private lives and family commitments requires a considerable effort that should be recognized as commendable. The very nature of the external delivery mode of our programs and the vast difference in the role of the lecturer for that, in our particular circumstances, means that life for our academic staff is certainly different too, and yes, it is not necessarily easy either.

While the online presentation of courses is a very positive and welcome improvement for both students and staff it is important that we do not overdo the automation, creating a truly virtual reality and risk becoming faceless, disconnected and depersonalised as a result.

Many changes are occurring in our building construction-based professions, and the service we are providing, is one in which we all need to demonstrate flexibility while retaining focus in how and what we do to enhance the learning experience for our students while retaining the support of our accreditation industries. Thus our future directions need to focus on matters such as those discussed by Orrell as key issues in the provision of quality work-integrated learning programs. The key issues are; management, teaching and supervision, assessment, legal and ethical matters and partnerships (2004, p1).

REFERENCES

Abeysekera, I. (2007). Issues relating to designing a work-integrated learning program in an undergraduate accounting degree program and its implications for the curriculum. *Asia-Pacific Journal of Cooperative Education*, *7*(1), 7–15.

Argyris, C., & Schön, D. A. (1976). *Theory in practice: Increasing professional effectiveness*. San Francisco, CA: Jossey-Bass Publishers.

Bates, M. (2003). The assessment of work integrated learning: Symptoms of personal change. *Journal of Criminal Justice Education*, *14*(2), 303–326. doi:10.1080/10511250300085811

Bloom, B. S. (1956). *Taxonomy of educational objectives, handbook 1: The cognitive domain*. New York, NY: David McKay Co.

Dewey, J. (1944). *Democracy and education*. New York, NY: The Free Press.

Field, D. M. (2007). *The world's greatest architecture past and present*. Rochester, NY: Grange Books.

Green, D. R. (1964). *Educational psychology*. New Jersey: Prentice-Hall, Inc.

Gregory, R. L. (Ed.). (1987). *The Oxford companion to the mind*. Oxford, UK: Oxford University Press.

Jancauskas, E., Atchison, M., Murphy, G., & Rose, P. (1999). *Unleashing the potential of work-integrated learning through professionally trained academic and industry supervisors*. Paper presented at the tenth World Conference on Cooperative Education. Washington, DC.

Linkon, S. L. (2005). *How can assessment work for us?* Academe Online, American Association of University Professors, July-August 2005. Retrieved July 26, 2006, from http://www.aaup.org/AAUP/pubsres/academe/2005/JA/Feat/link.htm

Orrell, J. (2004). *Work-integrated learning programmes: Management and educational quality.* Paper presented at the Australian Universities Quality Forum.

Osland, J. S., Kolb, D. A., Rubin, I. M., & Turner, M. E. (2007). *Organizational behaviour: An experiential approach* (8th ed.). New Jersey: Pearson Education.

Risebero, B. (1979). *The story of Western architecture. London, UK: The Herbert press*

Stokes, P. (2002). *Philosophy - 100 essential thinkers*. London, UK: Arcturus Publishing Limited.

Tagg, J. (2003). *The learning paradigm: College*. Bolton, MA: Anker Publishing Company, Inc.

Ulhbrich, A., Schier, P., Lindstaedt, S. N., & Görtz, M. (2006). *A context-model for supporting work-integrated learning.* APOSDLE project.

Wiredu, G. O. (2005). *Mobile computing in work-integrated learning: Problems of remotely-distributed activities and technology use.* Unpublished doctoral dissertation, University of London, United Kingdom.

Woodworth, R. S., & Marquis, D. G. (1949). *Psychology: A study of mental life*. London, UK: Methuen & Co Ltd.

Zimmerman, B. J. (2002). Becoming a self-regulated learner: An overview. *Theory into Practice*, *41*(2), 64–70. doi:10.1207/s15430421tip4102_2

KEY TERMS AND DEFINITIONS

External Studies: One term by which CQ University refers to its non-campus based students who are literally studying externally, the other relevant term is 'flexible' which was discussed above.

Flexible: which has two interpretations here, first; the literal flexibility or adaptability of what and how we present the learning experience, and second; the method of external delivery at CQ University is called 'flexible' delivery.

Learning while Earning: A term devised for its perceived memorability, due to the use of similar sounding words. While it may shift the reader's interpretation of that work-related term to the mere earning of money, it is used intentionally to deflect the reader's focus from the traditional notion or understanding of 'work experience' or 'work placement' during studies.

Online E-Courses: Should be self-evident as the website presence of a course, containing all or most matters related to the course and utilised throughout the term by enrolled students and relevant staff.

Self-Paced: Meaning that it is literally the student who determines how much he or she can study in each term (irrespective of recommended progression of studies schedules).

Strengths, Weaknesses and Opportunities: Are the first three terms (with self-evident meanings) from the typical SWOT analysis, which have been utilised in several parts of this chapter as a succinct way of summarising the preceding discussion.

Theory into Practice: In this context, means the demonstration of a student's ability to put the theory that he or she has learned into practice, which means to link that theory to hypothetical workplace situations as may have been devised for assessment items.

Chapter 11
Work Integrated Learning and Construction Project Management:
A Case Study of an Industry-Academia Partnership in Ireland

Ken Thomas
Waterford Institute of Technology, Ireland

John Wall
Waterford Institute of Technology, Ireland

Brian Graham
Waterford Institute of Technology, Ireland

Patrick Troy
BAM Contractors Ltd., Ireland

David Crowe
BAM Contractors Ltd., Ireland

Aidan O'Connell
BAM Contractors Ltd., Ireland

ABSTRACT

This chapter concerns the design, delivery and management of a unique part-time postgraduate MSc in Construction Project Management (MScCPM) programme through an industry-academia partnership in Ireland during the period 2007-2010. The partners are BAM Contractors, part of the wider Royal BAM Group based in The Netherlands, and Waterford Institute of Technology (WIT). There are many innovative Work Integrated Learning (WIL) aspects to this bespoken programme, including the blending of teaching and assessment by both WIT lecturers and senior BAM staff. There is also a blend of traditional classroom activities and e-learning technologies to suit the geographically dispersed participants. All stakeholders in this programme have benefited from their participation. These benefits and the associated lessons learned are described in the hope that they may be of use to those developing WIL postgraduate programmes in the future.

INTRODUCTION

This Chapter reviews the bespoke two-year part-time postgraduate MSc in Construction Project Management programme (MScCPM) for a leading construction company in Ireland. It is an example of a successful Industry-Academia partnership that incorporates 'Work Integrated Learning'. The two partners are Waterford Institute of Technol-

DOI: 10.4018/978-1-60960-547-6.ch011

ogy (WIT) and BAM Contractors, a member of the Royal BAM Group. The time-frame under review for this Case Study is 2007 to 2010 which essentially breaks down into three phases:

- **September 2007 to May 2008:** Programme design and approval;
- **June 2008 to May 2009:** Year 1 preparation, implementation and review;
- **June 2009 to September 2010:** Year 2 preparation and implementation.

The economic context to the past three years has been extremely difficult for many countries in the world and this has been particularly true for Ireland. The impact of the recession on the Irish Construction Industry and the Irish Higher Education Sector has been very significant. While all stakeholders are devising strategies to cope with the short-term challenges there are also other medium to long-term issues to be accommodated.

There are a number of challenges that are facing the construction industry and higher education. Many of these challenges are driven by the need for change. These changes are forced upon us often due to changes in government policies, society and culture, making it harder to strike the balance between the educational methods of delivery and meeting the demands of a highly competitive construction industry. This challenges higher educational providers and curricular developers to bridge the gap between academia and industry, which is often proven difficult, particularly in view of the rapid changes due to the arrival of new technologies and economic cycles which continue to challenge the construction industry.

All construction organizations will need to continuously improve if they are to survive and prosper and central to achieving such improvements is 'learning'. The combination of current economic and other influences is encouraging individuals to step-up their knowledge and qualifications. In Ireland and elsewhere there is a growing realization of the need for professional and managerial staff to have postgraduate qualifications. While learning doesn't necessarily have to involve qualifications there are many advantages to both employees and employers. Qualifications for an individual can fulfill a personal desire and in most cases enhance their career opportunities. The need for construction client groups to be assured that they are giving work to the best company is also a major factor in encouraging employers to enhance the qualifications of their employees. The challenge for enlightened companies is therefore to attract, retain and develop their employees' learning and qualifications, while continuing to use their talents on live projects.

Lifelong learning programmes aimed at people in the workplace must be suited to their particular needs (Davey et al., 2004). Within the construction industry the difficulty that exists with respect to undertaking post-graduate programmes has been somewhat addressed by the latest evolution/trend among educational providers to broaden access by offering subjects as modules or blocks that can be packaged up as short courses (Cole, 2004).

In parallel to companies improving their learning, the higher education sector globally in recent years, and certainly in Ireland, is actively seeking stronger links with industry. Many of these links are being encouraged by government, particularly with an emphasis on research and development. The influence in Europe of the EU's policies on 'lifelong learning' is also significant. 'Work Integrated Learning' is therefore a growing issue for both industry and academia and the boundaries between the two are blurring. Lees (2009) summarized the current situation nicely by stating *'the message is clear; engagement between industry and higher education in the built environment is not optimized and needs to be improved. The debate about whether to take action is over; now we must determine how to change.'*

The specific objectives of this Case Study chapter are to:

- Outline the economic and education context;
- Describe the formation and development of the BAM-WIT Industry-Academia partnership;
- Review the process of aligning the business objectives and the learning outcomes;
- Describe the participants and their selection process;
- Reflect on the operation of the programme;
- Consider the Return on Investment;
- Identify the innovative aspects and lessons learned.

ECONOMIC AND EDUCATION CONTEXT

Before considering the specifics of the Case Study under review in this Chapter it is important to appreciate the associated general economic and education context. This context is considered under the two headings:

- Irish Population, Economy and Construction Industry;
- Higher Education in Ireland and Lifelong Learning.

Irish Population, Economy and Construction Industry

The 2006 census calculated the population of Ireland at 4.235 million, the highest in more than 135 years (CS0, 2006). While there has not been another census since 2006, the subsequent annual estimates for the population figures indicate that the population has continued to increase to 4.559 million in 2009 (CSO, 2009). While the recent economic conditions have led to greater number of people emigrating and smaller number immigrating, the increasing birth-rate has underpinned this growth. Giannakouris (2008) predicted that the EU27 population *'will increase from 495 million on 1 January 2008 to 521 million in 2035, and thereafter gradually decline to 506 million in 2060.'* While this time period is beyond the scope of this Chapter, it is interesting to note the general trend and that the Ireland is predicted to have a strongest population growth at +53%, second only to Cyprus' +66%.

In their most recent review of the Irish economy the OECD (2009) stated that it is '*experiencing a severe recession as large domestic imbalances correct, but there are recent signs that the pace of contraction is slowing. Ireland should benefit from the world trade upswing along with restored competitiveness as a result of the decline in wages and prices. The ongoing domestic adjustment will nevertheless be prolonged, and the economic recovery weak*'. In stark contrast to the approximate average of 4% unemployment figures that Ireland had in the 2002-2006 period, the ESRI (2009) predicted an unemployment peak at close to 14% in the latter part of 2010. Figure 1 summarizes the Economic Growth figures for the 2000-2010 period, including the change in GNP from approximately +5% in 2007 to -10% in 2009. It is worth noting that although GDP is the international standard for economic comparison, economists regularly use the GNP figure for Ireland. While in other countries there is usually little difference between GNP and GDP, in Ireland GDP is significantly larger than GNP because of the big multinational activity which boosts GDP but nets out of GNP.

In 2006 the Construction Industry in Ireland accounted for close to a quarter of economic activity in the state as well as for one in every four jobs created in the economy (CSO, 2006). The subsequent decline in recent years of both the Irish economy and the Irish construction industry has been dramatic and unprecedented. While it is estimated that the overall global construction industry will grow by 1% there are significant differences in the various regions and countries. The construction output figure for Europe is likely to fall by between 2 and 4% with

Figure 1. Economic Growth 2000 to 2010 – Ireland & the Euro Area (adapted from epartment of Finance, 2010)

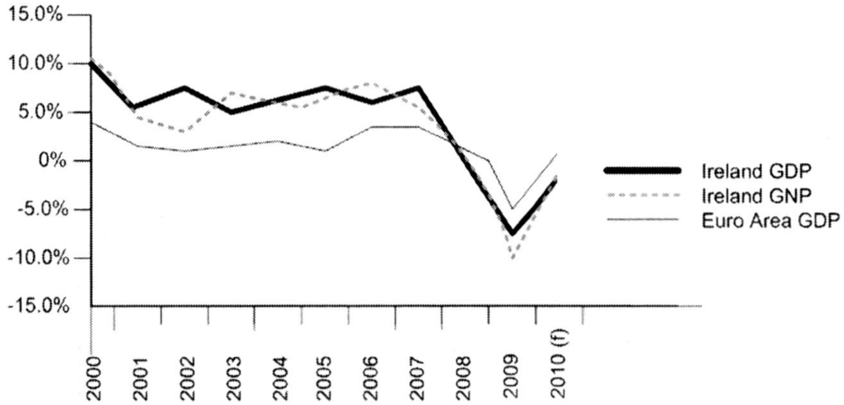

the hardest hit markets in Ireland, Spain and the UK (IHS Global Insight, 2010). Figure 2 below summarizes some of the key figures that show this decline in terms of overall output, share of GNP and direct employment.

The current and short-term context for the Irish construction industry is therefore very challenging. There have been a significant number of businesses that have ceased to operate and there are likely to be more in 2010. Those construction companies that want to survive and where possible develop their activities have to take a range of actions. Many of these actions are focused on reducing costs and increasing efficiencies throughout the supply-chain. Researching and developing business opportunities at home and overseas are also important factors. Those employees that remain with such companies become even more crucial to the success of the business. How the best are retained and developed is therefore a core business requirement.

Higher Education in Ireland and Lifelong Learning

The Higher Education Authority (HEA) is the statutory planning and policy development body for higher education and research in Ireland. It has identified that '*lifelong learning must become a guiding principle of Irish education*' and that '*lifelong learning will ensure that all individuals are facilitated in up-skilling and maintain a relationship with education throughout their lives*' (HEA, 2010). The HEA has wide advisory powers throughout the whole of the third-level education sector. In addition, it is the funding authority for the

Figure 2. Summary of Irish Construction Industry statistics for 2006-2008 and estimates for 2009-2010 (adapted from: DKM, 2009; Scully, 2009)*

	2006	**2007**	**2008**	**2009***	**2010***
Construction Output (€ billion)	37.6	38.4	32.0	19.9	12
Share of GNP (%)	24	23	20	14	12
Direct Employment (000s)	267	261	214	138	120

Figure 3. 2008/9 enrolments of students in Irish Higher Education institutions (adapted from HEA, 2010)

HEA Groups	Full-time Undergrad's	Part-time Undergrad's	Full-time Postgrad's	Part-time Postgrad's
University Sector [7 Uni's & 6 other institutions]	73,098	7,535	18,128	9,138
Institute of Technology Sector [14 IoTs & 1 other institution]	53,183	21,370	2,572	2,347

Universities, Institutes of Technology and a number of designated higher education institutions.

In the context of a developing a 'knowledge economy' and the pressure on government resources, the higher education system in Ireland is subject to much debate and regular review. The overall growth in the numbers of students in Irish Higher Education institutions in the past forty years is nothing short of phenomenal. While the statistics for part-time students are unavailable, the number of full-time students has increased from approximately 19,000 to 147,000 between 1966 and 2009. The HEA collate the annual students' statistics under two broad headings of the 'University Sector' and 'Institute of Technology Sector'. While the 2009/10 academic year statistics have not been published as yet, the figures for 2008/9 are outlined below in Figure 3.

The National Qualifications Authority of Ireland (NQAI) was established in February 2001. One of the key outputs from the NQAI has been the creation of the National Frame National Framework of Qualifications (NFQ) which is defined as follows (NQAI, 2003): *'The Framework comprises ten levels, with each level based on specified standards of knowledge, skill and competence. These standards define the outcomes to be achieved by learners seeking to gain awards at each level. A key aspect of the awards at different levels is that they are made on the basis of 'learning outcomes'*. The Institutes of Technology in Ireland are engaged in NFQ Levels 6,7,8,9 and 10 while the Universities focus exclusively on NFQ Levels 8,9 and 10.

In parallel with the development of the NFQ in Ireland has been the development of the European Qualifications Framework (EQF) for lifelong learning. '*The EQF is intended to act as a translation device to make national qualifications more readable across Europe, promoting workers' and learners' mobility between countries and facilitating their lifelong learning*' (European Commission, 2008). The core element of the EQF is a set of eight reference levels describing what a learner knows, understands and is able to do, i.e. their 'learning outcomes'. The EQF was adopted by the European Parliament and Council on 23 April 2008. While progress has varied throughout the EU, all countries are due to have their national qualifications systems or frameworks related to the EQF by 2010 and to ensure that all new qualifications issued from 2012 carry a reference to the appropriate EQF level. In parallel with the EQF initiative the EU Bologna process is attempting to harmonize all higher level education into a three-cycle structure (e.g. bachelor-master-doctorate). Figure 4 summarizes the relationship of the current Irish Higher Education Awards, their NFQ Level and their European Credit Transfer System (ECTS) Credits to the relevant Bologna Cycle and EQF Level.

There have been a number of factors encouraging the development of programmes at NFQ Level 9 for graduates and employees in recent years. Keithley and Redman (1997) predicted such a rise in the demand for bespoke management development programmes by organizations demanding the customization of programmes to meet their specific needs. Chivers (2007) considered *'the increased demand for professional services in modern society'* and concluded that *'this has led to an upsurge in professional pro-*

Figure 4. Summary of Awards and Levels in Irish Higher Education

NFQ Level	NFQ Award Title	ECTS Credits	Bologna	EQF Level
6	Higher Certificate	120	First Cycle	5
7	Ordinary Bachelor Degree	180	First Cycle	5
8	Honours Bachelor Degree	180-240	First Cycle	6
9	Masters Degree Postgraduate Diploma	60-120	Second Cycle	7
10	Doctoral Degree	180	Third Cycle	8

grammes of all kinds.' More recently Walker (2009) stated the *'need to promote and instil culture of training, lifelong learning and continuous professional development. This is essential if the managers and professionals of the future are to attain the necessary awareness and capabilities to deliver a built environment appropriate for the 21st century'*. In relation to the demand for postgraduate programmes from participants in the Irish construction industry Thomas et al (2007) identified that less that 5% of the approximate 10,000 students registered on construction related programmes in the Irish Higher Education sector in 2006 were at postgraduate level (i.e. NFQ Level 9 & 10).

There is an on-going debate in Ireland as to how well the higher education providers are responding to this demand. Constructing Excellence (Wolstenholme, 2009) in the UK for example reported that *'while clearly the industry itself carries a responsibility in this respect, the educational sector could do much more to understand the needs of the industry it supports.'* The same report also identified however that many staff development programmes are *'inferior'* and that *'leadership training at all levels of the industry is inadequate'*. The European Construction Institute Industry Futures Task Force (Goodier et al, 2008) also referred to the need for construction organisations to *'expand company alliances with universities'* and to *'influence the content and skills of educational programmes'*.

Given the current economic circumstances, the relatively large pool of graduates and the need for industry to increase learning, it would seem inevitable that the demand for postgraduate programmes will grow. This demand is putting the spotlight on the Higher Education providers in Ireland to facilitate such programmes, but ideally programmes that have a relevance to industry. Continuing Professional Development (CPD) can be considered as the planned acquisition of knowledge, experience and skills and the development of personal qualities necessary for the execution of professional and technical duties throughout a constructional professional life, encompassing both technical and non-technical matters (Wall and Ahmed, 2004). Academia has a key role in giving structure to learning and in mapping learning outcomes at the appropriate NFQ level. Acting in partnership with industry should ensure that the programmes are of benefit to the particpants as part of the lifelong learning continuum. The bespoke MSc CPM programme that is the subject of this Case Study chapter is at NFQ Level 9. It is intended to be interesting, relevant and useful to the individual participants, BAM Contractors and WIT.

BACKGROUND TO THE BAM-WIT PARTNERSHIP

BAM Contractors (previously known as Ascon-Rohcon) and WIT have had a long association that stretches back to 1990. The original links were developed for the purpose of providing internships to undergraduate students. The internships typically involved seven months work experience to students in their third year of four on their BSc [Honours] Construction Management and Engineering degree programme. A significant number of the graduates subsequently joined the company, many have remained and some of these are at senior management level within the organisation. The Managing Director of the company also acted as a WIT External Examiner for five years in the 1990s.

The process of developing the bespoke MSc CPM programme formally commenced in September 2007. By that time however the HRM Department in BAM Contractors had already developed the general objectives and indicative content of a Postgraduate Diploma programme that they wanted to implement. WIT was already running the MSc CPM programme and it was clear that there was significant overlap between this programme and the BAM concept. A series of meetings were subsequently arranged between senior staff in the company and WIT. The deliberations and the eventual programme are reviewed in the next section of this Chapter. However before considering how the company's objectives and the learning outcomes were aligned to everyone's satisfaction the following sections give further background information to each of the two main partners.

BAM Contractors

BAM Contractors is a wholly owned subsidiary of Royal BAM Group of the Netherlands and covers the entire spectrum of construction in Ireland. The 2008 turnover for BAM Contractors was €460 million which represented 5% of the overall Royal BAM Group turnover for that year. BAM Contractors is ranked in the top five construction companies in Ireland and the Royal BAM Group is in the top ten construction organizations in Europe. The Royal BAM Group generally and BAM Contractors in Ireland specifically believe that one of the key factors that contributes to current and future business success is the continuous development of its young professionals and middle management staff members. The company in Ireland has been delivering a structured graduate development programme to young professional staff since 1998. In 2007 BAM Contractors made a decision to further strengthen the development of talent by providing a postgraduate education opportunity to middle management staff members in the area of Construction Project Management. This learning and development initiative was conceived as a vehicle to encourage talent development after graduate training has been completed. It is expected that it will provide BAM with motivated and reliable project managers who can continue to deliver quality projects on time and within budget to satisfied clients.

In addition to the desire to develop the management talent within the company the programme was also seen as a mechanism to help retain the best employees. During 2006 the HRM Department identified that there was a significant problem with staff turnover, in particular that of project managers. Internal surveys indicate that in the 2002-2006 period 32% of the project managers left the company. When staff turnover was looked at from a length of service perspective in the 2002-2006 period 48% of the leavers left within two years of joining. While anecdotal evidence would suggest that such a high turnover rate was not uncommon in the leading Irish construction firms during the boom period, the associated disruption and costs were such that they need to be addressed. One of the mechanisms to do so was the proposed Postgraduate Diploma programme.

Waterford Institute of Technology

Waterford Institute of Technology (WIT) is one of the largest Institutes of Technology in Ireland. It was established in 1970 and now confers its own awards at NFQ Levels 6-10, subject to standards set and monitored by the Higher Education and Training Awards Council (HETAC). WIT has the highest number of students within the Irish Institute of Technology sector, outside of Dublin. There are currently approximately 6.500 registered full-time students and 3,500 registered part-time students on a range of programmes in Engineering, Architecture, Science, Informatics, Humanities, Health, Business and Education.

The School of Engineering at WIT has always worked closely with industry in the development and running of its education and research activities. The internship requirement on a number of the degree programmes has ensured that WIT students have gained valuable experience as part of their education. It has also ensured that the relevant staff has had constant interaction with the companies on a wide range of topics (inc. best practice in construction project management and technological advances) and direct access to the leading construction projects (mainly national but some international). The first postgraduate programme in the School was the MSc in European Construction Management. The setting-up and delivery of this one-year full-time programme in 1993 involved partnerships with educational institutions in the UK, France and Germany. This programme subsequently evolved into the MSc in International Construction Management and a total number of 190 participants from 24 different countries graduated between 1993 and 2007. The MSc in Construction Project Management commenced in 2005 and this was developed in response to a demand from Irish applicants. This programme is at NQF Level 9 and it has been designed for participants who want to develop their Construction Project Management knowledge and skills at postgraduate level without leaving their employment.

The Application of Blended Learning Approach at WIT

The capabilities and facilities that e-learning offers in the delivery of training are considerable. Technological advances and the Internet have opened up new opportunities for educational providers to configure traditional classroom and technology assisted learning to promote and encourage access to CPD learning opportunities (Klein and Ware, 2003). With the growing acceptance of e-learning technologies and the increasing need of access to distance learning opportunities, higher-level institutes face; (i) technological, (ii) organisational, (iii) pedagogical and (iv) cultural challenges in helping to integrate these changes (Howell et al, 2004).

The programme is delivered in the format outlined diagrammatically in Figure 5. Students attend for a specific period of formal lectures in Waterford for the relevant modules. Prior to / after or both prior to and after, handouts, other relevant material and websites will be uploaded on the Institute Learning Management System (LMS), Moodle. All assignments and other relevant discussions are posted on the LMS. In order to

Figure 5. Schematic of operational delivery of a module

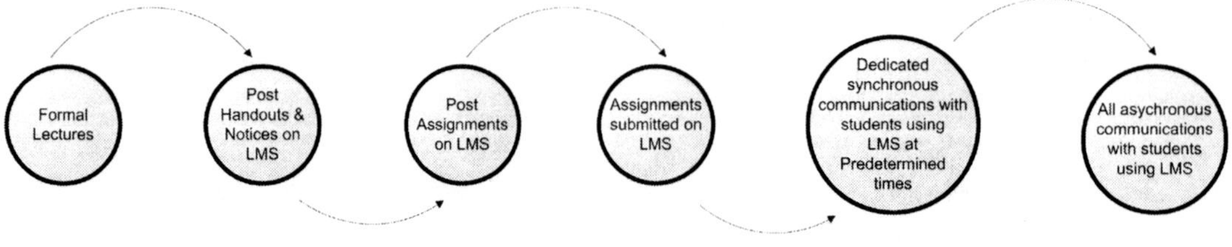

keep formal records and structure the programme appropriately all written assignment work is submitted for the individual modules using the LMS. Separately individual lecturers may require participants separately to submit a hard copy of the assignment. To encourage student participation between face to face lectures, lecturers may post topics for discussion on the LMS.

Whilst offering many advantages to students, a major consideration in delivering training through distance learning modes is maintaining a high level of motivation and commitment, avoiding the sense of isolation, and returning to the requirements and rigours of further education, especially for those who are returning to education after a number of years in industry. Through the use of this blended learning approach outlined earlier the use of both classroom and distance learning modes help to maintain a high level of commitment and remove the sense of isolation that students can often experience in traditional distance learning approaches.

ALIGNMENT OF BUSINESS OBJECTIVES AND LEARNING OUTCOMES

In any Industry-Academia partnership the need to clearly articulate the requirements of both parties is vital. The following sections outline the separate programme development activities in BAM and WIT as well as the subsequent development of the agreed MScCPM framework.

BAM Objectives and Supervisory Committee (2006-2007)

BAM formed an internal Supervisory Committee in 2006 consisting of senior personnel in order to devise the key objectives and indicative content for a management development programme. The overall aim established for the programme was to *'develop, retain and inspire project management talent to ensure that BAM Contractors continues to deliver successful projects to satisfied clients.'* The objectives derived from this aim were that the participants would on completion be able to:

- *manage projects in a contracting environment meeting the performance standards set out in their job descriptions;*
- *drive business improvements on their projects through innovative thinking and innovative action;*
- *generate, develop and maintain people relationships that demonstrate they are easy to do business with;*
- *evaluate and challenge personal strengths/weaknesses, beliefs, attitudes and knowledge to optimize personal effectiveness;*
- *adapt and readjust to change and uncertainty in project deadlines, information, client instructions etc. without any decrease in performance.*

The BAM Supervisory Committee identified the following three guiding 'pillars' that would underpin the proposed programme, namely *Operational Expertise*, *Financial Expertise* and *People Expertise*. In addition to the objectives and fundamental guiding pillars, it was also decided that the programme should ideally give the participants a recognized qualification when completed. A number of options were researched, discussed and examined by the Supervisory Committee. The strategy chosen was to pursue a 'partnership' with WIT, mainly because they were already running a similar postgraduate programme which could potentially be modified to suit the specific needs of BAM Contractors. The Supervisory Committee liked many aspects of the existing WIT MScCPM and the fact that there was a good working relationship between BAM and WIT was also a strong influence.

The Existing WIT MScCPM Programme (2005-2007)

The development of project management skills and knowledge in the construction domain continues to be the subject of significant debate. Large construction companies employ a variety of graduates form a variety of professional disciplines. The majority of these people either become project managers or become involve in some management function within the organisation. Programmes in the 'engineering' domain have generally focussed on traditional disciplines with strong professional institutions. However there are difficulties in defining 'engineering' in general and the various specific engineering disciplines, particularly when considering the future.

There has been much debate at national and international levels concerning the structure, content and delivery of successful engineering and construction education programmes of the future. Elmuti et al (2006) considered the role of education in the development of leadership skills. They concluded perhaps unsurprisingly that the many aspects of the leadership curriculum were not adequate and changes were required. These changes included the need for more in-depth training on interpersonal and conceptual skills allied to ethical training and global perspective.

In developing the MScCPM programme in 2005, the School of Engineering at WIT used the twelve years experience of running the related MSc programmes in European and International Construction Management. In addition to this experience however and taking into account a range of feedback from a variety of industry and academic sources the following list of generic requirements were used for the overall programme and the individual modules:

- *develop technical excellence appropriate to the relevant discipline;*
- *focus on projects and problem-solving;*
- *develop management and communication skills;*
- *encourage innovation and entrepreneurship.*
- *include integration with other disciplines and working in teams;*
- *include customer relations and market awareness;*
- *include environmental and sustainability awareness;*
- *include ethics and a global awareness;*
- *include industrial placement and strong links with industry;*
- *include the latest ICT and develop ICT skills.*

Agreed MSc CPM Framework and Bespoke BAM-WIT Programme (2007-2008)

The challenge of aligning the existing programme and the BAM objectives through an approved academic MScCPM framework was an interesting exercise for the partners. WIT were conscious of the need to develop a solution that would accommodate both the bespoke MScCPM programme (appropriate to the needs of one employer - BAM) and a general 'open' programme (appropriate to a variety of participants from a range of employment backgrounds).

While BAM through its Supervisory Council had strong views as to what was important and what needed to be included, there was an underlying assumption that the participants would automatically agree with all proposals. In their reflection on 'work-based learning' Hegarty and Johnston (2008) emphasized the potential conflict that could exist in a Graduate Development Programme (GDP) where the needs and benefits of the employer and the employee may not be aligned. Important issues for the participants include the perceived career benefits and especially their work-education-leisure balance for the duration of the programme. Cox and King (2006) referred to the needs of the three main stakeholders (employer, employee, academic institution) in any WIL course design, particularly the issues of employability. In the context of the Irish construction industry in the 2007-2010 period this issue was central to everyone's thoughts. The distinction between being employable and employed is key to addressing the future success of all stakeholders. These future requirements include the balance of both the requisite transferable skills and relevant subject skills.

Another interesting aspect to the alignment of the company's requirements with the WIT academic requirements was the required flexibility for the bespoke programme. Costley (2001) explored some of the tensions that existed between the employer organization and the individual employee on work-based programmes and concluded that '*partnership between and organization and a university or college means universities adhering to their level criteria and quality standards of process and assessment as well as being flexible enough to apply them to the needs of a particular organizational contexts.*' More recently Virolainen (2007) highlighted the potential problem of learning outcomes becoming '*compromised*' by too great an emphasis on the technical aspects of industry-academia agreements.

While the difficulties of successfully incorporating industry practitioners into an academically validated programme were acknowledged by all, the potential upside of including such experience and expertise was far greater. Apart from the various senior personnel that BAM wanted to be involved in the programme, the company also wanted separate specialist training firm to play a major role in the area of personal effectiveness. Tranfield et al (2004) referred to the gap between academic knowledge and its relevance for practice as an 'enduring problem'. They stated that this was particularly the case in academic management research and it relationship to management practice. However a potential solution to this problem that they proposed was a '*co-production model*' which included a synthesis of academic research, practitioner experience and professional practice.

How in practice these or any such agreed principles of 'co-production' are incorporated into an education programme is always a challenge. The associated debate can include issues such as ownership and academic standards. The balance of setting and demonstrably achieving the required education standards while also having the necessary flexibility for such programmes to be successful was a key challenge. The logical approach to addressing this balance was the development of an academic framework which identified the general learning outcomes for the overall programme as well as the individual constituent modules. Within this framework there was and is scope to deliver associated content and assess student performance in a range of approaches.

The framework went through the WIT CE3 evaluation process in May 2008. This CE3 process is the WIT Academic Council mechanism by which proposed changes to existing frameworks and programmes are ratified. Apart from preparing the relevant documentation demonstrating that the learning outcomes are at the required level the MScCPM team was questioned by an independent panel including two external experts about all aspects. The panel approved the framework subject to some minor alterations. The bespoke version of the MScCPM programme for BAM and the revised 'open' programme commenced in September 2008. Both programmes typically consists of thirty six tutor contact days that take place over a two-year period. The training sessions take place in a dedicated CPM training centre in WIT. The planned schedule consists of 6 x 3-day sessions (Thursday, Friday and Saturday) in each academic year with three sessions in the September-December period and the other three in the January-May period. The programme material is delivered through a blend of traditional methods (lectures, workshops, seminars, tutorials, projects) and e-learning. Students also use the Institute's

Learning Management System 'Moodle' and engage in online interaction.

Each module has a WIT Module Leader (ML) who is assigned to co-ordinate the various activities involved in each module. One of their key roles is to ensure that the assessments are carried out in an appropriate manner and checked to ensure that the stated module learning outcomes have been achieved. When the original framework was being developed the full extent of non-WIT staff involved in the delivery of the bespoke MSc-CPM programme was not known. However the general principle and the associated flexibility to allow a mix of WIT and non-WIT personnel in the delivery of each module were incorporated. The different context for the delivery of each module (i.e. 'bespoke' or 'open') was also specifically referred to in the framework. This meant that the emphasis on certain aspects within each module will vary for the bespoke programme to the open programme (e.g. the assessment approach) but that the overall learning outcomes for each module will be consistent.

Figure 6. Flowchart of developing and approving the MScCPM Framework

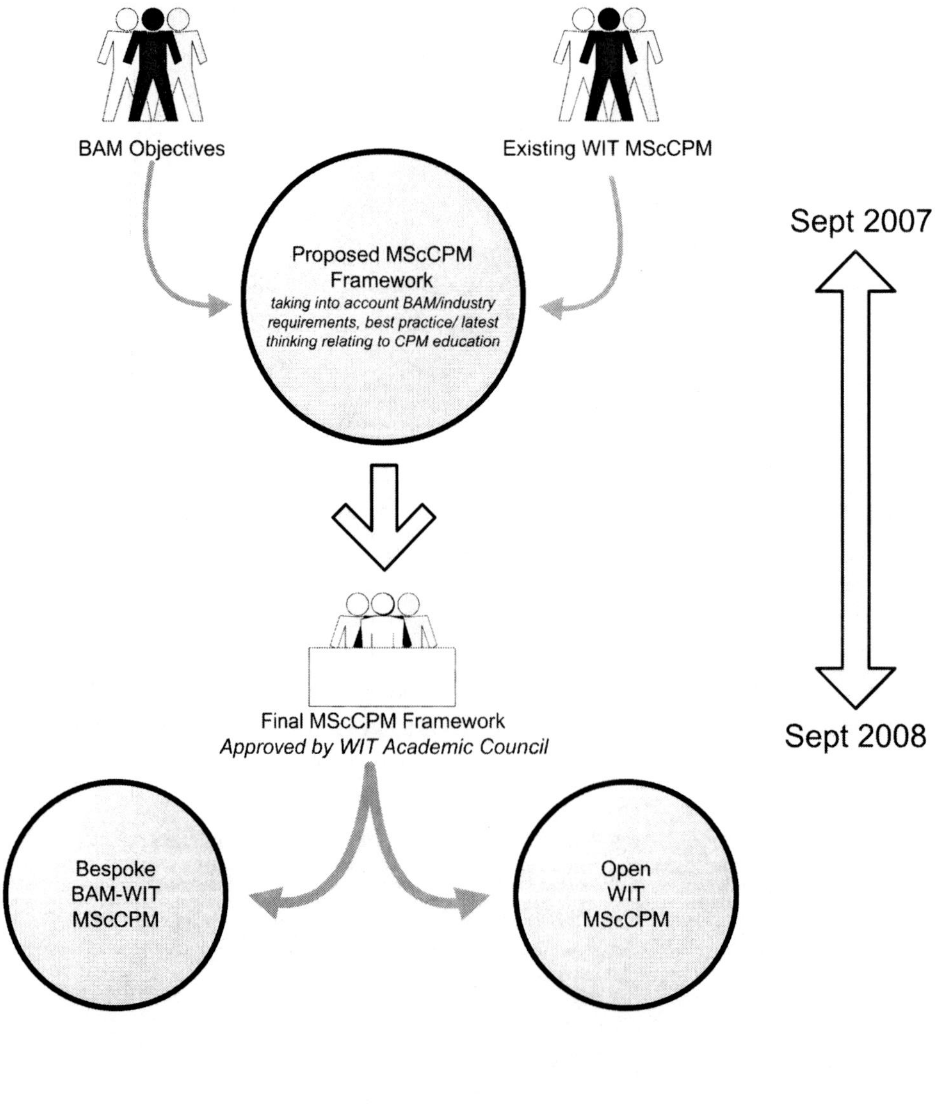

While the official Module Descriptors for each of the modules are consistent for both the open and the bespoke programme, there are differences between programmes. For example BAM wanted a greater emphasis on 'Commercial Management' and 15 ECTS were accordingly allocated and agreed under the framework. The open programme however only covers two-thirds of this module (i.e. 10 ECTS) and the balance of 5 ECTS is devoted to a separate module called 'Sustainability & the Built Environment'. The approach taken to the 'Personal Effectiveness' module was also substantially different between the bespoke and open programmes. The specialists that BAM wanted involved in the delivery of this module also contributed to the overall communications and presentation skills on the other modules on the bespoke programme. The role of the WIT Module Leader for the bespoke 'Personal Effectiveness' module was almost exclusively to ensure that the content and assessments were in accordance with the agreed framework.

LEARNERS, EXPERTS, FACILITATORS AND MENTORS

The programme was advertised during May 2008 throughout all of the BAM divisions and regional offices in Ireland. Many of those that subsequently applied were encouraged to do so by their Director or Senior Manager. The eventual 20 participants selected by the company in consultation with WIT during June/July 2008 were a mix of civil engineers, quantity surveyors and construction managers. The ranged in age from 25 to 38, they had been with the company for at least three years. There were more applicants than places and central to BAM Supervisory Committee's decision to make offers to the chosen 20 was that they were generally expected to be the future senior managers and leaders within BAM Contractors. The only exception to the criteria was the one participant who joined the group from a separate company, Suir Engineering. This arrangement suited everyone as the particular individual was anxious to complete the programme and there was a good relationship between BAM and Suir Engineering. The government backed 'Skillnets' initiative who were financially supporting the programme were also in favour.

The participants in the agreed bespoke programme not only get the opportunity to review theory and practice relating to Construction Project Management (Irish & global) but they also are encouraged to reflect on their role with the company. The need to maximise the use of everyone's time became increasingly obvious throughout the implementation of the programme. The need for the appropriate balance of the participant's:

- contact time with the relevant WIT Module Leaders and Facilitators;
- contact time with the BAM Experts and Mentors;
- time for independent learning, reflection and completion of assessments;
- 'full-time' employment;
- personal and family time.

The overall management of the programme was shared between the School of Engineering at WIT and the HRM Department at BAM. This required regular and frequent contact to ensure that all of the various elements (including the participants and indeed the various module facilitators, experts and mentors) were organized and co-ordinated. Figure 7 attempts to show the range and relationships of the various groups involved in the management and delivery of the bespoke programme.

PROGRAMME STRUCTURE, ASSESSEMENT AND RESEARCH

The bespoke programme was officially launched in WIT on 16th September 2009. This event included a number of speeches from BAM, Skillnets and

Figure 7. List of MScCPM Modules on the bespoke BAM programme, the associated academic credits (ECTS) and their mapping against the three BAM pillars of expertise

WIT Module Title	ECTS	BAM Operational Expertise	BAM Financial Expertise	BAM People Expertise
Management of Projects 1	10	✓	✓	✓
Management of Projects 2	10	✓	✓	✓
Commercial Management	15	✓	✓	
Construction Law	5	✓		
Strategic Management	5	✓	✓	✓
ICT for Construction	5	✓		
Management of People	5			✓
Personal Effectiveness	5			✓
Postgraduate Diploma CPM	60 ECTS			
Research Methodology	5			✓
Research Thesis	25	✓	✓	✓
MSc CPM	90 ECTS			

WIT representatives. Perhaps the most important statement on the day was by Mr. Brendan Barrett, MD of BAM Contractors who said that

the WIT BAM MSc in Construction Project Management (CPM) is a first in construction executive education in Ireland and will inspire our project management talent, ensuring that we continue to deliver successful projects to our clients. Industry, academia and management science have fused to produce this precedent setting innovation in learning. It represents a true investment in the belief that people are our most important asset. This is a good news story; it is an investment in innovation, in people, in the industry and ultimately in the future of Ireland.

The following sections outline the Programme Structure, the Assessments for each of the modules and the approach taken to the Research Thesis.

Programme Structure

The dates for the various sessions and their sequence are shown in Figure 8. In line with the agreed MScCPM framework the participants were scheduled to attend classes at WIT on six number three-day occasions (Thursday/Friday/

Figure 8. Key groups in the management and delivery of the bespoke MScCPM programme

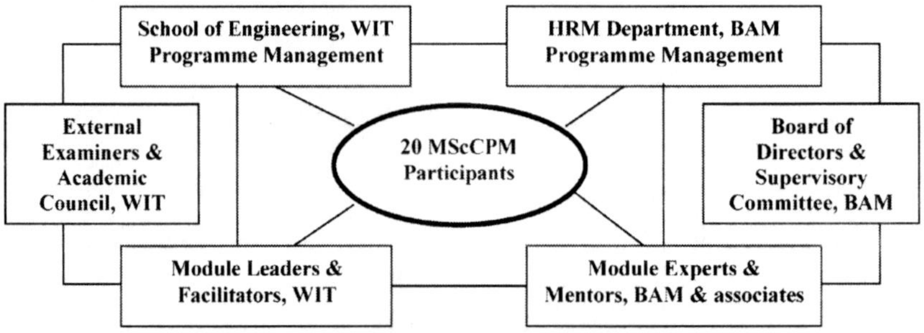

Saturday) throughout each academic year. However an additional session was held in the Royal BAM Group headquarters during the second year of the programme. This bonus session was added by agreement between BAM and WIT and it served a number of important purposes. The main objective was for the participants to realize that they work for an Irish company that is part of a major european construction group. By physically going to the headquarters in Bunnik, Netherlands and meeting many of the key staff, including the Royal BAM Group Managing Director Mr. Joop van Oosten, the three days had a profound effect on the group. The impact of the Irish participants on their hosts was also significant and many, including the in-house BAM Business School, were both curious and impressed that BAM in Ireland were running such an MScCPM programme. Other benefits from the visit included the team bonding aspects and indeed the insight that the traveling WIT staff got in relation to the Royal BAM Group activities.

Assessment

The approved MScCPM, framework at WIT specifies that almost all of the associated assessment work does not include formal written exams. In fact the only such examination is relation to Financial Management which forms part of the overall assessment for the Commercial Management module on the bespoke programme. The thinking behind this approach is to try and encourage the participants, as individuals and in groups, to improve their report writing and oral presentation skills.

The overall volume of assessment on the bespoke MScCPM programme for the BAM Group is similar to that on the open programme. However while a number of the assessments are the same, the majority of those on the bespoke programme relate specifically to BAM business activities. BAM saw the benefits of getting the participant either individually or in groups to investigate issues that are of importance to the company. The advantages of such an approach from the participants' perspective are also supported by the academic literature. Thurairajah (2009:49) advise that

when designing work-based assessment it is always useful to review the work students are currently doing in their jobs. If you make opportunities to assess real work, this will provide an efficient use of time for students and their employees.

Each assessment was designed and agreed by the relevant WIT Module Leader and the associated BAM Expert in line with the agreed MScCPM framework. In a number of cases there were two assessments the first of which concerned a general topic (e.g. Contract Law) and the second was more company specific (e.g. the use of the New Forms of Government Contracts by BAM). The processes of aligning the required programme assessments with topics and problems of direct interest to the company may not always be straightforward, but the potential befits are clear. '*The 'learning by doing' approach is an important factor in the successful research knowledge transfer from the company's perspective but also enhances the academic research and teaching activities'* (Sas, 2009). An example of this approach is in relation to the ICT in Construction module. In this case the WIT Module Leader agreed the topics and approach with the ICT Manager for BAM and his team. Essentially the group was divided up into 6 groups of three peoples and the other two participants were given individual projects (see Figure 9). The topics selected were of interest to both BAM and WIT and it was expected that the results would be of direct benefit to both parties.

Research Thesis

By far the largest single element of assessment on the MScCPM programme is the Research Thesis. While the BAM Supervisory Committee

Figure 9. Key groups in the management and delivery of the bespoke MScCPM programme

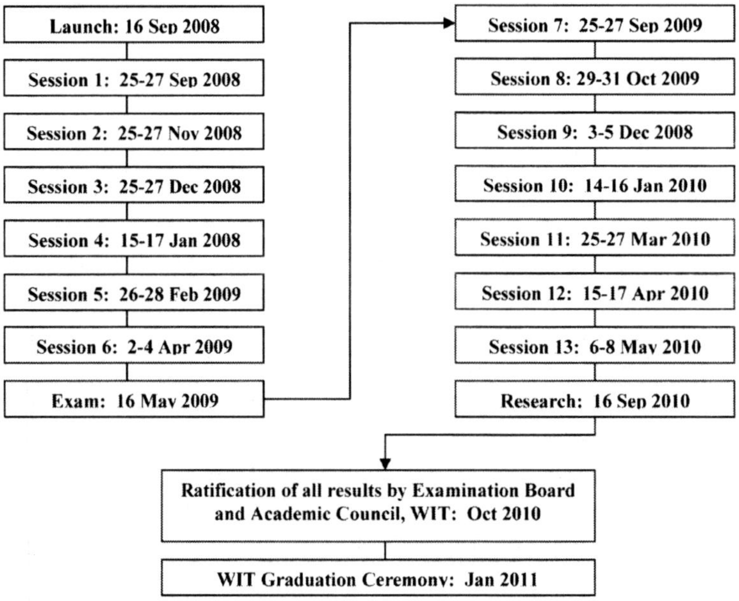

did not include this element in their original plans for the bespoke programme, the opportunity for the participants to take complete this work was recognized by all. The company could potentially gain significant benefits from a series of comprehensive studies on a range of relevant topics, the participants also had the carrot of gaining an MSc qualification instead of the Postgraduate Diploma qualification by successfully completing the research.

The academic literature relating to the role of research in an Industry-Acdemia Partnership related includes Elmuti (2005) who reviewed such strategic alliances in the US. He highlighted the associated benefits, particularly in relation to research activities. Among the advantages for the academic partners were the increased funding opportunities which the industry partners benefited from technology transfer opportunities that effect competitiveness. Wynn et al (2008) describe a '*Knowledge Transfer Partnership (KTP)*' which highlights how academics can work productively in a commercial construction environment with industry partners and that '*such a partnership can be used to develop new business ideas and initiatives*'. Martin et al (2010) also support the shared benefits of such partnerships, particularly in relation to the joint development of research activities. In their recent review of the strategic collaboration between the international civil engineering consultancy Halcrow Group Ltd. and Cardiff University they say the collaboration

has been mutually beneficial. Since its inception in 2001 the collaboration has been of direct benefit for the two main parties and also the wider industry-academia community, specifically in relation to the promotion of innovation and the exploitation of industry relevant research.

In order to gain the 30 ECTS for the research element of the MScCPM programme, participants must first complete the Research Methods module. This module is essentially the vehicle to give the participants grounding in the purpose of postgraduate research (NFQ Level 9) and the associated methodologies. By the end of the module each participant should have a clear proposal for the specific research topic that they intended to complete. This proposal includes the associated aim, objectives and hypothesis as well as an initial review of the relevant literature and a outline of the selected research methodology. In the case of the bespoke programme BAM had the view that each research topic should not only meet the academic criteria described in the MScCPM framework, it also should be directly relevant to the company. The specific guidelines given to the BAM group were that:

- *a successful thesis should ultimately lead to increased turnover greater efficiencies, problem solving, and development of relevant expertise; and*
- *the research results should form the basis of improvements in the company and how it conducts its business*

After much deliberation three of the group decided that they were not in a position to undertake the research element of the programme, due mainly to the required time commitment. However the other seventeen did complete the 'Research Methods' module and submit their proposals to both WIT and to the BAM Board of Directors for approval. After some further discussions and some alterations, the proposals were approved by WIT and BAM. In September 2009 each individual was assigned a WIT Academic Supervisor (who had an expertise in the relevant subject area) and a BAM Mentor (typically the Director or Senior Manager who had the company responsibility for the relevant subject area). In order to encourage progress in the September to December 2009 period, the Literature Reviews were to be submitted by 4th January 2010. Most achieved this deadline but a number did not, mainly due to work commitments (e.g. construction projects coming to a conclusion).

RETURN ON INVESTMENT

When reflecting on the Return on Investment (ROI) for the programme the primary stakeholder that had an interest in this topic was the company. Given the fact that BAM Contractors were paying the direct costs as well a number of indirect costs (e.g. time away from work for the participants, senior staff time involvement) they had a requirement to undertake an ROI study. This process is described below and is followed by a section on the ROI for the other main stakeholder groups, i.e. the participants and WIT.

ROI for BAM

Given the scale of the investment in this programme the HRM Department in BAM Contractors was asked by the company's Board of Directors to carry out a thorough analysis of the benefit of the programme to our business. The 'ROI Methodology' used was developed by Dr. Jack J. Phillips in the 1970s, refined through application and use in the 1980s and implemented globally during the 1990s (ROI Institute, 2010). The Phillips model of evaluates learning & development interventions across the following levels:

Level 1: Reaction to the learning.
Level 2: Measurement of learning.
Level 3: Application of the learning in the workplace.
Level 4: Business impact as a result of the application of the learning.
Level 5: Return on Investment (financial return based on overall business impact).

The potentially massive administrative workload that would be involved in trying to carry out the evaluation and ROI exercise via manual coordination of hard copy surveys etc. was quickly recognized. Given this challenge the HRM Department reviewed and selected a software application developed by Gaelstorm Ltd., which has been developed specifically to automate the ROI process for Levels 1 and 2 of the Phillips Model. After each 3-day session in WIT the participants received an email which links them to an online evaluation document. The participants complete the electronic survey, which is structured around each content module that they recently experienced on the postgraduate programme.

They answer questions regarding the programme content and the trainers, for example:

- *Clarity of learning objectives?*
- *Quality of training materials and presentations?*
- *Skill and knowledge of each trainer, i.e. pace of delivery, ability to meet learning objectives, openness to answer questions etc.?*
- *Relevance of content to their current roles?*
- *Increase in learning and levels of understanding of programme content?*
- *Intention to apply the learning back on the job?*

Regarding the Level 3 'application of learning in the workplace', a separate survey was developed by the HRM Department to periodically measure the broad range of relevant skills and competences of the programme participants. This survey document is shown in Figure 10 and it was intended to be distributed to the participants on five occasions. The first survey was carried out in November 2008 to give a baseline measurement. Further 'live' measures were then carried out in May 2009 and December 2009. There will be two more measures taken in May 2010 and December 2010. It was expected that the results would demonstrate a steady and visible increase in overall skill and ability from both the participants' and their managers perspectives as a result of the application of learning from the postgraduate programme.

With regard to Levels 4 and 5 of the ROI model, the plan is to use the research thesis element of the MScCPM programme to facilitate an internal review and performance improvement initiative for a range of areas in our business where senior management would like to see operational, technical, innovation and cost-reduction improvements in the business operations. The commercial benefit and ultimately the return on investment from this postgraduate programme will be calculated in the short to medium term from the efficiencies and savings that will result from the findings and implementation of the theses recommendations.

Figure 10. 'ICT in Construction' assessment topics

Group	ICT in Construction Topic
G1	**COINS** - A review of current and future use of COINS for BAM construction projects
G2	**SharePoint** – An investigation into the current and potential use of SharePoint within BAM
G3	**Building Information Modeling (BIM)** – The potential use of BIM by BAM for tendering and project management
G4	**Project Planning Software** – An investigation into the current use of Project Planning software in BAM
G5	**ICT on Sites** - An investigation into the best approach to ICT services on each BAM construction site
G6	**Mobile ICT Devices** - An investigation into the current and potential use of mobile devices by BAM
I7	**ICT in BAM UK** - A review of the ICT facilities and activities in BAM UK and the lessons for BAM Contractors, Ireland.
I8	**ICT in Suir Engineering** – A review of the ICT facilities and activities in Suir Engineering, including the ICT links with BAM.

Further ROI Aspects and Perspectives

While the financial investment by the company in the bespoke programme has been significant, each individual participant has had to invest a lot of their own personal time. Whether they feel they have had a return on this investment may be best reflected upon at the end of the programme. However the fact that all participants are still in employment while a significant number of others in BAM and indeed the Irish Construction Industry in general have become unemployed is a return in itself. Assuming that continues to be the case at the end of the programme it will be interesting to see how the careers of each participant develop in the short to medium term. The learning and the qualification gained from the programme should result in significant opportunities both within BAM. Even allowing for the difficult current economic context this likely development is supported by Osborne (2008) who pointed out that

leaders who are promoted from within an organisation already have a good understanding of the organisations culture and working methods, and will have been nurtured and trained by the organisation to have the exact suite of skills and knowledge required to take on a new role.

While WIT has received fees from BAM under the agreement to deliver the bespoke MScCPM programme, these fees have not covered all of the direct costs. The management and co-ordination activities required to make the programme operate successfully have been at a greater level than

Figure 11. BAM Survey Questionnaire for Level 3 ROI, Application of the learning in the workplace

This is an anonymous questionnaire. Please take a few minutes to think about your current level of skills and provide responses to the following questions. Please indicate your level of skills using the following scale by checking the appropriate response:
VS = Very Strong (completely competent), St = Strong, Sa = Satisfactory – could improve, W = Weak, VW = Very Weak (limited or no skills)

	VS	St	Sa	W	VW
My ability to manage my time at work is	O	O	O	O	O
My ability to plan and organize my work is	O	O	O	O	O
My ability to manage the work and time of my team and others is	O	O	O	O	O
My ability to identify and analyze problems and develop effective solutions is	O	O	O	O	O
My ability to organize and schedule resources in a productive and cost effective manner is	O	O	O	O	O
My ability to manage the flow of information and documentation is	O	O	O	O	O
My ability to arrange and run meetings is	O	O	O	O	O
My ability to delegate effectively to others is	O	O	O	O	O
My ability to manage finances is	O	O	O	O	O
My ability to prepare reports in a timely manner is	O	O	O	O	O
My ability to relate to and negotiate effectively with subcontractors and suppliers is	O	O	O	O	O
My ability to implement the Company's Quality Management System [QMS] is	O	O	O	O	O
My ability to maintain the trust of others is	O	O	O	O	O
My ability to manage & lead those staff assigned to me is	O	O	O	O	O
My ability to implement excellent cost control measures is	O	O	O	O	O
My ability to maximize Income, though not to the detriment of repeat business, is	O	O	O	O	O
My ability to represent the Company with professionalism and integrity is	O	O	O	O	O
My ability to consult with architects, engineers, and other technical stakeholders is	O	O	O	O	O

originally envisaged. However the overall WIT learning experience of the Industry-Academia partnership and the associated knowledge benefits are also far greater. This discovery is in keeping with Ryan (2009) who explored the increasing number of university-corporate education partnerships which included quantitative research from North America, UK and Australia. The primary reasons for such partnerships was the desire from the employer organizations to have a recognized university award for their programme and also the 'strength and credibility' that a university can add to such programmes. Ryan conclusions supported the 'win-win' argument for both the company and the university. '*Rather than compromising the content of a university's programmes, the closer the working relationship between a university and it corporate education partner, the more a university can evolve programmes.*'

REFLECTION: INNOVATION AND LESSONS LEARNED

The journey of developing and delivering this bespoke postgraduate programme in the 2007-2010 period has generated a number innovations and lessons learned for all stakeholders. Although it fulfils all the academic requirements under the approved MScCPM framework at WIT, the partnership approach has involved strong co-operation and commitment through:

- joint programme management;
- a mix of academic and industry facilitators, experts and mentors;
- a blend of traditional classroom interaction, e-Learning technologies and work activities; and
- assessments that address relevant BAM business topics at the required academic standard.

The dramatic economic developments during 2007-2010 period and the associated collapse in the output of the Irish construction industry over that period was something that was unforeseen by most stakeholders, including BAM and WIT. This context was foremost in everyone's mind throughout the delivery of the programme. The need for BAM to be more competitive in all aspects of its operations gave a particular emphasis to much of the content delivery and the associated assessment activity. While it might have been expected that the global and national economic difficulties put a strain on the BAM-WIT partnership, in reality it probably contributed to strengthening the relationship. These economic difficulties led to a greater focus by the company on the programme's content and assessments. This naturally required even greater interaction between the relevant BAM and WIT personnel. Elmuti et al (2005) highlighted the risk and problems facing strategic alliances and these included cultural differences, differences in objectives, dealing with the unexpected and a break down in trust. They wisely pointed out that '*as long as the value gained from the partnership exceeds the costs of both partners, the basis of the alliance is set.*' However this basis '*must be supported by continuous learning and restructuring processes to overcome the differences between the partners.*' Williams and Lees (2009) emphasise the need to continually work on such partnership *'as with other successful relationships, a relationship with an employer moves through phases and is built upon trust.'*

There is always a debate as to how much training a company should facilitate and pay for in spite of the multitude of 'our people are our greatest assets' statements. Research findings from Germany, Ireland, Sweden, the UK and the US show that the vast majority of job-related training appears to be employer paid, at least partially (O'Connell, 2009). The underlying assumption of this human-capital approach is that *'individual workers undertake training, and employers invest in training, on the basis of their estimates of future returns (including employment prospects, wages and productivity gains).'* While the full ROI evaluation has yet to be completed the indications to date would suggest that the investment by BAM Contractors has been worthwhile. Butcher and Sheehan (2010) recently considered the issue of being an excellent contractor in the UK construction industry from the customer's perspective. Among the key behaviors they identified in being excellent was '*demonstrating the desire to learn and share learning'*. The Board of Directors and specifically the HRM Department at BAM Contractors have demonstrated such a desire through this bespoke MScCPM programme.

In addition to the organizational learning the participants have also benefited significantly from their participation in the programme. While it has been a challenge for them to complete the programme in addition to their employment responsibilities, particularly in the context of an industry in turmoil, their own personal development has been obvious. Their communication skills have improved substantially, they have become far more aware of the factors that contribute to a successful construction company and they have a much better insight into all aspects of their own organization in Ireland as well as the wider Royal BAM Group. Their assessment work has forced them to reflect on their company's act ivities but often in a flexible manner to incorporate their own individual role as espoused by Williams and Thurairajah (2009:29). The need for engineering education programmes to focus more on innovation and innovative thinking has been supported by a number of industry experts and academics (including Armitt & Baldwin, 2008; MacLeod, 2010). The bespoke BAM-WIT MScCPM programme has incorporated many of the task-based approaches that MacLeod has suggested are central to developing the participants' innovative thinking.

The main lessons learned by the stakeholders are as follows:

- Setting up an Industry-Academic Partnership is a challenge but maintaining the partnership is a greater challenge, particularly in changing economic circumstances. The key to a successful Industry-Academic partnership is trust and this is achieved through frequent contact including weekly phone calls, emails and regular meetings.
- The link between the Academic Module Leader who has overall responsibility for ensuring that the academic requirements for their individual module and the associated Industry Expert/Mentor should be strong. The successful integration of the general/non-company content and the specific company content can be achieved through good planning and co-ordination.
- The nature and timing of the assessments are key to the programme. These assessments not only demonstrate that the participants have achieved the required learning outcomes but they can also to lead to tangible benefits for the company. Apart from the personal development of the individual staff members and being able to justify 'our people are our greatest asset claims', the various reports and presentations can lead to positive changes within the company in the short to medium term.
- Plan all aspects of the programme in as much detail as possible before commencing the programme, particularly the teaching and assessment schedule, but incorporate some flexibility to cope with the inevitable changes. These changes will include the non-availability at certain times of the facilitators, experts, mentors and participants. Other changes are likely to relate to the business environment and the incorporation of new technologies and practices.
- Participants should be interviewed and informed (warned!) in advance of com-

mencing a postgraduate programme of the significant requirements, particularly the likely time commitment. While it is always good to emphasise the positive aspects of completing such a programme (including 'personal development' and the enhanced career prospects) there is no point in hiding the significant additional workload to people who are already busy in their work and private lives.

CONCLUSION

Developing a framework to bridge the gap between industry and academia in the delivery of Work Integrated Learning (WIL) is a necessary challenge for continuous improvement. Provided that there is trust between the principal stakeholders (academic institution, company and participants) and a recognition that the required learning outcomes must be achieved in order for the qualifications to be awarded, then WIL postgraduate programmes can be very successful.

The Case Study described in this chapter has demonstrated not only value to both of the Industry-Academia partners, i.e. BAM Contractors and WIT, but also the individual participants. Keithley and Redman (1997) emphasized the need to avoid of *'rose-tinted glasses'* view that can be used in relation to industry-academia alliances. They wisely identified that *'the acid test for successful industry-academic partnerships is whether they can prosper over the long term and accommodate new contexts in a turbulent and highly competitive business environment'*. Turbulent and highly competitive do not do justice to the Irish Construction Industry in the 2007-2010 period and the fact that there are advanced plans to run the BAM-WIT MSc in Construction Project Management programme again in 2010-2012 with an new cohort emphasizes the perceived success of the bespoke programme from everyone's perspective.

REFERENCES

Alam, M., Gale, A., Brown, M., & Kidd, C. (2008). The development and delivery of an industry led project management professional development programme: A case study in project management education and success management. *International Journal of Project Management*, *26*(3), 223–238.

Armitt, J., & Baldwin, R. (Eds.). (2008). *Innovation process in the construction sector*. UK: CICE & University of Loughborough.

Barrett, A., Kearney, I., Goggin, J., & Conefrey, T. (2009). *Quarterly economic commentary – Winter 2009*. Dublin, Ireland: ESRI.

Butcher, D. C. A., & Sheehan, M. (2010). Excellent contractor performance in the UK construction industry. *Engineering, Construction, and Architectural Management*, *17*(1), 35–45. doi:10.1108/09699981011011302

Chivers, G. (2007). Professional competence enhancement via postgraduate post-experience learning and development. *Journal of European Industrial Training*, *31*(8), 639–659. doi:10.1108/03090590710833688

Cole, M. (2004). Opening university. *New Civil Engineer, March*, 38-40

Costley, C. (2001). Organizational and employee interests in programs of work based learning. *The Learning Organization*, *8*(2), 58–63. doi:10.1108/09696470110387991

Cox, S., & King, D. (2006). Skill sets: An approach to embed employability in course design. *Education + Training*, *48*(4), 262-274.

CSO. (2006). *2006 statistical yearbook*. Dublin, Ireland: Central Statistics Office.

CSO. (2009). *2009 statistical yearbook*. Dublin, Ireland: Central Statistics Office.

Davey, C., Powell, J., Cooper, I., & Powell, J. (2004). Innovation, construction SMEs and action learning. *Engineering, Construction, and Architectural Management, 11*(4), 230–237. doi:10.1108/09699980410547586

Department of Finance. (2010). *Monthly economic bulletin – February 2010*. Ireland: Government Publications.

DKM. (2009). *Review of the construction industry 2008 and outlook 2009-2011*. Ireland: Government Publications.

Drake, J., Blake, J., & Swallow W. (2009). Employer engagement: The critical role of employee commitment. *Education + Training, 51*(1), 23-43.

Elmuti, D., Abebe, M., & Nicolosi, M. (2005). An overview of strategic alliances between universities and corporations. *Journal of Workplace Learning, 17*(1/2), 115–129. doi:10.1108/13665620510574504

Elmuti, D., Minnis, W., & Abebe, M. (2005). Does education have a role in developing leadership skills? *Management Decision, 43*(7/8), 1018–1031. doi:10.1108/00251740510610017

Gale, A., & Brown, M. (2003). Project management professional development-an industry led programme. *Journal of Management Development, 22*(5), 410–425. doi:10.1108/02621710310474769

Giannakouris, K. (2008). *Ageing characterises the demographic perspectives of the European societies*. Eurostat, European Commission.

Global Insight, I. H. S. (2009). *Global construction outlook*. IHS Global Insight.

Goodier, C., Austin, S., Guthrie, W., & Metzdorf, C. (Eds.). (2008). *Anticipating tomorrow, the future of the European construction industry – final report of the ECI industry futures task force*. UK: European Construction Institute.

HEA. (2010). *Student statistics–2008/2009: HEA funded institutions*. The Higher Education Authority. Retrieved February 17, 2010, from http://www.hea.ie/en/statistics

Hegarty, C., & Johnston, J. (2008). Graduate training: Evidence from FUSION projects in Ireland. *Education + Training, 50*(5), 391-405.

Howell, S. L., Saba, F., Lindsay, N. K., & Williams, P. B. (2004). Seven strategies for enabling faculty success in distance education. *The Internet and Higher Education, 7*, 33–49. doi:10.1016/j.iheduc.2003.11.005

Institute, R. O. I. (2010). *About the ROI Institute*. Retrieved February 12, 2010, from http://www.roiinstitute.net/about/

Keithley, D., & Redman, T. (1997). University-industry partnerships in management development: A case study of a world-class company. *Journal of Management Development, 16*(3), 154–166. doi:10.1108/02621719710164319

Klein, D., & Ware, M. (2003). E-learning: New opportunities in continuing professional development. *Learned Publishing, 16*(1), 34–46. doi:10.1087/095315103320995078

Kwak, Y. H., & Anbari, F. T. (2009). Analysing project management research: Perspectives from top management journals. *International Journal of Project Management, 27*(5), 435–447. doi:10.1016/j.ijproman.2008.08.004

Lees, M. (2009). Foreword. In Williams, A., & Lees, T. (Eds.), *Knowing your market: Strategy, communication and relationships – employer engagement series two, engaging employers*. Salford, UK: University of Salford.

MacLeod, I. A. (2010). The education of innovative engineers. *Engineering, Construction, and Architectural Management, 17*(1), 21–34. doi:10.1108/09699981011011294

Martin, J. C., Akande, K., & Falconer, R. A. (2010). Innovations from the Halcrow-Cardiff University strategic collaboration. *Engineering, Construction, and Architectural Management, 17*(1), 46–64. doi:10.1108/09699981011011311

Mitchell, P. (2009). Borrowed time. In Cantwell, T. (Ed.), *Irish construction overview 2010 – opinion, analysis and the year ahead in construction.* Ireland: Commercial Publications Ltd.

National Research Council. (2009). *Advancing the competitiveness and efficiency of the US construction industry.* Washington, DC: The National Academies Press.

NQAI. (2003). *National framework of qualifications - policies, actions and procedures for access, transfer and progression of learners.* Ireland: The National Qualifications Authority of Ireland.

O'Connell, P. (2009). *What do we know about training at work?* ESRI Research Bulletin 2009/1/4. Retrieved February 17, 2010, from http://www.esri.ie/publications/ latest_publications/

OECD. (2009). *Economic outlook – Ireland.* (Economic Outlook No. 86, OECD). Retrieved January 28, 2010, from http://www.oecd.org/document/18/

Ryan, L. (2009). Exploring the growing phenomenon of university-corporate education partnerships. *Management Decision, 47*(8), 1313–1322. doi:10.1108/00251740910984569

Sas, C. (2009). Research knowledge transfer through business-driven student assignment. *Education + Training, 51*(8/9), 707-717.

Scully, D. (2009). *Prospects for the Irish construction industry.* Paper presented at the National Construction Conference 2009, Croke Park, Dublin, Ireland.

Thomas, K., Graham, B., & Wall, J. (2007). Learning about IT and learning using IT-a review of current practice on Higher Education AEC programmes in Ireland. In D. Rebolj (Ed.), *Bringing ITC knowledge to work - 24th W78 Conference Maribor & 5th ITCEDU Workshop & 14th EG-ICE Workshop* (pp. 723-729). University of Maribor, Slovenia.

Tranfield, D., Denyer, D., Marcos, J., & Burr, M. (2004). Co-producing management knowledge. *Management Decision, 42*(3/4), 375–386. doi:10.1108/00251740410518895

Virolainen, M. (2007). Workplace learning and higher education in Finland: Reflections on current practice. *Education + Training, 49*(4), 290-309.

Wall, J., & Ahmed, V. (2004). *E-learning and pedagogical challenges in construction management: Bridging the gap between academia and industry.* Paper presented at the 20th Annual Conference Association of Researchers in Construction Management ARCOM, (pp. 603 – 613).

Williams, A., & Lees, T. (Eds.). (2009). *Knowing your market: Strategy, communication and relationships – employer engagement series two, engaging employers.* Salford, UK: University of Salford.

Williams, A., & Thurairajah, T. (Eds.). (2009). *Working the curriculum: Approval, delivery & assessment–employer engagement series two, work-based learning.* Salford, UK: University of Salford.

Wolstenholme, A. (Ed.). (2009). *Never waste a good crisis–a review of progress since rethinking construction and thoughts for our future.* UK: Constructing Excellence.

Wynn, M., Jones, P., Roberts, C., & Little, E. (2008). Innovation in the construction and property management industries - case studies of the knowledge transfer partnership scheme. *Property Management, 26*(1), 66–78. doi:10.1108/02637470810848903

Chapter 12
An Archetype of WIL in Information Technology at Baden-Württemberg Cooperative State University Ravensburg, Germany

Karin Reinhard
Baden-Württemberg State University of Cooperative Education, Germany

Shalini Singh
Durban University of Technology, Republic of South Africa

ABSTRACT

The chapter provides an overview of Work Integrated Learning (WIL) into the Information Technology (IT) programme offered at the Baden-Württemberg Cooperative State University in Ravensburg, Germany. The opinions and debates of leading role-players in WIL are featured. The university's pose and the operations adopted in managing this programme will be presented. These include the structure of the IT programme, its accreditation process, strengths, and weaknesses. The chapter concludes with the programmes direction for the future.

INTRODUCTION

This chapter will provide a holistic perspective of Work Integrated Learning (WIL) and its integration in the Information Technology (IT) programme offered in the School of Information Technology at the Baden-Württemberg Cooperative State University in Ravensburg, Germany. The university is also known as the Duale Hochschule Baden-Württemberg (DHBW) University. This chapter will present the views and discussions of leading role-players in WIL, the university's

DOI: 10.4018/978-1-60960-547-6.ch012

position on WIL and the operations adopted in managing the programme at this university.

The operations will cover the main focus of the chapter which is the structure of the IT programme, its accreditation, strengths and weaknesses and problems and controversies. The chapter will close on the viability of the programme and its direction for the future

BACKGROUND

The policymakers of state of Baden-Württemberg, Germany, determined that there was a need for a new method of teaching that would directly impact the students who were attempting to learn the duties and responsibilities that industry desired in today's highly technical and increasingly global world. Therefore in 1972, three world-renowned organisations: Bosch, Daimler Benz and SEL initiated, in cooperation with the Baden-Württemberg Chamber of Commerce, an innovative and highly successful system of academic study at university level. Their intention was to establish a model programme for WIL.

Shortly thereafter, the DHBW was founded in 1974 in Stuttgart, the capital of the state of Baden-Württemberg. Today, eight main locations and three branch campuses with their areas of responsibility and close networking with the regional organisations form the pillars of the DHBW group of universities (Müller, 2009).

On 1 March 2009 the German state of Baden-Württemberg changed the status of these former universities. Among other changes, one was to include a more "robust research component" (Reinhard, Osburg and Townsend 2010, p.2). Therefore the university was encouraged to change from a diploma-only university to one that offers a variety of bachelor degree programmes in the fields of business, engineering, social work and research. Executive master programmes will also be initiated by the year 2011.

There are number of definitions or models of WIL. Martin (1997) tabulates a variety of these which are widely used globally and range from pre-course experience, sandwich courses, co-operative programmes, cognitive apprenticeship or job shadowing, joint industry courses, new traineeship and apprenticeships, placements or practicum, fieldwork and post course internship. The DHBW University models WIL according to Jarvis and Wilson (1999) where the university and the related industries work closely together to provide programmes where the students alternate between attending classes at university and industry. There is an overlap between what Martin (1997) refers to as a co-operative and sandwich programme models at the DHBW University.

The DHBW University allows students to alternate work and industrial experience several times before graduation. Students alternate every three months between industry and university. The objective of this teaching method is to provide a strong correlation between theoretical and practical phases of studies to allow students to experience in reality the principles mentioned in the classroom.

The DHBW University currently works with over 10,000 organisations of varied sizes representing diverse disciplines. These organisations comprise both national and subsidiaries of international organisations and work with the University as partners. The majority of these organisations are located in the state of Baden-Württemberg.

Heinemann (1988) states that in 1984 the Cooperative Education Incorporating Internships Association commissioned an investigation regarding the position of WIL in higher education. Although the growth and success of WIL were undisputed, the concern of the association was that the programmes being offered would not be seen as intellectual and did not provide the level of standard expected of a university. Du Pre (2009) concurs with Heinemann (1988) and is of the view that in some instances it was thought that WIL programmes were equivalent to high school curricula.

Du Pre (2009) is of the view that WIL programmes will achieve its rightful place in education when there is a shift in the mind-set of the core academic functions of universities. In the past, the focus was on academic quality, which was determined by the intake of students, while faculty quality was measured by resources and facilities. He believes that this focus should be on student performance and productivity. Perhaps he sees this because measuring student outcomes according to industry standards is not regarded by traditional academics as intellectual. Zegwaard, Coll and Hodges (2003) argue that the content taught at co-operative universities focuses more on the discipline's technical content at the expense of theoretical principles. From the literature above, it can be inferred that the qualifications from co-operative universities can be underrated because WIL does not prepare the student for work life. Could this be because curricula were dictated by industrial needs and not necessarily by what traditional academics deem as relevant to the discipline? Designers of WIL programmes are mindful whether professors, employers, and particularly students, in this globalised world want to know if they are able to apply their knowledge when working in a global market (Dawson, 1989). This is evident in certain departments at DHBW Universities, where students tend to select programmes according to whether the programmes are being internationalised. Other views are that education authorities do not regard WIL as a learning tool, the design of WIL programmes are vague and inappropriate, and industrial practitioners do not want to see themselves as educators (Eames & Cates, 2004). There are perceptions that the student's soft skills around behavioural development and culture have been neglected.

On the contrary, Harvey, Geall and Moon (1998) are of the view that WIL academic programmes can be very beneficial when they are strategically planned, organised and presented by an accredited institution. Zegwaard, Coll and Hodges (2003) indicate that combining the connection of student`s technical and behavioural skills is necessary. In addition, according to Eames (2000) the WIL curriculum should include content that students should learn about cultural norms, i.e. norms of practice and norms of behaviour at their workplace as part of their assimilation into a new community of practice. Students believe that intercultural competencies should also be included in assessments at the university and at the partnering organisations.

WIL programmes should develop learning outcomes of the programme and include the views and needs of students, employers, lecturers and professors. This is not evident as yet at the DHBW University. Perhaps the reason for this is that the Bologna Process (which will be discussed later) is still in its infancy (Zegwaard, Coll & Hodges, 2003). Research reports have identified employer's views on graduate competencies (Coll, Zegwaard & Hodges 2002). These views are that soft skills are as important as the content taught in IT and will influence the competencies required for effective performance. These reports show that according to employers graduates should have both cognitive (content) and behavioural (social skills) competencies

The IT programme was introduced in 1998 by the Department of Information Technology as an undergraduate offering. Students must be in a possession of a matriculation pass before attempting to register for this qualification. They must also sign a contract with standardised clauses with an organisation. This is a pre-requisite for enrolment at the university. The student will graduate with a Bachelor of Science degree.

The programme uses information system technologies to support, optimise and apply processes in different disciplines such as production planning, and controlling, image processing in traffic monitoring, business management and engineering.

In the past, the university did not have to adhere to a structured programme plan as presented in Figure 1 but was given broad learning disciplines of study such as law, economics and IT, to name a few. The head of department had the academic freedom to select content considered suitable. No approval from "curriculum designing committees" was required from accreditation organisations. Perhaps, this gave the IT Department at DHBW University a competitive edge as they were free to change content as the IT discipline rapidly evolved. Sometimes subject content taught was dependent on the software available or sponsored by partnering organisations or major software industries.

However, according to the Bologna Process, the programme had to be restructured because the European Union decided to standardise all Bachelor and Master Programmes in all member countries by the year 2010. The purpose of the Bologna Process (or Bologna Accords) was to make academic degrees and quality assurance standards more comparable and compatible throughout Europe. This was in accordance with the Lisbon Recognition Convention and the signing in 1999 of the Bologna declaration by Ministers of Education from 29 European countries.

The Bolgna Process required a structured curriculum that presented a detailed description of the learning module outcomes, the credits achieved for each learning module as well as the number of contact hours with students and their workload. The academic qualification of the lecturers as well as their experience in industry is also specified. The students also need to complete two projects in the prescribed format that is given to them by the university. This system has been fully implemented by the DBHW.

It is acceptable that organisations from other parts of Germany choose to contract and educate students at the DHBW University in the state of Baden-Württemberg. Organisations and social institutions play an essential role within the DHBW educational system. They are equal partners to the state-run DHBW universities on all decision-making committees. All partnering organisations that pay for the education of the students exert a strong, ongoing influence on curricula in the programme. For organisations to be accepted as partners of the co-operative learning system, they must fulfil a number of requirements such as size of the organisation, number of qualified personnel and accessibility of state-of-the-art training facilities. When they have fulfilled these requirements, they are certified as training organisations. The curriculum of the DHBW University's courses is matched to the skills required by industry. This match makes the students very attractive and marketable for employment after graduation (Reinhard, 2006a).

This full study co-operative education model benefits the organisation, the university and the student. The university cooperates with various organisations. This cooperation also provides the organisations with a competitive advantage in search for good employees. It shows the attractiveness of the university in the market place. Starting to lay the groundwork of cooperation with future employers significantly raises the profile of the university (Waterkamp & Reinhard, 2006).

Although the programme has numerous strengths, some of the weaknesses are: increased workloads for students and too intense curricula, thereby making the theoretical phase cramped. Practitioners and professors present lectures in the face- to-face teaching style which is contrary to teaching in the preferred problem solving style. Also, although interdisciplinary learning was possible in the past, the learning modules are now so rigidly defined that there is no room for interdisciplinary learning and teaching. In some instances, the projects conducted in industry cannot be completed within the time required. To overcome these challenges the university has adopted the following:

- Due to the compact workloads, topics are selected strategically between the lecturers

Figure 1. Structure of programmes

Structure of the programme

The structure programme is as follows:

Group of Modules	1st Academic Year		2nd Academic Year		3rd Academic Year		CP*
Practicals in Company	**Practical Module I**		**Practical Module II**		**Practical Module III**		60
	Students learn first working and problem-solving methods based on IT systems. They get an overview of the IT infrastructure in the company and learn to work in teams.		Students apply tools, techniques and methods of the the study content to real-life problems in the company. They learn to solve conflicts in teams.		Students learn to analyse and manage the complex relationships between business processes and IT systems. They plan and manage smaller IT projects in the company.		
General Study Contents							**125**
Methods of Business Informatics	Introduction to Business Informatics	System Analysis	Case Study System Analysis	Project Management	Management of Business Processes Project	E-Business CRM Business Intelligence	25
Information Technology	Introduction to Computer Systems	Communication Systems	Computer Architecture and Operating Systems	Database Technology	IT-Management/Security/ITIL		13
System Engineering	Programming I	Programming II Algorithms and Data Structures	Database Design and Programming	Web Engineering Media Design	Service-Orientated Architecture	Knowledge-Based Systems	24
Mathematics Statistics	Mathematics	Logistics and	Statistics	Operations Research			8

and the head of department to ensure that only value-adding content is delivered

- Lecturers are assessed at the end of every trimester. This assessment focuses mainly on the teaching style of the lecturer and his/her ability to present lectures with an interdisciplinary approach. Lecturers who are unable to present in the problem-solving or interdisciplinary style are removed from his/her position.
- Students, who have not completed their projects in the designated time, carry their work to the next semester. During this semester their projects are completed outside their normal working hours.

Strengths of the programme are as follows: Exposure to industry has increased the maturity and professionalism of the students. The students have improved their social skills, are more adaptable to changing situations, are very conscious of adhering to time-frames, have exposure to multi-cultures and languages and have opportunities to work overseas.

Additional benefits to these students are first-class practical education and networking opportunities. Almost 90% of the students are employed before they leave the university and the partnering organisation that facilitates the student's employment is highly regarded among industry. In addition, the university also improves the quality of teaching by offering lectures from experts from the partnering organisations. This might be considered beneficial to students and sometimes to lecturers and professors who are constantly confronted and challenged with new developments outside their schooling. From a financial perspective, the university and students also receive direct funding for necessary expenses, that is, material, travel or services needed from the State of Baden Württemberg. Funding obtained from various organisations also helps to maintain projects that require additional external costs (Reinhard, Osburg & Townsend, 2008).

Students experience additional academic benefits such as improved learning, motivation and problem solving. Furthermore, they will have a greater understanding of suitable career choices in terms of clarity, decision-making, planning and salaries. Progress in the student's career is fast due to the work experience they were exposed to. Personal benefits such as increased autonomy (salary), enhanced self confidence (have a business card as a student), improved organisational learning, increased ability to set plans and priorities can perhaps be attributed to the tight schedule during their theoretical phase and the requirements which they have to fulfil within a 3 month period during their practical phase. In addition, work skill development focuses on understanding the demands of industry, learning job requirements and assuming a functional role which often leads to an international focus. Partnering organisations also satisfy their corporate social responsibility aims by employing students.

The DHBW acknowledges that the programmes are demanding on the students, but does not want to change their stance. This perhaps is due the 210 credit points which are assigned to this degree as opposed to the 180 credits points assigned to the "traditional" bachelors' degrees.

To improve the viability of the programme, the IT Department has introduced learning outcomes to all the internal and external lecturers to help strengthen the programme and standardise work practices. These outcomes are designed to enhance the professional, methodological, social and self competence of the student. The IT Department nurtures professional competence in solving problems, in acting intellectually and in being physically self-organised. In this way problems are dealt with and solved, knowledge is classified and evaluated and solutions are found and implemented.

The graduates of the IT Department acquire knowledge in informatics and business administration. They analyse business scenarios and processes from the point of view of business and define adequate IT concepts to transform information systems. As a result of their academic and practical experiences, students have a broad, solution-oriented knowledge. This enables them to step up to the challenges of international business as employees or consultants.

The IT Department graduates have a good command of analysing, modelling and conceptualizing methods of IT. As a consequence of the high speed of innovations in the IT discipline, these graduates have to work on developing new methods and concepts. This realization is practiced and enhanced during project work in which students develop the necessary knowledge in project management and the soft skills such as the ability to communicate, the ability to work in a team and the ability to motivate for the success of the project. The graduates also gain a broad knowledge in scientific methods and procedures. They have gained the ability to obtain and to retain the information efficiently. Moreover, they get the ability to develop practice-oriented solutions and methods regarding the problem. They gather mathematical and statistical knowledge as well as analytical abilities which allows them to solve complex duties efficiently. The graduates are able to communicate in foreign languages, both linguistically and written. Moreover, they are able to work together with professionals of other areas.

The graduates develop social competence by the ability to act in a communicative and cooperative way. They are able to work as a team and handle intended or unintended actions with high emotional competence. The graduates are characterised by a significant personal maturity, professional expertise, colleagueship and the ability to make decisions. They are aware of social responsibilities, are prepared to assume personal responsibility, and are able to adapt and accept criticism. They demonstrate self-competence in their ability to evaluate and act in a reflective (self-evaluating) manner appropriate to situations which emerge. Their exposure to a working environment helped them to develop their own abilities to foster a productive position, values, motives, self-perceptions, achievements and to continuously improve them.

Problems and Controversies to WIL and Solutions and Recommendations

In times when most universities are confronted with a decline in funding and new technologies are emerging that demand a substantially higher investment, educational institutions are being forced to find new financing solutions (Reinhard, Osburg & Townsend, 2007). This subject is even more relevant for WIL programmes, where the university has to balance the conflicting demands of industry and its pedagogical needs. Particularly WIL IT programmes are in a controversy when trying to meet the requirements of industry.

Problems occur when the partnering organisations of the university sponsor hard and software. In return, the organisations want to restrict the curriculum to meet their own criteria instead of those of the academic institution and its learners. Students cannot afford to be restricted in terms of their employment opportunities; therefore, the IT curricula need to facilitate the entry of the learners into a variety of industries, some of which may be using different software as that of the sponsoring organisations (Reinhard, 2006b). Therefore, the new structure of the IT curriculum could be regarded as advantageous because it prevents the dominance of one organisation in the programme. The university needs to state clearly to the sponsoring organisations that they are bound by the curriculum. In addition, academic staff may be reluctant to design work integrated curricula which only answer the immediate needs of industry without developing the breadth of a traditional university education.

One way to overcome financial problems, according to Reinhard, Osburg and Townsend (2010), is to increase industrial sponsorships with a view for organisations to satisfy their corporate social responsibility.

Adherence to the curriculum is monitored. The programme was accredited in 2006 by the department of education and has since undergone several reviews annually. Reviews are conducted by the accreditation body, the education department and selected industries in Germany. Zentrale Evaluations- und Akkreditierungsagentur (ZEvA) is the accreditation body recognized to accredit programmes. This accreditation body has been in existence since 1998. There are three accreditation options. The first option called the "Quality Audit" evaluates the entire system of the university. The second option is a programme evaluation. In this instance only the programme gets evaluated. Third option evaluates the delivering a topic. The DHBW complies with option one only. The last review was conducted in October 2009. The following improvements were suggested as recommendations from this review. The accreditation body suggested that the student assessments should demonstrate a transfer of knowledge gained from both the class-room and industry. The body also commended and encouraged the presentation of lectures and assessments in the English language as this embraced globalisation. The department of education advised that the number of lecture hours should be reduced and that students should spend more time engaged in self-study. They also requested that the students be exposed to more international lecturers and that students conducting research should be supervised by promoters with doctorate qualifications. Management from industry suggested that students be taught by lecturers who have previous industrial experience. Some of these recommendations have been implemented while the others will be introduced in due course.

Lectures are presented in a variety of formats such as face-to-face, group work, presentations, self-study and field projects. An industrial and academic mentor is assigned to monitor each student. In addition, guest lecturers with expert knowledge and international lecturers conduct classes at various times on the programme. Lectures are facilitated to present both theoretical and practical aspects of the subject. Students are exposed to innovative teaching practices such as game-theory, simulations, dvd's, videos, role playing, inter-university competitions, exhibitions and social media. Furthermore, classes are also supported by the information technology department's Learning Management System (ELI), WIKIS, forums and instant messaging for communication and collaboration with students. Topics range from Project Management, Information Management, Business Administration, Intercultural Communications and Management, Mathematics, Statistics and Languages. Students are also required to conduct a project in the third year of their studies. This project is related to a specific problem in the organisation, is conducted in collaboration with that organisation and is presented as a dissertation.

The programme is supported by German students mainly from the Ravensburg area and to a smaller extent from surrounding areas such as Stuttgart and Bavaria. There is an eighty percent male to twenty percent female complement in the programme. The programme does not currently have any students with disabilities. However, a challenge posed was that industry preferred male students to female students. This choice cannot be generalised to all departments in the University. Preferences from organisations are usually based on language skills and grades in high school. Often when it comes to employing IT students, Human Resource recruiters believe that male students have better IT skills than female students. Male students tend to apply for these jobs whereas female students tend to apply for programmes where language skills are needed. Some departments prefer to be part of the recruiting process with organisations.

Another controversy arises when students are asked to conduct project work on behalf of their organisations, which is outside the scope of the academic requirements. In some instances the students request for remuneration. To create a win-win situation for both students and the organisation, the University suggests that the student be issued with a certificate to demonstrate competence in additional work. Additional work can be beneficial if it helps the students develop other skills and self management. Cases have been noted where students were asked to conduct project work for their organisations which was outside the scope of their content skills, and in some instances students could not attend the compulsory academic session. It is suggested that when these requests arise, expert academics from the relevant departments assess the new project and identify and match the activities to the outcomes of the curriculum.

In many instances, students feel that they have not been given sufficient responsibility for learning to have taken place. This has been rectified by DHBW University by evaluating the learning within the partnering organisation. Therefore the DHBW University has designed a questionnaire to evaluate the learning outcome of the student's practical phase.

FUTURE RESEARCH DIRECTIONS

The success of the university has led to a number of improvement strategies, such as offering executive master's degrees, expanding on co-operative research projects and making industry partners' part of the DHBW University board.

The focus of the DHBW University has expanded to include research at higher levels. It is in the process of introducing and encouraging research at higher academic levels, such as the executive masters. There are two points of departure about research at the University. Firstly, the academics are embarking on extensive research projects to improve their knowledge in their disciplines. Secondly, the University is conducting research on the subject of WIL with a view to provide the best possible service. Bartkus and Stull (2004), Wilson (1998), Ricks et.al, (1990) and Weaver (1993) are of the view that a move towards higher levels of research is bound to keep WIL programmes highly acclaimed by higher education authorities. Developing higher levels of research can be regarded as innovation which Reinhard, Osburg and Townsend (2008) believe will enable the sustainability of WIL institutions. It is observed that more traditional universities from South Africa, Australia, United States of America, Canada and South America who are knowledge partners to the DHBW University are not only introducing WIL programmes, but are also benchmarking their programmes against the University.

Research in co-operative education focuses on the problem inherent in bridging different cultures, namely the academic culture and the business culture. The perceived shortcoming of the academic culture is related to its focus on theory instead of practice. Students in WIL education must learn to differentiate between the various values, behaviours and attitudes of the two cultures. The significance of this transition is highlighted by Eames (2003). Such issues become even more prominent when the internationalization of industry and bringing in cultural learning outcomes in WIL programmes are considered (Reinhard, Satow & Sisco, 2007).

The DHBW University offers the world to the students in the form of international lecturers, academic exchange programmes and educational partnerships which link them to the global market. Information Technologies are important methodical means to support and optimise processes in different fields and for the development of new technical methods. The applied IT technologies comprise both software (user and system programs) and interconnected systems. Therefore, it is necessary for students in this field to deal intensively with applications and computer science, programming, databases, networks, software and system technology.

The learning experience of students at the DHBW is different from other IT programs in that they are exposed to e-learning. There is a learning platform as well as a WIKI. These are designed to form a virtual community of practice where students have an option to decide for example, what a guest professor from a foreign country should teach them in accordance to their interests or trends. They would discuss this among the group of students and after a consensus is reached, the head of department is informed of their choice.

Project-based learning is also part of our curriculum which is largely oriented towards competence-based learning. Because of the practical environment of the programmes offered at the University, students gain a deeper appreciation of theory which they can use to improve their professional skills. Thus, disciplinary and professional skills are gained concurrently. Together with organisations, students are assessed on their skills, capabilities, and abilities. Each semester, after spending 3 months at the University, students return to their respective organisations where they are able to immediately apply their theoretical knowledge. These out-of-the-classroom programmes expand the learning opportunities for students and build relationships across the various subjects of the curriculum.

Tests and examinations are the main forms of assessment. Assignments, presentations, projects, group projects and a work related dissertation are the other forms of assessment. These may be written or oral and may be assessed by the academic lecturer or the mentor from industry. Assessments are moderated by any of or all of the following: university lecturer, external lecturer or open-university before and after being administered.

There is a perception of an overload of information in the academic component of the programmes. This overload of information can be overcome by introducing more project work and simulations that covers a large module. Another option to reduce the volume of work is to replace theory via a student excursion to a related organisation and to acquire the knowledge by studying its processes. Various organisations operate differently and some students get a better insight of outcomes than at others. Also, some experts out of industry are not always available for teaching and do not like to contract with the DHBW; therefore, it may be better to visit them at their workplace. The length of the programme should not be extended to accommodate for the volume of work. Students would lose their advantage of "being so young and available on the labour market". An increase in the length of the programme will have financial implications for students, parents and organisations.

CONCLUSION

Although students have developed a good sense of professional, methods, social and self confidence in the past, the main focus of the Information Technology Department is to continually strive to improve its performance. The orientation in export as well as the use of new information and communication appliances within of the German economy, the increasing Europeanization of Germany and the neighbouring countries as well as the global economic integration leads to extreme changes in the working field of the organisation. Both large and medium sized industries and service organisations need more practice-oriented graduates with the focus on IT. This was the rationale in introducing an IT department at the DHBW University in Ravensburg.

The contents of the information technology programme have been supplemented by special features of rhetoric and presentation techniques as well as consulting, conference management and languages. The main focus is on establishing the necessary abilities like expertise, methodological competence, linguistic skills, soft skills and intercultural competence. This includes dispositive skills such as planning, control, communication, mobility and intercultural flexibility.

After graduation, the graduates of the IT Department have the ability to support the management of their international oriented organisations. Moreover, they are able to assume responsible duties, to act as an adviser, free-lance or become an independent corporate.

REFERENCES

Bartkus, K. R., & Stull, W. A. (2004). Research in cooperative education. In Coll, R., & Eames, C. (Eds.), *International handbook for cooperative education* (pp. 67–78). Boston, MA: WACE Int. Secretariat.

Coll, R. K., Zegwaard, K., & Hodges, D. (2002). Science and technology stakeholders' ranking of graduate competencies, part 1: Employers' perspective. *Asia Pacific Journal of Cooperative Education*, *3*(2), 19–28.

Dawson, J. D. (1989). Educational outcomes for students in cooperative education. *Journal of Cooperative Education*, *25*(2), 6–13.

Du Pre, R. (2009). *The place and role of universities of technology in South Africa*. Bloemfontein, South Africa: Durban University of Technology.

Eames, C. (2000). A comprehensive database of New Zealand cooperative programs. *Asia Pacific Journal of Cooperative Education*, *1*(1), 1–2.

Eames, C. (2003). Learning to work: Becoming a research scientist through work experience placements. *Asia Pacific Journal of Cooperative Education*, *4*(2), 7–15.

Eames, C., & Cates, C. (2004). Theories of learning in cooperative education. In Coll, R., & Eames, C. (Eds.), *International handbook for cooperative education* (pp. 37–47). Boston, MA: WACE Int. Secretariat.

Harvey, L., Moon, V., & Geall, S. (1997). *Graduates work: Organisational change and students' attributes*. Birmingham, UK: Centre for Research into Quality (CRQ) and Association of Graduate Recruiters (AGR).

Heinemann, H. (1988). Cooperative education and the academy. *Asia Pacific Journal of Cooperative Education*, *24*(2-3), 109–119.

Jarvis, P., & Wilson, A. L. (1999). *International dictionary of adult and continuing education* (2nd ed.). London, UK: Kogan Page.

Martin, E. (1997). *The effectiveness of different models of work-based university education*. Department of Employment, Education, Training and Youth Affairs. J. S. McMillan Printing Group.

Müller, B. (2009). *Cooperative state university campus* (p. 3). Ravensburger Stadtmagazin.

Reinhard, K. (2006a). Berufsakademie in Germany: Can this work-integrated learning program serve as an effective model in higher education institutions in western and eastern countries? *Asia Pacific Journal of Cooperative Education*, *7*(1), 16–21.

Reinhard, K. (2006b). The role of business and industry in supporting education: Case studies from the United States and Germany in educational sponsorship–an option for IT-educational funding. *Journal for Global Business Education*, *6*, 43–50.

Reinhard, K., Osburg, T., & Townsend, R. (2008). The sponsoring by industry of universities of cooperative education: Case study in Germany. *Asia Pacific Journal of Cooperative Education*, *9*(1), 1–13.

Reinhard, K., Osburg, T., & Townsend, R. (2010). Introducing corporate social responsibility as component in cooperative education: Results from a student research project in Germany and the United States supported by Intel Corporation. *Asia Pacific Journal of Cooperative Education*, *11*(1), 1–12.

Reinhard, K., Satow, L., & Sisco, L. (2007). Internationalizing cooperative education: An introduction to the Berufsakademie Ravensburg's work integrated partnership with the multinational company SAP Incorporation. *Asia Pacific Journal of Cooperative Education*, *8*(2), 109–119.

Ricks, F., van Gyn, G., Branton, G., Cut, J., Loken, M., & Ney, T. (1990). Theory and research in cooperative education: Practice implications. *Asia Pacific Journal of Cooperative Education*, *27*(1), 7–20.

Waterkamp, D., & Reinhard, K. (2006). System of industry sponsorship in education. In Bagusat, A., & Hermanns, A. (Eds.), *Management-handbuch Bildungssponsoring* (pp. 111–126). Berlin, Germany: Erich Schmidt Verlag.

Weaver, K. A. (1993). Research and employer oriented learning objectives. *Journal of Cooperative Education*, *28*(3), 7–11.

Wilson, J. W. (1988). Research in cooperative education. *Journal of Cooperative Education*, *24*(2-3), 77–89.

Zegwaard, K., Coll, R. K., & Hodges, D. (2003). Assessment of workplace learning: A framework. *Asia Pacific Journal of Cooperative Education*, *4*(1), 9–18.

Compilation of References

<eref>Clark, D. (1999). *Constructivism.* Retrieved April 4, 2006, from http://www.nwlink.com/~donclark /hrd/ history/ constructivism.html</eref>

<eref>James, R., & Beckett, D. (2001). *The changing expectations of university students and the implications for learning*. 1-6. Retrieved from http://www.cshe.unimelb.edu.au/ people/staff_pages/ James/James&Beckett= Singapore.pdf</eref>

Abbott, J. (2004). *Report to James Goldston, Faculty of Engineering and Physical Systems: A strategic review of the postgraduate maintenance management courses.* Unpublished consultancy report.

Abeysekera, I. (2007). Issues relating to designing a work-integrated learning program in an undergraduate accounting degree program and its implications for the curriculum. *Asia-Pacific Journal of Cooperative Education, 7*(1), 7–15.

ACCED-I. (1994). *Statement of professional standards and ethics.* Retrieved May 2, 2006, from http://accedi.colostate.edu/ imis_web/staticcontent/3/ standardsandethics/standards.htm

ACCI/BCA. (2005). *Employability skills for the future.* Retrieved from http://www.acci.asn.au/text_files/ issues_papers/Employ_Educ /ee21.pdf

ACER. (2001). *Graduate skills assessment: Summary report.* Canberra, Australia: Department of Education, Training and Youth Affairs.

ACS. (2007). *About us.* Retrieved April 9, 2007, from http://www.acs.org.au/index.cfm ?action=show&conID=aboutacs

ACS. (2010). *About us.* Retrieved April 9, 2010, from http://www.acs.org.au /index.cfm?action=show &conID=aboutacs

Ahrens, C. (2000). European Civil/ Construction Engineering Management (ECEM/ ECM) as an example for integrated European engineering curriculum. *Proceedings of the International Conference of SEFI.* Paris.

Alam, M., Gale, A., Brown, M., & Kidd, C. (2008). The development and delivery of an industry led project management professional development programme: A case study in project management education and success management. *International Journal of Project Management, 26*(3), 223–238.

Albrecht, A. (2006). *Enriching students learning through blended learning.* Educause Centre for Applied Research. *Research Bulletin (Sun Chiwawitthaya Thang Thale Phuket), 2006*(12). Retrieved from http://net.educause.edu/ir/ library/pdf/ERB0612.pdf.

Alexander, F. K. (2000). The changing face of accountability. *The Journal of Higher Education, 17*(4), 411–431. doi:10.2307/2649146

Allen, I. E., Seaman, J., & Garrett, R. (2007). Blending in: Extent and promise of blended learning in the United States. Retrieved July 24, 2010, from http://www.sloan-consortium.org/sites /default/files/Blending_In.pdf

Anderson, M. J. (2006). Degree of fit: University students in paid employment, service delivery and technology. *Australasian Journal of Educational Technology, 22*(1), 88–103.

Andresen, L., Boud, D., & Cohen, R. (2000). Experience-based learning. In Foley, G. (Ed.), *Understanding adult education and training* (2nd ed., pp. 225–239). Allen & Unwin.

ANECA. (2006). *Criterios e indicadores de calidad para la autoevaluación de las unidades de empleo que gestionan las universidades*. Agencia Nacional Para la Evaluación de la Calidad. Retrieved February 01, 2010, from http://www.aneca.es/media/ 158226/estu_oil_fase3b_criteriosindicadores.pdf

ANECA. (2007). *Reflex. Informe ejecutivo. El profesional flexible en la sociedad del conocimiento*. Agencia Nacional Para la Evaluación de la Calidad. Retrieved February 01, 2010, from http://www.aneca.es/media/ 158162/informeejecutivoaneca_ jornadasreflexv20.pdf

Anthes, G. (2004). Quality model mania. *Computerworld Management*(March).

Argy, P. (2006). Professionalism is a matter of practice not just attitude. *Information Age*, 2.

Argyris, C., & Schön, D. A. (1978). *Organizational learning: A theory of action perspective*. Reading, MA: Addison-Wesley.

Argyris, C., & Schön, D. A. (1976). *Theory in practice: Increasing professional effectiveness*. San Francisco, CA: Jossey-Bass Publishers.

Armitt, J., & Baldwin, R. (Eds.). (2008). *Innovation process in the construction sector*. UK: CICE & University of Loughborough.

Armsby, P., Costley, C., & Garnett, J. (2006). The legitimacy of knowledge: A work based learning perspective of APEL. *Journal of Lifelong Education, 25*(4).

Asavasanti, S., Bhumitratana, S., & Chayawattana, W. (2005). *FEPS: A successful story of constructionism learning approach in Thailand*. Paper presented at ICLORD2005: 19th International Conference, Bangkok, Thailand.

Assiter, A. (Ed.). (1995). *Transferable skills in higher education*. London, UK: Kogan Page.

Atherton, J. (1999). Resistance to learning: A discussion based on participants in in-service professional training programmes. *Journal of Vocational Education and Training, 51*(1), 77–90. doi:10.1080/13636829900200070

Australian Association of Graduate Employers. (2008). *Employer survey*. Melbourne, Australia: AAGE.

Australian Library Association. (2008). *What is PD?* Retrieved February 18, 2010, from http://www.alia.org.au/education /pd/description.html

Australian Physiotherapy Association. (2005). *The health workforce*. Melbourne, Australia: Australian Physiotherapy Association.

Back, A., Bendel, O., & Stoller-Schai, D. (2001). *E-learning in Unternehmen. Grundlagen - Strategien - Methoden - Technologien*. Zurich, Switzerland: Orell Flüssli.

Balanced scorecard Institute. (2010). *Official website for BSC*. Retrieved from http://www.balancedscorecard.org /basics/bsc1.html

Barrett, A., Kearney, I., Goggin, J., & Conefrey, T. (2009). *Quarterly economic commentary – Winter 2009*. Dublin, Ireland: ESRI.

Bartkus, K. R., & Stull, W. A. (2004). Research in cooperative education. In Coll, R., & Eames, C. (Eds.), *International handbook for cooperative education* (pp. 67–78). Boston, MA: WACE Int. Secretariat.

Bartlett, A., & Mercer, G. (Eds.). (2001). *Postgraduate research supervision: Transforming (r)elations*. New York: Peter Lang.

Bates, A. (2000). *Managing technological change: Strategies for college and university leaders*. San Francisco, CA: Jossey-Bass.

Bates, M. (2003). The assessment of work integrated learning: Symptoms of personal change. *Journal of Criminal Justice Education, 14*(2), 303–326. doi:10.1080/10511250300085811

Beach, K. (1999). Consequential transitions: A sociocultural expedition beyond transfer in education. In Iran-Nejad, A., & Pearson, P. D. (Eds.), *Review of research in education* (*Vol. 24*, pp. 101–140). Washington, D.C.: American Educational Research Association.

Beard, C., & Wilson, J. P. (2002). *The power of experiential learning: A handbook for trainers and educators*. London, UK: Kogan Page.

Becher, T. (1996). The learning professions. *Studies in Higher Education, 21*(1), 43–55. doi:10.1080/03075079612331381447

Bellinger, G. (2004). Knowledge management-emerging perspectives. Retrieved March 5, 2007, from http://www.systems-thinking.org /kmgmt/kmgmt.htm

Bendrich, G., & Pugsley, T. S. (1998). The practical side of engineering at the University of New Brunswick. *Chemical Engineering Education, 32*(3), 208–213.

Berger, A. A. (1998). *Media research techniques* (2nd ed.). Thousand Oaks, CA: Sage.

Berinato, S. (2004, December 16). Are you on board with enterprise risk management? You had better be. It's the future of how businesses will be run. *CIO*.

BIHECC. (2007). *Graduate employability skills*. Canberra, Australia: DEST.

Billett, S. (2001). *Participation and continuity at work: A critique of current workplace learning discourses. Context, power and perspective: Confronting the challenges to improving attainment in learning at work.* Paper presented at the Joint Network/SKOPE/TLRP International workshop, Northhampton.

Bioinformatic. (2010). *About Bioinformatic*. Retrieved February 6, 2010, from http://www.bioinformatics.kmutt.ac.th/about.php

Birenbaum, M. (1996). Assessment 2000: Towards a pluralistic approach to assessment. In Birenbaum, M., & Dochy, F. J. R. C. (Eds.), *Alternatives in assessment of achievements, learning processes and prior knowledge* (pp. 3–29). Boston, MA: Kluwer Academic Publishers.

Birenbaum, M. (2003). New insights into learning and teaching and the implications for assessment. In Segers, M., Dochy, F. J. R. C., & Cascallar, E. (Eds.), *Optimising new modes of assessment: In search of qualities and standards*. Dordrecht, The Netherlands: Kluwer Academic Publishers. doi:10.1007/0-306-48125-1_2

Birla Institute of Technology. (2010). In *Wikipedia, the free encyclopedia*. Retrieved February 13, 2010, from http://en.wikipedia.org/w/index.php ?title=Birla_Institute_of_Technology &oldid=342663877

Blakely, P. N., & Tomlin, A. H. (Eds.). (2008). *Adult education: Issues and developments*. New York, NY: Nova Science Publishers, Inc.

Bloom, B. S. (1956). *Taxonomy of educational objectives, handbook 1: The cognitive domain*. New York, NY: David McKay Co.

Boadle, M. (2005). Adult learning. *ACSLearn* Retrieved April 7, 2006, from http://www.acs.openlab.net.au/ content.php?article.82

Bologna Process. (1999 - 2010). *The Bologna process and Declaration: Trends in learning structures in higher education.* Retrieved from http://www.ond.vlaanderen.be/ hogeronderwijs/bologna/ search/index.htm?cx=013531861688426506363%3 Afnpyfltxhbu&cof=FORID%3A11&q=Trends+in+Learning+Structures+in+Higher+Education&sa= Zoeken&siteurl=www.ond.vlaanderen. be%2Fhogeronderwijs%2Fbologna%2Fsearch%2F#1664

Boud, D., & Middleton, H. (2003). Learning from others at work: Communities of practice and informal learning. *Journal of Workplace Learning, 15*(5), 194–202. doi:10.1108/13665620310483895

Boud, D., & Lee, A. (Eds.). (2009). *Changing practices of doctoral education*. London, UK: Routledge.

Boud, D., & Falchikov, N. (2005). Redesigning assessment for learning beyond higher education. [Sydney, Australia: HERDSA.]. *Proceedings of the HERDSA Conference, 2005*, 34–41.

Boud, D., & Solomon, N. (Eds.). (2001). *Work-based learning: A new higher education?*Buckingham, UK: SRHE and the Open University Press.

Bourner, T., Katz, T., & Watson, D. (Eds.). (2000). *New directions in professional higher education*. Buckingham, UK: SRHE and Open University Press.

Bova, B. M., & Philips, R. E. (1981). *The mentor relationship, a study of mentors and proteges in business and academia*. (ED 208 233)

Bradney, A. (1992). Ivory towers and Satantic mills: Choices for university law schools. *Studies in Higher Education, 17*(1), 5–20. doi:10.1080/03075079212331382726

Brown, J. S., Collins, A., & Duguid, P. (1989). Situated learning and the culture of learning. *Education Researcher, 18*(1), 32–42.

Brown, J. S., & Duguid, P. (1991). Organizational learning and communities of practice: Toward a unified view of working, learning, and innovation. *Organization Science, 2*(1), 40–57. doi:10.1287/orsc.2.1.40

Bruner, J. (1996). *The culture of education* (pp. 224 + xvi). Cambridge, MA: Harvard University Press.

Buarod, P. (2009). *An application of statistical models to the significance of admission criteria for the chemical engineering practice school (ChEPS)*. Unpublished master's thesis, King Mongkut's University of Technology Thonburi, Bangkok.

Butcher, D. C. A., & Sheehan, M. (2010). Excellent contractor performance in the UK construction industry. *Engineering, Construction, and Architectural Management, 17*(1), 35–45. doi:10.1108/09699981011011302

California State Department of Education. (1983). *California mentor teacher program, program advisory.* (ED 241 473).

Calway, B. A., & Murphy, G. A. (2007). The educational imperatives for a work-integrated learning philosophy. *Journal of Cooperative Education and Internships, 41*(2), 95–109.

Calway, B. A. (2005). *Rethinking a learning environment strategy.* Deakin, Australia: Geelong.

Calway, B. A., & Murphy, G. A. (2006). *WIL theory or philosophy? - A grounded study*. Swinburne University of Technology, Working Paper.

Cámara, J. M., Sánchez, P. L., & Represa, C. (2009). Las prácticas en empresa en el marco del EEES. *Proceedings of the XVII Congreso Universitario de Innovación Educativa en las Enseñanzas Técnicas*. Valencia.

CAMM2010. (2010). *Proceedings of the Conference on Asset and Maintenance Management 2010*, Gladstone, Australia. Retrieved 29 July, 2010, from http://research.cqu.edu.au/ FCWViewer/getFile.do?id=32434

CAPPE. (2006). *Professionalism, ethics and integrity systems: The promotion of professional ethical standards and the protection of clients and consumers*. Melbourne, Australia: Centre for Applied Philosophy and Public Ethics.

Catts, R. (2004). *Lifelong learning and higher education-reflections and prospects.* Keynote address, 3rd International Lifelong Learning Conference, Yeppoon, June 2004.

CEED Program Qld. (2010). *Company-based student training.* Retrieved February 13, 2010, from http://www.corptech.com.au/ home/

Chappell, C., Solomon, N., Rhodes, C., Tennant, M., & Yates, L. (2003). *Reconstructing the lifelong learner: Pedagogy and identity in individual, organisational and social change*. London, UK: Routledge. doi:10.4324/9780203464410

Châtelet, J.-M., & Nuninger, W. (2011). *Quality Seminar for WIL leading to Value Management and Innovation.* XIIIème Congrès de la Société Française de Génie des Procédés, SFGP 2011.

Châtelet., et al. (2006). *The social and educational challenge of work based learning in European higher education and training. Results of a pilot experience. Chapter 5: Quality control*, (p. 153), (Ed. Reinhard Schmidt), Firenze.

Chem, E. (2010). *Practice school-history*. Retrieved February 13, 2010, from http://web.mit.edu/cheme/academics/practice/ history.html

Chivers, G. (2007). Professional competence enhancement via postgraduate post-experience learning and development. *Journal of European Industrial Training, 31*(8), 639–659. doi:10.1108/03090590710833688

Chivers, G. (2006). *Informal learning by professionals in the UK.* Retrieved October 14, 2008, from www.leeds.ac.uk/medicine/ meu/lifelong06/papers/ P_GeoffChivers.pdf

Christl, C., Ghidini, C., Guss, J., Lindstaedt, S., Pammer, V., & Rospocher, M. ... Serafini, L.(2008). Deploying Semantic Web technologies for work integrated learning in industry. A comparison: SME vs. large sized company. In A. P. Sheth, S. Staab, M. Dean, M. Paolucci, D. Maynard, T. Finin, K. Thirunarayan (Eds.), *The Semantic Web - ISWC 2008. Proceedings of ISWS 2009*. (LNCS 5318), (pp. 709–722). Berlin, Germany: Springer.

CHSRF (Canadian Health Services Research Foundation). (2007). *Knowledge transfer and exchange at work*. Retrieved February 28, 2007, from http://www.chsrf.ca/home_e.php

CICA. (2005). *The CA competency map*. Retrieved March 23, 2006, from http://www.cica.ca/index.cfm/ci_id/17150/la_id/1.htm

CIO Executive Council. (2006). *Situation analysis. Council initiatives-Australia*. Retrieved March 15, 2006, from http://www.cioexecutivecouncil.com/ public/au/mentoring.html

Clark, D. (1999). *Andragogy*. Retrieved May 15, 2006, from http://www.nwlink.com/~donclark/ hrd/history/andragogy.html

Clark, D. (2000). *John Dewey (1867-1949)*. Retrieved April 23, 2006, from http://www.nwlink.com/~donclark/hrd/history/history.html

Clough, G. W. (2002). *Defining the technological research university of the 21st Century: The strategic plan of Georgia Tech*. Atlanta, GA: Georgia Institute of Technology, Institute Communications & Public Affairs.

Clough, G. W. (2005). *Strengthening the global competence and research experiences of undergraduate students: A quality enhancement plan*. Atlanta, GA: Georgia Institute of Technology, Institute Communications & Public Affairs.

Cole, M. (2004). Opening university. *New Civil Engineer, March*, 38-40

Coll, R. K., Zegwaard, K., & Hodges, D. (2002). Science and technology stakeholders' ranking of graduate competencies, part 1: Employers' perspective. *Asia Pacific Journal of Cooperative Education*, *3*(2), 19–28.

Coll, R. K., & Eames, C. (2004). Current issues in cooperative education. In Coll, R. K., & Eames, C. (Eds.), *International handbook for cooperative education*. Boston, MA: The World Association for Cooperative Education.

Colley, H., Hodkinson, P., & Malcolm, J. (2002). *Non-formal learning: Mapping the conceptual terrain. A consultation report*. University of Leeds Lifelong Learning Institute. Leeds. Retrieved August 18, 2008, from http://www.infed.org/archives/ e-texts/colley_informal_learning.htm

Collins, A., Brown, J. S., & Holum, A. (1991). Cognitive apprenticeship: Making thinking visible. *American Educator*(Winter).

Connor, H., & MacFarlane, K. (2007). *Work Related Learning (WRL) in HE-a scoping study*. Glasgow, UK: Caledonian University.

Costley, C. (2001). Organizational and employee interests in programs of work based learning. *The Learning Organization*, *8*(2), 58–63. doi:10.1108/09696470110387991

Costley, C. (2000), The boundaries and frontiers of work based knowledge. In D. Portwood & C. Costley (Eds.), *WBL and the university: New perspectives and practices*. Birmingham, UK: SEDA Paper 109.

Cox, S., & King, D. (2006). Skill sets: An approach to embed employability in course design. *Education + Training*, *48*(4), 262-274.

Crebert, G. (1995). *Links between higher education and industry: Workplace-based learning programs in Australia: Practices and issues*. Paper presented at the Keynote address, Workplace-based learning conference.

Crequis, A., Leroy, N., Nuninger, W., Berthoux, A., & Pauwels, J.-P. (2010). *3 partners, 3 frameworks, 3 goals: The WIL to achieve the success of all*. International Conference on Work Integrated Learning, Wace 2010, Hong-Kong, 3-5 February, 5 pages.

Cross, J. (2007). *Informal learning*. John Wiley & Sons.

CSO. (2006). *2006 statistical yearbook*. Dublin, Ireland: Central Statistics Office.

CSO. (2009). *2009 statistical yearbook*. Dublin, Ireland: Central Statistics Office.

CTI. (2010). *Presentation of the commission for accredited diploma*. Retrieved from http://www.cti-commission.fr/spip.php?page=sommaire-en

Dadashzadeh, M., Saber, A., & Saber, S. (Eds.). (2002). *Information technology education in the new millennium*. Hershey, PA: IRM Press.

Dadds, M. (2000). Continuing professional development: Nurturing the expert within. In Solar, J., Craft, A., & Burgess, H. (Eds.), *Teacher development: Exploring our own practice*. London, UK: Paul Chapman Publishing.

Daley, B. J. (2001). Learning and professional practice: A study of four professions. *Adult Education Quarterly*, *52*(39), 39–54.

Daniels, M., Carbone, A., Hauer, A., & Moore, D. (2007). *Panel-Ill-structured problem solving in engineering education*. 37th annual Frontiers in Education Conference, Global Engineering: Knowledge Without Borders, Opportunities Without Passports. Oct. 10-13.

Davenport, T. H., & Prusak, L. (1998). *Working knowledge*. Boston, MA: Harvard Business School.

Davey, C., Powell, J., Cooper, I., & Powell, J. (2004). Innovation, construction SMEs and action learning. *Engineering, Construction, and Architectural Management*, *11*(4), 230–237. doi:10.1108/09699980410547586

David, M. E., & Bathmaker, A. M. (Eds.). (2010). *Improving learning by widening participation in higher education*. Abingdon, UK: Routledge.

Dawson, J. D. (1989). Educational outcomes for students in cooperative education. *Journal of Cooperative Education*, *25*(2), 6–13.

Dearing, R. (1997). *Report of the National Committee into Higher Education (Dearing report)*. London, UK: NCIHE.

Department of Education South Africa. (1995). *White Paper on education and training*. Cape Town.

Department of Finance. (2010). *Monthly economic bulletin – February 2010*. Ireland: Government Publications.

DEST. (2002). *Striving for quality: Learning, teaching and scholarship*. Canberra, Australia: Department of Education, Science and Training.

DEST. (2007). *Graduate employability skills*. Canberra, Australia: BIHECC.

Dewblam Project. (2006), *The social and educational challenge of Work based Learning in European higher education and training: results of a pilot experience*. Ed. Reinhard Schmidt, Firenze, December 2006, 352 p.

Dewey, J. (1938). *Experience and education*. New York, NY: Macmillan.

Dewey, J. (1944). *Democracy and education*. New York, NY: The Free Press.

DiBella, A. J., & Nevis, E. C. (1998). *How organizations learn: An integrated strategy for building learning capacity*. San Francisco, CA: Jossey-Bass.

Dimopoulos, N., & Walker, D. (2005). Lifelong learning challenges for educators, governments, firms and individuals. *CEDA Australian Chief Executive*(August).

Direction de l'information légale et administrative (French State informative center). (2005). *Apprenticeship in France, chronology*. Retrieved from http://www.vie-publique.fr/ politiques-publiques/ apprentissage-enseignement -professionnel/chronologie/

DKM. (2009). *Review of the construction industry 2008 and outlook 2009-2011*. Ireland: Government Publications.

Domínguez, U. (2002). Spanish engineering curricula. Moving towards the European higher education space? *Proceedings of 30th SEFI Annual Conference*, Florence.

Domínguez, U., & Magdaleno, J. (2003). Industrial training in education. Towards an international approach in engineering curricula. *Proceedings of the International Conference of SEFI*. Porto.

Domínguez, U., Martin, M. A., García, J. M., & Magdaleno, J. (2006). Engineering education in Spain. The teacher's point of view on innovation. In U. Domínguez (Ed.), *Innovative teaching and learning in engineering education* (pp. 219-236). Valladolid, Spain: Escuela Universitaria Politécnica. El- Sayed, J. (2001). Industrial-academic integration takes learning out of the classroom. *Proceedings of the SEFI Annual Conference*. Copenhagen

Dotan, A., Maiden, N., Lichtner, V., & Germanovich, L. (2009). *Designing with only four people in mind? - A case study of using personas to redesign a work integrated learning support system*. In. INTERACT 2009 12th IFIP TC13 Conference in Human-Computer Interaction. Uppsala, Sweden.

Doyle, D. (2002). *Putting learning to work: The distance learner and transfer of learning.* Retrieved March 23, 2010, from http://www.aare.edu.au/ 03pap/doy03366.pdf

Drake, J., Blake, J., & Swallow W. (2009). Employer engagement: The critical role of employee commitment. *Education + Training, 51*(1), 23-43.

Driscoll, A. (1985). Designing a mentor system for beginning teachers. *Journal of Staff Development, 6*(2).

Du Pre, R. (2009). *The place and role of universities of technology in South Africa.* Bloemfontein, South Africa: Durban University of Technology.

Eames, C. (2000). A comprehensive database of New Zealand cooperative programs. *Asia Pacific Journal of Cooperative Education, 1*(1), 1–2.

Eames, C. (2003). Learning to work: Becoming a research scientist through work experience placements. *Asia Pacific Journal of Cooperative Education, 4*(2), 7–15.

Eames, C., & Cates, C. (2004). Theories of learning in cooperative education. In Coll, R., & Eames, C. (Eds.), *International handbook for cooperative education* (pp. 37–47). Boston, MA: WACE Int. Secretariat.

Ehrenberg, R. G., & Kuh, C. V. (Eds.). (2009). *Doctoral education and the faculty of the future*. Ithaca, NY: Cornell University Press.

Elmuti, D., Abebe, M., & Nicolosi, M. (2005). An overview of strategic alliances between universities and corporations. *Journal of Workplace Learning, 17*(1/2), 115–129. doi:10.1108/13665620510574504

Elmuti, D., Minnis, W., & Abebe, M. (2005). Does education have a role in developing leadership skills? *Management Decision, 43*(7/8), 1018–1031. doi:10.1108/00251740510610017

Engeström, Y. (2001). Expansive learning at work: Toward an activity theoretical reconceptualization. *Journal of Education and Work, 14*(1), 133–156.

Engineers Australia. (2008). *Continuing professional development.* Retrieved February 18, 2008, from http:// www.engineersaustralia.org.au/ education/continuing-professional- development/continuing-professional -development.cfm

ENMM20012. (2010). Maintenance organisation course profile, term 2. Retrieved July 15, 2010, from http:// content.cqu.edu.au/ CourseProfile/deliverPDF.jsp? docId=34902&course Code=ENMM20012

EPC. (2000). *The EPC engineering graduate output standard.* Interim Report of the EPC Output Standards Project. The Engineering Professor's Council. Warwick. Retrieved February 24, 2010, from http://www.epc.ac.uk/ publications /standards/index.php

Eraut, M. (1994). *Developing professional knowledge and competence*. London, UK: Falmer Press.

Eraut, M. (2004). Informal learning in the workplace. *Studies in Continuing Education, 26*(2), 247–274. doi:10.1080/158037042000225245

Eraut, M. (1998). Concepts of competence. *Journal of Interprofessional Care, 12*(2), 127–139. doi:10.3109/13561829809014100

Eraut, M., & Hirsh, W. (2007). *The significance of workplace learning for individuals, groups and organisations. (SKOPE Monograph 9)*. Oxford University, Department of Economics.

Etzkowitz, H. (1999). Bridging the gap: The evolution of industry-university links in the United States. In Branscomb, L. M., Kodama, F., & Florida, R. (Eds.), *Industrializing knowledge: University-industry linkages in Japan and the United States* (pp. 203–234). Cambridge, MA: MIT Press.

Evenson, J. S. (1982). *Workplace mentoring, far West laboratory for educational research and development.* (ED 246 182).

Everett, J., & Van Dessel, M. (2005). Using cases in a mixed learning environment: Meeting the challenges of higher education reform. *International Journal of Case Method Research and Application, 17*(3), 341–350.

Fernandez, A. (2008). *Les nouveaux tableaux de bord pour les managers.* 4ème édition Septembre

Field, D. M. (2007). *The world's greatest architecture past and present*. Rochester, NY: Grange Books.

Fook, J., Ryan, M., & Hawkins, L. (2000). *Professional expertise: Practice, theory and education for working with uncertainty*. London, UK: Whiting & Birth Ltd.

French Agency for Research and Higher Education Assessment. (2006). *Official website*. Retrieved from http://www.aeres-evaluation.fr

French Government. (2002). *Loi de modernisation sociale*. (Loi n°2002-73 du 17.01.2002, et décret d'application du 24.04.2002, Version consolidée au 01 juillet 2007).

French Government. (2006). *AERES establishment law*. (n ° 2006-450 of April 18, 2006 followed by the decree No. 2006-1334 of November 3, 2006 version 2009, February 6 defining AERES organization and operation).

Friedman, A., Davis, K., Durkin, C., & Phillips, M. (2000). *Continuing professional development in the UK: Policies and programmes*. Bristol, UK: PARN.

Friedman, A., & Mason, J. (2007). *Distinguishing Australian professional bodies*. Bristol, UK: PARN.

Friedman, A., & Phillips, M. (2002). The role of mentoring in the CPD programmes of professional associations. *International Journal of Lifelong Learning*, *21*(5), 269–284.

Friedman, A., & Phillips, M. (2004). Continuing professional development: Developing a vision. *Journal of Education and Work*, *17*(3), 361–376. doi:10.1080/1363908042000267432

Friedman, A., & Phillips, M. (2001). *Leaping the CPD hurdle: A study of the barriers and drivers to participation in continuing professional development.* Paper presented at the British Educational Research Association, Leeds.

Funding Agency to advance French research (ANR). (2010). *Official website and ANR presentation*. Retrieved from http://www.agence-nationale -recherche.fr/Intl

Gale, A., & Brown, M. (2003). Project management professional development-an industry led programme. *Journal of Management Development*, *22*(5), 410–425. doi:10.1108/02621710310474769

Gardner, H., & Shulman, L. S. (2005). The professions in America today: Crucial but fragile. *Daedalus*, *134*(3), 13–18. doi:10.1162/0011526054622132

Ghidini, C., Kump, B., Lindstaedt, S., Mahbub, N., Pammer, V., Rospocher, M., & Serafini, L. (2009). MoKi: The enterprise modelling wiki. In Sure, Y., & Domingue, J. (Eds.), *The Semantic Web: Research and applications. (LNCS 5554)* (pp. 831–835). Berlin, Germany: Springer. doi:10.1007/978-3-642-02121-3_65

Ghidini, C., Rospocher, M., Serafini, L., Kump, B., Pammer, V., Faatz, A., & Guss, J. (2007). *Integrated Modelling Methodology (IMM)*. version 1. Retrieved from http://www.aposdle.tugraz.at /media/multimedia/¯les/integrated modelling methodology version 1

Giannakouris, K. (2008). *Ageing characterises the demographic perspectives of the European societies*. Eurostat, European Commission.

Gielen, S., Dochy, F., & Dierick, S. (2003). Evaluating the consequential validity of new modes of assessment: The influence of assessment on learning, including pre-, post- and true assessment effects. In Segers, M., Dochy, F., & Cascallar, E. (Eds.), *Optimising new modes of assessment: In search of quality and standards*. Dordrecht, The Netherlands: Kluwer Academic Publishers. doi:10.1007/0-306-48125-1_3

Global Insight, I. H. S. (2009). *Global construction outlook*. IHS Global Insight.

Goodier, C., Austin, S., Guthrie, W., & Metzdorf, C. (Eds.). (2008). *Anticipating tomorrow, the future of the European construction industry – final report of the ECI industry futures task force*. UK: European Construction Institute.

Graduate Careers Australia. (2007). *Snapshot: University and beyond*. Carlton.

Green, D. R. (1964). *Educational psychology*. New Jersey: Prentice-Hall, Inc.

Gregory, R. L. (Ed.). (1987). *The Oxford companion to the mind*. Oxford, UK: Oxford University Press.

Greve, D., Schonk, P., & van der Aa, P. (2009). Matching, monitoring and assessing learning outcomes of students in practical assignments. *Proceedings of 37th SEFI Annual Conference,* Rotterdam.

Griffin, G. A. (1984). *Crossing the bridge: The first years of teaching*. Paper prepared for the National Commission on Excellence in Teacher Education. (ED 250 292).

Griffith University. (2006). *Introduction to Work Integrated Learning (WIL)*. Retrieved June 11, 2010 from http://www62.gu.edu.au/ policylibrary.nsf/xmainsearch/ a0d60f04626b7f144a257180 0063dd56?opendocument

Groenewald, T. (2004). Towards a definition for cooperative education. In Coll, R. K., & Eames, C. (Eds.), *International handbook for cooperative education: An international perspective of the theory, research and practice of work-integrated learning* (pp. 17–25). Boston, MA: World Association for Cooperative Education.

Gulikers, J., Bastiaens, T., & Kirschner, P. (2004). A five-dimension framework for authentic assessment. *Educational Technology Research and Development*, *52*(3), 67–86. doi:10.1007/BF02504676

Gumport, P., Nozzi, M., Hamah, S., & Zemsky, R. (1997). Trends in United States higher education from massification to post massification. Retrieved 27 July, 2010, from http://www.stanford.edu/group/ ncpi/documents/pdfs/1-04 _massification.pdf

Guri-Rosenblit, S., Sebkova, H., & Teichler, U. (2007). Massification and diversity of higher education systems: Interplay of complex dimensions. *Higher Education Policy*, *20*, 373–389. doi:10.1057/palgrave.hep.8300158

Haines, C., Scott, K., & Lincoln, R. (2006). *Australian blueprint for career development*. Retrieved June 11, 2010, from http://www.dest.gov.au/NR/ rdonlyres/B2B0AACE -D719-48AB-8A42-D78203FD4F0D/ 10560/Blueprint-Trial Version.pdf

Hamid, A. (2002). E-learning is it the 'e' or the learning that matters? *The Internet and Higher Education*, *4*, 311–316. doi:10.1016/S1096-7516(01)00072-0

Harris, S. (2003). An andragogical model: Learning through life experiences. *Kappa Delta Pi Record*, 38–41.

Harvey, L., Moon, V., & Geall, S. (1997). *Graduates work: Organisational change and students' attributes*. Birmingham, UK: Centre for Research into Quality (CRQ) and Association of Graduate Recruiters (AGR).

Haselgrove, S. (Ed.). (1994). *The student experience*. Buckingham, UK: Society for Research into Higher Education, Open University Press.

Haskell, R. E. (1998). *Reengineering corporate training. Intellectual capital and transfer of learning*. Westport, CO: Quorum Books.

HEA. (2010). *Student statistics–2008/2009: HEA funded institutions*. The Higher Education Authority. Retrieved February 17, 2010, from http://www.hea.ie/en/statistics

Hegarty, C., & Johnston, J. (2008). Graduate training: Evidence from FUSION projects in Ireland. *Education + Training, 50*(5), 391-405.

Heinemann, H. (1988). Cooperative education and the academy. *Asia Pacific Journal of Cooperative Education*, *24*(2-3), 109–119.

Heinz, W., & Taylor, A. (2005). Learning and work transition policies in a comparative perspective: Canada and Germany. In Leithwood, D. L. K., Cumming, A., Bascia, N., & Datnow, A. (Eds.), *International handbook of educational policy* (*Vol. 2*). New York, NY: Kluwer. doi:10.1007/1-4020-3201-3_45

Heitmann, G. (2003). Employability through innovative curricula. In Borri, C., & Maffili, F. (Eds.), *E4 thematic network* (*Vol. C*). Firenze, Italy: Firenze University Press.

Heshmati, A., Sohn, Y. B., & Kim, Y. R. (Eds.). (2007). *Commercialization and transfer of technology: Major country case studies*. New York: Nova Science Publishers.

Hessels, L., & Van Lente, H. (2008). Re-thinking new knowledge production: A literature review and a research agenda. *Research Policy*, *37*, 740–760. doi:10.1016/j.respol.2008.01.008

Hodgson, I. (1993). Learning contracts for the industrial experience of hospitality management students. In Stephenson, J., & Laycock, M. (Eds.), *Using learning contracts in higher education* (pp. 107–114). London, UK: Kogan Page.

Houssaye, J. (2000). *Le triangle pédagogique*. Edition Peter Lang.

Howell, S. L., Saba, F., Lindsay, N. K., & Williams, P. B. (2004). Seven strategies for enabling faculty success in distance education. *The Internet and Higher Education*, *7*, 33–49. doi:10.1016/j.iheduc.2003.11.005

HRSDC. (2005). *Transfer of learning planning workplace education programs: HRSDC*. Government of Canada.

Huba, M. E., & Freed, J. E. (2000). *Learner-centered assessment on college campuses*. Needham Heights, MA: Allyn & Bacon.

Huerta, M. M., Portela, J. M., Díaz, J. E., & Pastor, A. (2009). Seguimiento de los alumnos en las Prácticas de Empresa. *Proceedings of the XVII Congreso Universitario de Innovación Educativa en las Enseñanzas Técnicas*. Valencia.

Huffman, G., & Leak, S. (1986). Beginning teachers' perceptions of mentors. *Journal of Teacher Education*, *37*(1), 22–25. doi:10.1177/002248718603700105

Hyland, T. (1998). Work-based experience and higher professional learning in British universities. In James, J. J. F. (Ed.), *University teaching: International perspectives* (pp. 305–324). New York: Garland Publishers.

IBM. (2004). *Services science: A new academic discipline*. Retrieved February 22, 2010, from http://www.almaden.ibm.com /asr/resources/facsummit.pdf

IESP Departement. (2009). *English presentation booklet of the IESP department and training*. Polytech'Lille.

Institute of International Education. (2009). *Open doors web site report on international education exchange, table 22*. Retrieved February 1, 2010, from http://opendoors.iienetwork.org /?p=150836

Institute, R. O. I. (2010). *About the ROI Institute*. Retrieved February 12, 2010, from http://www.roiinstitute.net/about/

International Council for Open and Distance Education. (2009). *Global trends in higher education, adult and distance learning*. Retrieved 27 July, 2010, from http://www.icde.org/filestore/ Resources/Reports/FINALICDE ENVIRNOMENTALSCAN05.02.pdf

Ireland, R. D., Kuratko, D. F., & Morris, M. H. (2006). A health audit for corporate entrepreneurship: Innovation at all levels: Part I. *The Journal of Business Strategy*, *27*(1), 10–17. doi:10.1108/02756660610640137

Iribarne, P. (2004). *Balanced Scorecard et Management Qualité - AFNOR*. Retrieved from http://www.itii.fr/img/ itii_charte_2009.pdf

ITII. (2009), *ITII Engineer charter*, Retrieved from http:// www.itii.fr/img/ itii_charte_2009.pdf (on line, french), Institute of Technical Engineering of Industry of North-Pas-de-Calais, created 2006 updated 2009

Jackson, S. E., & Schuler, R. S. (2000). *Managing human resources: A partnership perspective*. Cincinnati, OH: South-Western College Publishing.

James, P. (2000). *Lessons in learning the lore of the library: Theories and strategies for reader education*. Paper presented at the CAVAL Reader Education Conference: Librarians behaving badly - taking reader education to the people, Melbourne.

Jancauskas, E., Atchison, M., Murphy, G. A., & Rose, P. (1997). *Unleashing the potential of work-integrated-learning through professionally trained academic and industry supervisors*. Paper presented at the WACE Conference, Cape Town 1997.

Jancauskas, E., Atchison, M., Murphy, G., & Rose, P. (1999). *Unleashing the potential of work-integrated learning through professionally trained academic and industry supervisors*. Paper presented at the tenth World Conference on Cooperative Education. Washington, DC.

Jarvis, P., & Wilson, A. L. (1999). *International dictionary of adult and continuing education* (2nd ed.). London, UK: Kogan Page.

Johnston, G. (Ed.). (1983). *The Australian pocket Oxford dictionary* (5th ed.). Melbourne, Australia: Oxford University Press.

Jones, E. A., Voorhees, R. A., & Paulson, K. (2002). *Defining and assessing learning: Exploring competency-based initiatives*. Washington, DC: Council of the National Postsecondary Education Cooperative (NPEC), U.S. Department of Education National Center for Education Statistics.

Kaplan, R. S., & Norton, D. P. (1992). The balanced scorecard: Measures that drive performance. *Harvard Business Review*, (Jan – Feb): 71–80.

Kapp, A. (1883). *Die angragogik ober bildung im mannlichen alter. Platons Erziehungslehre, als Padagogik fur die Einzelen und als Staatspafagogik*. Germany: Minden and Leipzig.

Katula, R. A., & Threnhauser, E. (1999). Experiential education in the undergraduate curriculum. *Communication Education, 48*, 238–255. doi:10.1080/03634529909379172

Katz, R. L. (1974). Skills of an effective administrator. *Harvard Business Review*, 51.

Keithley, D., & Redman, T. (1997). University-industry partnerships in management development: A case study of a world-class company. *Journal of Management Development, 16*(3), 154–166. doi:10.1108/02621719710164319

Keleher, P., Wong, K., & Howard, P. (2006). The nexus: Ivory towers and muddy trenches-studying maintenance management by distance education at postgraduate level at central Queensland University, Australia. *The Journal of Cooperative Education and Internships, 40*(2), 26–32.

Keleher, P., Moxham, L., & Shakespeare, P. (2007). Work integrated learning: A comparative analysis of an Australian and UK experience. In *Proceedings 15th World Conference on Cooperative Education, Transcending Boundaries: Work-Integrated Learning for the Global Economy*. Suntec International Convention & Exhibition Centre. Singapore.

Kemp, D. (1998). *Strategic developments in higher education, address to the OECD thematic review seminar on the first years of higher education*. Sydney. Retrieved June 15, 2010, from http://www.dest.gov.au/ archive/ministers/kemp /ks210498.htm

Kerr, C. (1995). *The uses of the university* (3rd ed., pp. 1–34). Cambridge, MA: Harvard University Press.

Ketteridge, S., Marshall, S., & Fry, H. (Eds.). (2002). *The effective academic: A handbook for enhanced academic practice*. London, UK: Kogan Page.

King, P. M., & Kitchener, K. S. (1994). *Developing reflective judgment: Understanding and promoting intellectual growth and critical thinking in adolescents and adults*. San Francisco, CA: Jossey-Bass.

Klein, D., & Ware, M. (2003). E-learning: New opportunities in continuing professional development. *Learned Publishing, 16*(1), 34–46. doi:10.1087/095315103320995078

Knapper, C. (2004). *Research on college teaching and learning: Applying what we know*. Paper presented at the Teaching Professor Conference, Philadelphia.

Knight, A. (2005). *Australian integrated compact dictionary & thesaurus*. Melbourne, Australia: Oxford University Press.

Knowles, M. (1975). *Self-directed learning-a guide for learners and teachers*. New York, NY: Association Press.

Knowles, M. (1978). *The adult learner: A neglected species* (2nd ed.). Houston, TX: Gulf Publishing Company.

Knowles, M. S. (1980). *The modern practice of adult education*. New York, NY: Association Press & Cambridge Book Publishers.

Kohler, J. (2004). *The Bologna process and employabilty: The impact of employability on curricular development*. Paper presented at the A Key Objective of Academic Studies and for Academic Institutions, Bled, Slovenia.

Kolb, D. (1983). *Experiential learning: Experience as the source of learning and development*. Englewood Cliffs, NJ: Prentice Hall.

Kolb, D. A. (1984). *Experimental learning experience as a source of learning and development*. New Jersey: Prentice Hall.

Kooken, J., Ley, T., & de Hoog, R. (2007). How do people learn at the workplace. Investigating four workplace learning assumptions. In E. Duval, R. Klamma & M. Wolpers (Eds.), *Creating new learning experiences on a global scale* (pp. 158–171). Second European Conference on Technoloy Enhanced Learning, EC-TEL 2007, Crete, Greece, September 2007. Berlin, Germany: Springer.

Kounkou, A., Lichtner, V., Dotan, A., & Maiden, N. (2008). *APOSDLE use scenarios and requirements for 3rd prototype*, Deliverable D6.4. Retrieved June 2010, from http://www.aposdle.tugraz.at/ media/multimedia/files/ aposdle_use_scenarios_requirements_for_3rd_prototypes

Krupp, J. A. (1984). *Mentor and protege perceptions of mentoring relationships in an elementary and secondary school in Connecticut*. Annual meeting of the American Educational Research Association. (ED 245 004).

Ku, H. M., Thonglek, S., & Bhumiratana, S. (2005). A graduate-level chemical engineering practice school model in Thailand. In Pudlowski, Z. J. (Ed.), *Monash engineering education series: 4th Asia-Pacific Forum on Engineering and Technology Education* (pp. 43–46). Melbourne, Australia: Monash University.

Ku, H. M., Sriwatanapongse, W., & Thonglek, S. (2007). Operating a chemical engineering practice school. In Z. J. Pudlowski (Ed.), *Monash engineering education series: 10th UICEE Annual Conference on Engineering Education* (pp. 251-254). Melbourne, Australia: Monash University.

Ku, H. M., Thonglek, S., Thipayarat, A., & Cheevadhanarak, S. (2005). *A graduate-level science and engineering practice school model in Thailand.* Paper presented at the 2005 Regional Conference on Engineering Education, Jahor, Malaysia.

Kump, B. (2008). A validation framework for formal models in adaptive technology-enhanced work-integrated learning. In W. Nejdl, J. Kay, P. Pu, & E. Herder (Eds.), *Adaptive hypermedia and adaptive Web based systems,* (pp. 416-420). 5th International Conference, AH 2008, Hannover, Germany, July 29 - August 1, 2008. Berlin, Germany: Springer.

Kwak, Y. H., & Anbari, F. T. (2009). Analysing project management research: Perspectives from top management journals. *International Journal of Project Management, 27*(5), 435–447. doi:10.1016/j.ijproman.2008.08.004

Lamancusa, J. S., George, R. T., & Soyster, A. L. (1997). *Industry-based projects in academia-what works and what doesn't.* Paper presented at 1997 ASEE Annual Conference, Milwaukee, WI.

Lamancusa, J. S., Jorgensen, J. E., Zayas-Castro, J. L., & Ratner, J. R. (1995). *The learning factory-a new approach to integrating design and manufacturing into the engineering curriculum.* Paper presented at 1995 ASEE Annual Conference, Anaheim, CA.

Lambe, P. (2002). *The autism of knowledge management.* Retrieved June 28, 2006, from http://greenchameleon.com/ thoughtpieces/autism.pdf

Lave, J., & Wenger, E. (1991). *Situated learning: Legitimate peripheral interaction.* Cambridge, UK: Cambridge University Press. (Original publication IRL report 90-0013, Palo Alto, CA.: Institute for Research on Learning) (1990).

Leemkuil, H., de Hoog, R., Lindstaedt, S., Ley, T., Scheir, P., Ulbrich, A., et al. Pammer, V.(2007). *Conceptual framework & architecture version 1.* Retrieved August 1, 2009, from http://www.aposdle.tugraz.at/ media/ multimedia/¯les/conceptual framework architecture version 1

Lees, M. (2009). Foreword. In Williams, A., & Lees, T. (Eds.), *Knowing your market: Strategy, communication and relationships – employer engagement series two, engaging employers.* Salford, UK: University of Salford.

Leung, T. P., & Siu, K. W. M. (2005). Design for new urban needs: Balanced consideration in design. In H. I. Lei & D. C. Dai (Eds.), *Selected papers from the international symposium on design in urban spaces* (Part I, pp. 42-48). Macao: Macao Polytechnic Institute.

Ley, T., Kump, B., & Albert, D. (2010). A methodology for eliciting, modelling, and evaluating expert knowledge for an adaptive work-integrated learning system. *International Journal of Human-Computer Studies, 68*(4), 185–208. doi:10.1016/j.ijhcs.2009.12.001

Lindstaed, S., Beham, G., Aehnelt, M., Hambach, S., Musielak, M., Kooken, J., et al. (2009). *The APOSDLE approach to self-directed work-integrated learning.* Retrieved April 1, 2009 from http://www.aposdle.tugraz.at/ media/multimedia/files/ the_aposdle_approach_to_self_ directed_work_integrated_learning

Lindstaedt, S. N., Ley, T., Scheir, P., & Ulbrich, A. (2008). Applying scruffy methods to enable work-integrated learning. *The European Journal of the Informatics Professional, 9*(3), 44–50.

Lindstaedt, S., Beham, G., Kump, B., & Ley, T. (2009). Getting to know your user–unobtrusive user model maintenance within work-integrated learning environments. In Cress, U., Dimitrova, V., & Specht, M. (Eds.), *Learning in the synergy of multiple disciplines* (pp. 73–87). Heidelberg, Germany: Springer. doi:10.1007/978-3-642-04636-0_9

Lindstaedt, S. N., & Mayer, H. (2006). A storyboard of the APOSDLE vision. In W. Neidl & K. Tochtermann (Eds.). *Proceedings of the 1st European Conference on Technology-enhanced Learning* (LNCS 4227). (pp.628–633). Heidelberg, Germany: Springer.

Lindstaedt, S. N., Kump, B., Beham, G., Pammer, V., & Ley, T. dotan, A., & de Hoog, R. (2010). *Providing varying degrees of guidance for work-integrated learning.* To appear in Proceedings of ECTEL 2010, Lecture Notes in Computer Science. Heidleberg, Germany: Springer

Linkon, S. L. (2005). *How can assessment work for us?* Academe Online, American Association of University Professors, July-August 2005. Retrieved July 26, 2006, from http://www.aaup.org/AAUP/ pubsres/academe/2005/JA/ Feat/link.htm

Lock, G., Bullock, K., Gould, V., & Hejmadi, M. (2009). Exploring the industrial placement experience for mechanical engineering undergraduates. *English Education, 4*(1), 42–51.

M. E. (2009). *Datos y cifras del sistema universitario Español.* Curso 2009/2010. Ministerio de Educación. Secretaría General Técnica. Retrieved February 24, 2010, from http://www.educacion.es/ educacion/universidades/ estadisticas-informes/ datos-generales.html

MacKinnnon-Slaney, F. (1994). The adult persistence in learning model: A road map to counseling services for adult learners. *Journal of Counseling and Development, 73*(2), 268–275.

MacLeod, I. A. (2010). The education of innovative engineers. *Engineering, Construction, and Architectural Management, 17*(1), 21–34. doi:10.1108/09699981011011294

Magdaleno, J., & Domínguez, U. (2006). Industrial training in engineering curricula. Proposals for spanish degrees. In Domínguez, U. (Ed.), *Innovative teaching and learning in engineering education* (pp. 145–153). Valladolid, Spain: Escuela Universitaria Politécnica.

Magdaleno, J., Domínguez, U., & González, J. (2002). La integración obligatoria de las prácticas en empresas en los planes de estudio de ingeniería. *Proceedings of the X Congreso Universitario de Innovación Educativa en las Enseñanzas Técnicas.* Valencia.

Maki, P. (2002). *Using multiple assessment methods to explore student learning and development inside and outside of the classroom.* NASPA's Net Results. Retrieved January 10, 2010, from http://virtual2.yosemite.cc.ca.us /mjcinstruction/CAI/ Summer2005Institute/ Training-MaterialsUsed/ NASPA's%20NetResults%20Using%20 Multiple%20Assessment% 20Methods.doc

Manipal Institute of Technology. (2010). In *Wikipedia, the free encyclopedia.* Retrieved February 4, 2010, from http://en.wikipedia.org/w/ index.php?title=Manipal_Institute _of_Technology&oldid =341735957

Marden, B. (2003). High aims for professional standards legislation. *Law Institute Journal, 77*(11), 28.

Martin, J. C., Akande, K., & Falconer, R. A. (2010). Innovations from the Halcrow-Cardiff University strategic collaboration. *Engineering, Construction, and Architectural Management, 17*(1), 46–64. doi:10.1108/09699981011011311

Martin, E. (1997). *The effectiveness of different models of work-based university education.* Department of Employment, Education, Training and Youth Affairs. J. S. McMillan Printing Group.

Marton, F., & Säljö, R. (1997). Approaches to learning. In Marton, F., Hounsell, D., & Entwistle, N. (Eds.), *The experience of learning* (2nd ed.). Edinburgh, UK: Scottish Academic Press.

Martyn, M. (2005). *The hybrid online model: Good practice, a hybrid approach to online learning offers important lessons for institutions entering the online arena.* Retrieved 24 July, 2010, from http://net.educause.edu/ir/library /pdf/EQM0313.pdf

Mattill, J. (1991). *The flagship: The M.I.T. School of Chemical Engineering Practice, 1916-1991.* Cambridge, MA: MIT Press.

McInnes, C., James, R., & Hartley, R. (2000). *Trends in the first year experience in Australian universities.* Canberra, Australia: Department of Education, Training and Youth Affairs.

Melotte, B. (1996). *The role of qualitative research in identifying praxis and hermeneutics for professional planning education in Australia. Different approaches: Theory and practice in higher education.* Paper presented at the HERDSA Conference, Perth, Western Australia.

Meyers, C., & Jones, T. B. (1993). *Developing and assessing instructional expertise: Promoting active learning.* San Francisco, CA: Jossey-Bass.

Michel, M. (2005). La démarche inductive en pédagogie. *Le Portique, 9.* Retrieved from http://leportique.revues.org/ document182.html

Mintzberg, H. (1973). *The nature of managerial work.* Harpercollins College Division.

Mirvis, P. H., & Hall, D. T. (1996). New organisational forms and the new career. In Hall, D. T. (Ed.), *The career is dead: Long live the career.* San Francisco, CA: Jossey-Bass.

MIT|TPD (2010). *Technology and development program.* Retrieved February 3, 2010, from http://web.mit.edu/ mit-tdp /www/thailand.html

Mitchell, P. (2009). Borrowed time. In Cantwell, T. (Ed.), *Irish construction overview 2010 – opinion, analysis and the year ahead in construction.* Ireland: Commercial Publications Ltd.

Monyarch, J. (2005). La adquisición de competencias de los ingenieros a través de las prácticas tuteladas a la empresa. *Proceedings of the XIII Congreso Universitario de Innovación Educativa en las Enseñanzas Técnicas.* Maspalomas.

Moore, D. (1999). Toward a theory of work-based learning (IEE brief no.23). *IEE brief.* Retrieved March 21, 2010, from http://www.tc.columbia.edu/ centers/iee/BRIEFS/ Brief23.htm

Morell, L., Zayas-Castro, J., & Velz-Arocho, J. (1998). *Some assessment tools for evaluating curricular innovations outcomes.* Paper presented at 1998 ASEE Annual Conference, Seattle, Washington.

Morell, L., Zayas-Castro, J., Velz-Arocho, J., & Torres, M. (1998). *Curricular innovations outcomes assessment and ABET 2000.* Paper presented at International Conference on Engineering Education, Rio de Janiero, Brazil.

Morris, P. W. G. (1999). *Body building.* Retrieved April 2, 2007, from http://www.pmforum.org/library/ papers/1999/crmpbody.pdf

Müller, B. (2009). *Cooperative state university campus* (p. 3). Ravensburger Stadtmagazin.

Murphy, G. A., & Calway, B. A. (2008). Deep learning for professional development of professionals. *Australian Journal of Adult Learning, 48*(3), 424–444.

Murphy, G. A., & Calway, B. A. (2007). *Professional development-the case for deep learning.* Paper presented at the BERA 2007 Conference, London.

Murphy, G. A., & Calway, B. A. (2007a). *Education for professionals through work-integrated learning.* Paper presented at the AARE 2006 Conference, Adelaide.

Murphy, G. A., & Calway, B. A. (2007b). *Professional standards legislation and supporting learning cultures.* Paper presented at the HERDSA 2007, Adelaide.

Murphy, G. A., Murphy, R. E., & Calway, B. A. (1997). *Career progression of cooperative education graduates.* Paper presented at the 10th World Conference on Cooperative Education, Cape Town.

Musielak, M., Hambach, S., & Christl, C. (2008). *APOSDLE contextualized cooperation.* In: ACM SIGCHI: ACM Conference on Computer Supported Cooperative Work 2008. Electronic Proceedings: CSCW 08. New York, NY: ACM Press.

National Research Council. (2009). *Advancing the competitiveness and efficiency of the US construction industry.* Washington, DC: The National Academies Press.

Nelson, B. (2002). *Higher education at the crossroads: An overview paper.* Retrieved June 15, 2010, from http://www.dest.gov.au/ ministers/nelson/apr02/ n55_260402.htm

Noe, R. A. (2008). *Human resource management: Gaining a competitive advantage* (6th ed.). Boston, MA: McGraw-Hill/Irwin.

NQAI. (2003). *National framework of qualifications - policies, actions and procedures for access, transfer and progression of learners.* Ireland: The National Qualifications Authority of Ireland.

Nuninger, W., & Chatelet, J.-M. (2011). *Quality WIL based model for IVT and CVT*. 17th World Conference on Cooperative Education and Work Integrated Learning, Wace 2011, Philadelphia, 14-17 june, 5 pages.

Nuninger, W., Chatelet, J.-M., & Abt, A. (2009). *Feedback on work based learning for engineers as a reply to industrial demands*. 16th World Conference on Cooperative Education and Work Integrated Learning, Wace 2009, Vancouver, 23-26 june, 5 pages.

O'Connell, P. (2009). *What do we know about training at work?* ESRI Research Bulletin 2009/1/4. Retrieved February 17, 2010, from http://www.esri.ie/publications/latest_publications/

O'Connor, A. J., Kandas, A. W., Natori, Y., & Hatton, T. A. (1999). Internationalizing practical ChE education: The M.I.T. practice school in Japan. *Chemical Engineering Education, 33*(2), 162–165.

OECD. (2002). *Education policy analysis.* Retrieved May 6, 2010 from http://www.oecd.org/document/42/0,3343,en_2649_ 39263231_1841194 _1_1_1_1,00.html

OECD. (2003). *Education policy analysis.* Retrieved May 6, 2010, from https://www.oecd.org/ document/14/0,2340,en_2649_ 34511_17735886_ 1_1_1_1,00.html

OECD. (2004). *Education policy analysis.* Retrieved May 6, 2010 from www.oecd.org/.../ 0,3343,en_2649_ 39263231_ 34989090_1_1_1 _1,00.html

OECD. (2005-06). *Education policy analysis.* Retrieved May 6, 2010 from www.oecd.org/.../ 0,3343,en_2649_39263238_ 37693017_ 1_1_1_1,00.html

OECD. (2009). *Economic outlook – Ireland.* (Economic Outlook No. 86, OECD). Retrieved January 28, 2010, from http://www.oecd.org/ document/18/

Order. (2009a). *Orden CIN/351/2009, de 9 de febrero, por la que se establecen los requisitos para la verificación de los títulos universitarios oficiales que habiliten para el ejercicio de la profesión de ingeniero técnico industrial.* Retrieved February 24, 2010, from http://www.boe.es/boe/dias/ 2009/02/20/pdfs/ BOE-A-2009-2893.pdf

Order. (2009b). *Orden CIN/311/2009, de 9 de febrero, por la que se establecen los requisitos para la verificación de los títulos universitarios oficiales que habiliten para el ejercicio de la profesión de ingeniero industrial.* Retrieved February 24, 2010, from http://www.boe.es/boe/dias/ 2009/02/18/pdfs/ BOE-A-2009-2740.pdf

Orr, J. (1990). Sharing knowledge, celebrating identity: War stories and community memory in a service culture. In Middleton, D., & Edwards, D. (Eds.), *Collective remembering: Memory in society*. Beverly Hills, CA: Sage Publications.

Orrel, J. (2004). Work-integrated learning programmes: Management and educational quality. *Proceedings of the Australian Universities Quality Forum 2004*. AUQA Occasional Publication. Retrieved January 10, 2010, from http://www.auqa.org.au/ auqf/pastfora/2004/program/papers/Orrell.pdf

Osland, J. S., Kolb, D. A., Rubin, I. M., & Turner, M. E. (2007). *Organizational behaviour: An experiential approach* (8th ed.). New Jersey: Pearson Education.

Osman, S. A., et al. (2008). The importance of industrial training: Students' perception in civil engineering sector. *Proceedings of the 7th WSEAS Conf. on Education and Educational Technology.* Retrieved January 10, 2010, from http://pkukmweb.ukm.my/ ~upak/paper/importance.pdf

Painter, J. (2006). *Limited returns: Why management development isn't fully developed.* Retrieved June 1, 2010, from http://www.ceda.com.au/ research/previous-topics/research/2005/09/28/ dnetto_management.aspx

Pammer, V., Kump, B., Ghidini, C., Rospocher, M., Serafini, L., & Lindstaedt, S. N. (2009). *Revision support for modeling tasks, topics and skills*. In A. Paschke, H. Weigand, W. Behrendt, K. Tochtermann, & T. Pellegrini, (Eds.), *Proceedings of I-SEMANTICS '09, 5th International Conference on Semantic Systems*, (pp. 501-508).

Patkó, G., Szentirmai, L., & Váradi, A. (2009). Adventures in engineering wonderland. *Proceedings of 37th SEFI Annual Conference,* Rotterdam.

Paul, M. (2004). L'accompagnement: Une posture professionnelle spécifique. In L. Schlesinger & J. Heskett (Eds.), Breaking the cycle of failure in services. *Sloan Management Review, 31*, 17-28.

Perkins, D. N., & Salomon, G. (1992). Transfer of learning. In Tuijnman, A. (Ed.), *International encyclopedia of education* (2nd ed.). Oxford, UK: Pergamon Press.

Petraglia, J. (1998). *Reality by design: The rhetoric and technology of authenticity in education.* Mahwah, NJ: Lawrence Erlbaum Associates Publishers.

Phelan, P. (2002). The medical colleges: Issues at the turn of the century. *eMJA -The Medical Journal of Australia.*

PIPS. (2010). *The Professional Engineering Placement Scholarship at The University of Queensland.* Retrieved February 13, 2010, from http://www.uq.edu.au/ peps/pips

Polytech-Lille. (2010). *Official site*. Retrieved from http://www.polytech-lille.fr/iesp

Poppins, P., & Singh, M. (2005). Work integrated learning in information technology education. In van Weert, T. J., & Tatnall, A. (Eds.), *Information and communication technologies and real-life learning: New education for the knowledge society* (pp. 223–230). New York: Springer. doi:10.1007/0-387-25997-X_25

Professions Australia. (2006). *Blueprint for national registration of the professions.* Retrieved June 12, 2010, from http://www.professions.com.au/ natreg.html

PSC. (2006). Professional Standards Council consultation paper: *Approval of schemes under the professional standards legislation for associations with firms only as members* Retrieved March 5, 2007, from http://www.lawlink.nsw.gov.au/ lawlink/professional_standards_council /psc_ll.nsf/vwFiles/PSC%20 Consultation%20Firm%20 Only_23 _Nov_06.doc/$file/PSC%20 Consultation%20 Firm%20Only _23_Nov_06.doc

PSC. (2007). *Application guidelines-guidelines for occupational associations to prepare an application to the Professional Standards Council for the approval or renewal of a scheme in New South Wales.* Retrieved May 21, 2007, from http://www.lawlink.nsw.gov.au/ lawlink/ professional_standards_ council/psc_ll.nsf/vwFiles/ Application%20Guidelines%20 (NSW).doc/$file/Application% 20Guidelines%20(NSW).doc

PSC. (2009). *Application guidelines for approval of a scheme.* Retrieved June 29, 2010, from http://www.lawlink.nsw.gov.au/ lawlink/psc/ll_psc.nsf/vwFiles/ Application_Guidelines.pdf/$file /Application_Guidelines.pdf

R. D. (1987). *Real Decreto 1497/1987, por el que se establecen directrices generales comunes de los planes de estudio de los títulos universitarios con carácter oficial y validez en todo el territorio nacional.* Retrieved February 24, 2010, from http://www.boe.es/aeboe/ consultas/ bases_datos/doc.php ?id=BOE-A-1987-27707

R. D. (2003). *Real Decreto 1125/2003, por el que se establece el sistema europeo de créditos y el sistema de calificaciones en las titulaciones universitarias de carácter oficial y validez en todo el territorio nacional.* Retrieved February 24, 2010, from http://www.boe.es/aeboe/ consultas/bases_datos/ doc.php?id=BOE-A-2003-17643

R. D. (2007). *Real Decreto 1393/2007, por el que se establece la ordenación de las enseñanzas universitarias oficiales.* Retrieved February 24, 2010, from http://www.boe.es/aeboe/ consultas/bases_datos/ doc.php?id=BOE -A-2007-18770

Ramirez, L. M., & Beauchamp, G. (1995). *Integration of skills development across the engineering curriculum.* Paper presented at 1995 ASEE Annual Conference, Anaheim, CA.

Ramsden, P. (1992). *Learning to teach in higher education.* London, UK: Routledge. doi:10.4324/9780203413937

Ramsden, P. (2003). *Learning to teach in higher education* (2nd ed.). London, UK: Taylor and Francis, Inc.

Randall, N., Roberts, T., & Rogers, S. (2005). *Connected assessment: Getting real with authentic assessment online.* Presented at the ETUG Spring Workshop on Educational Technologies, June.

Raybould, B. (2002). *Performance support engineering part one: Key concepts.* Ariel PSE Technology.

REACU. (2009). *Recomendaciones para el seguimiento de los títulos oficiales.* Red Española de Agencias de Calidad Universitaria. Retrieved February 24, 2010, from http://aplicaciones.acsucyl.com/ acsucyl/export/ system/modules/ org.opencms.module.acucyl/ elements/ galleries/galeria_descargas /Seguimiento_titulaciones REACU.pdf

Reeders, E. (2000). Scholarly practice in work-based learning: Fitting the glass slipper. *Higher Education Research & Development, 19*(2), 205–220. doi:10.1080/072943600445655

Reiko, Y. (2001). University reform in the post-massification era in Japan: Analysis of government education policy for the 21st century. *Higher Education Policy*, *14*(4), 277–291. doi:10.1016/S0952-8733(01)00022-8

Reinhard, K. (2006a). Berufsakademie in Germany: Can this work-integrated learning program serve as an effective model in higher education institutions in western and eastern countries? *Asia Pacific Journal of Cooperative Education*, *7*(1), 16–21.

Reinhard, K. (2006b). The role of business and industry in supporting education: Case studies from the United States and Germany in educational sponsorship–an option for IT-educational funding. *Journal for Global Business Education*, *6*, 43–50.

Reinhard, K., Osburg, T., & Townsend, R. (2008). The sponsoring by industry of universities of cooperative education: Case study in Germany. *Asia Pacific Journal of Cooperative Education*, *9*(1), 1–13.

Reinhard, K., Osburg, T., & Townsend, R. (2010). Introducing corporate social responsibility as component in cooperative education: Results from a student research project in Germany and the United States supported by Intel Corporation. *Asia Pacific Journal of Cooperative Education*, *11*(1), 1–12.

Reinhard, K., Satow, L., & Sisco, L. (2007). Internationalizing cooperative education: An introduction to the Berufsakademie Ravensburg's work integrated partnership with the multinational company SAP Incorporation. *Asia Pacific Journal of Cooperative Education*, *8*(2), 109–119.

Revans, R. W. (1991). Action learning: Its origins and practice. In Pedlar, M. (Ed.), *Action learning in practice* (2nd ed.). New York, NY: Gower Press.

Reynolds, A. J. (1997). Interdisciplinary programs in community and applied research. In O'Donnell, C. R., & Ferrari, J. R. (Eds.), *Education in community psychology: Models for graduate and undergraduate programs* (pp. 65–82). New York: Haworth Press.

Richardson, J. T. E., & King, E. (1998). Adult students in higher education: Burden or boon? *The Journal of Higher Education*, *69*(1), 65–88. doi:10.2307/2649182

Ricks, F., van Gyn, G., Branton, G., Cut, J., Loken, M., & Ney, T. (1990). Theory and research in cooperative education: Practice implications. *Asia Pacific Journal of Cooperative Education*, *27*(1), 7–20.

Risebero, B. (1979). *The story of Western architecture. London, UK: The Herbert press*

Robinson, D. G. (2003). *Skill and performance: They are not equal*. Apartment Professional Magazine.

Roelofsen, W. R. (2009). Engineering education: More fun with action learning. *Proceedings of 37th SEFI Annual Conference,* Rotterdam.

Rojter, J. (2007). *PBL as means to better engineering education?* Paper presented at International Conference on Engineering Education and Research, Melbourne, Australia.

Rothwell, A., & Herbert, I. (2007). Accounting professionals and CPD: Attitudes and engagement-some survey evidence. *Research in Post-Compulsory Education*, *12*(1), 121–138. doi:10.1080/13596740601155587

Rowe, P. (2005). *Survey of graduate programs with cooperative and internship*. Waterloo.

Ryan, L. (2009). Exploring the growing phenomenon of university-corporate education partnerships. *Management Decision*, *47*(8), 1313–1322. doi:10.1108/00251740910984569

Sas, C. (2009). Research knowledge transfer through business-driven student assignment. *Education + Training*, *51*(8/9), 707-717.

Scheir, P., Ghidini, C., & Lindstaedt, S. N. (2007). Improving search on the semantic desktop using associative retrieval techniques. *Proceedings of I-MEDIA 2007 and I-SEMANTICS 2007*, Graz, Austria, September 5-7, 2007, (pp. 221-228).

Schenk-Brill, D., & Boots, P. (2001). Innovative engineering. Experiences with industrial coeducation. *Proceedings of the SEFI Annual Conference*. Copenhagen.

Scottish Executive Health Department Management Executive. (1999). *Learning together* Retrieved May 2, 2010, from http://www.scotland.gov.uk/ health/mel/mellt-00.asp

Scottish Executive Health Department Management Executive. (1999). *Learning together.* Retrieved May 2, 2010, from http://www.scotland.gov.uk/ health/mel/ mellt-00.asp

Scully, D. (2009). *Prospects for the Irish construction industry*. Paper presented at the National Construction Conference 2009, Croke Park, Dublin, Ireland.

Segal, N. (2003). *International jobs: Where they are, how to get them* (6th ed.). New York: Basic Books.

Sharma, R. (Ed.). (1994). *Development and implementation of strategies for assuring teaching quality and performance*. Melbourne, Australia: Swinburne University of Technology.

Sharp, J. (1997). *Communities of practice: A review of the literature.* Retrieved March 6, 2007, from http:// www.nickols.us/

Shaw, J. E. (1992). *An evaluation of the cooperative education program in Information Technology*. Canberra, Australia: DEET.

Shields, J. (2004). *Corporate learning strategies: Politics, the new order and teachers work*. Paper presented at the International Education Research Conference. Retrieved from www.aare.edu.au/04pap/ shi04532.pdf

Shulman, L. (2002). *Making differences: A table of learning.* Retrieved June 14, 2010, from http://www.carnegiefoundation.org/ elibrary/making-differences-table-learning

Simon, H. (1973). The structure of ill-structured problems. *Artificial Intelligence*, *4*, 181–201. doi:10.1016/0004-3702(73)90011-8

Siu, K. W. M. (2009a). Review on the development of design education in Hong Kong: The need to nurture the problem finding capability of design students. *Educational Research Journal*, *23*(2), 179–202.

Siu, K. W. M. (2009b). New trends of research in postgraduate design education in China. *US-China Education Review*, *6*(9), 16–21.

Siu, K. W. M. (2005). Learning opportunity for product designers and engineers. In Acland, T., Samuel, J., Saunders, D., & Storan, J. (Eds.), *Access, retention and employability: Transforming higher education* (pp. 108–114). London, UK: Forum for the Advancement of Continuing Education, University of East London.

Siu, K. W. M. (2010). Meeting the new needs: Design research education in China. *Research in Higher Education Journal, 6*(March), 157-173.

Smith, S. W. (1999). The industrial perspective on university-industry relationships in Japan and the United States. In Branscomb, L. M., Kodama, F., & Florida, R. (Eds.), *Industrializing knowledge: University-industry linkages in Japan and the United States* (pp. 579–588). Cambridge, MA: MIT Press.

Smith, D. (2006). Measuring the impact of informal learning. Retrieved December 12, 2006, from http://www.trainingzone.co.uk/ cgi-bin/item.cgi?id=163069

Smith, M. K. (1996b, 1999). *Praxis, the encyclopedia of informal education.* Retrieved May 2, 2006, from http:// www.infed.org/ biblio/b-praxis.htm

Smith, M. K. (2001). David A. Kolb on experiential learning. In *The encyclopedia of informal education.* Retrieved June 16, 2010, from http://www.infed.org/ biblio/b-explrn.htm

Smith, M. K. (2005). Competence and competency. In *The encyclopedia of informal education.* Retrieved June 14, 2010, from http://www.infed.org/ biblio/b-comp.htm

Smith, M. K. (2005). Competence and competency. In *The encyclopedia of informal education.* Retrieved June 14, 2010, from http://www.infed.org/ biblio/b-comp.htm

Smith, P. A. C., & McLaughlin, M. (2004). Knowledge management: People are important! *Journal of Knowledge Management Practice*.

Spiro, R. J., Feltovich, P. J., & Jacobson, M. J. (1996). *Cognitive flexibility, constructivism, and hypertext*, 2nd ed. Retrieved April 3, 2006, from http://phoenix.sce.fct.unl.pt/ simposio/Rand_Spiro.htm

Spiro, R. J., Feltovich, P. J., & Jacobson, M. J. (1996). Cognitive flexibility, constructivism, and hypertext, 2nd ed. Retrieved April 3, 2006, from http://phoenix.sce.fct.unl.pt/ simposio/Rand_Spiro.htm

Standards Australia. (2007). *What is a standard?* Retrieved March 12, 2010, from http://standards.org.au/ DevelopingStandards/ WhatisaStandard.aspx

Stewart, R. A., & Chen, L. (2009). Developing a framework for work integrated research higher degree studies in an Australian engineering context. *European Journal of Engineering Education*, *34*(2), 155–169. doi:10.1080/03043790902833325

Stokes, P. (2002). *Philosophy - 100 essential thinkers*. London, UK: Arcturus Publishing Limited.

Symes, C., & McIntyre, J. (Eds.). (2000). *Working knowledge: The new vocationalism and higher education*. Buckingham, UK: SRHE/Open University Press.

Tagg, J. (2003). *The learning paradigm: College*. Bolton, MA: Anker Publishing Company, Inc.

Teicher, U. (2003). The future of higher education and the future of higher education research. *Tertiary Education and Management*, *9*(3), 171–185.

Tennant, M. (2000). Learning to work, working to learn: Theories of situational education. In Symes, C., & McIntyre, J. (Eds.), *Working knowledge: The new vocationalism and higher education*. Buckingham, UK: SRHE/Open University Press.

Thalheimer, W. (2007). *Measuring learning results: Creating fair and valid assessments by considering findings from fundamental learning research.* Retrieved May 30, 2007, from http://www.work-learning.com/ catalog/

The Cambridge-MIT Institute. (2010). *Home page information.* Retrieved February 10, 2010, from http://www.cmi.cam.ac.uk

The Chamber of Minerals and Energy. Western Australia. (2005). *Fly in/fly out: A sustainability perspective sustainable, a discussion of the triple bottom line impact of fly in/ fly out operations in Western Australia's resources sector.* Retrieved July 24, 2010, from http://www.peopleforthefuture.com.au /files/files/20_FIFO_Report.pdf

Thomas, K., Graham, B., & Wall, J. (2007). Learning about IT and learning using IT-a review of current practice on Higher Education AEC programmes in Ireland. In D. Rebolj (Ed.), *Bringing ITC knowledge to work - 24th W78 Conference Maribor & 5th ITCEDU Workshop & 14th EG-ICE Workshop* (pp. 723-729). University of Maribor, Slovenia.

Tranfield, D., Denyer, D., Marcos, J., & Burr, M. (2004). Co-producing management knowledge. *Management Decision*, *42*(3/4), 375–386. doi:10.1108/00251740410518895

Treby, E., & Shah, A. (2005). Bridging the gap between academia and practitioners: Training coastal zone managers. *Planet*, *14*, 16–17.

Tuning Project. (2007). *Tuning general brochure of Tuning Educational Structure in Europe*. European Commission, Socrates & Tempus. Retrieved from http:// tuning.unideusto.org/ tuningeu/images/stories/ template/ General_Brochure _final_version.pdf

Ulhbrich, A., Schier, P., Lindstaedt, S. N., & Görtz, M. (2006). *A context-model for supporting work-integrated learning.* APOSDLE project.

University. (2001). *Studying in e-espace and other challenges for learning. Learning by working. University and Industry co-operation as a resource for learning.* E4 - Activity 5 & Best Symposium. Helsinki University of Technology. Helsinki.

Usher, R., & Solomon, N. (1999). Experiential learning and the shaping of subjectivity in the Workplace. *Studies in the Education of Adults*, *31*(2), 155–163.

Valiathan, P. (2002). *Blended learning models*. Retrieved July 24, 2010, from http://www.astd.org/LC/2002 /0802_valiathan.htm

van Geenhuizen, M. S. (Ed.). (2009). *Value-added partnering and innovation in a changing world.* West Lafayett, IN: Purdue University Press.

Van Gyn, G. (1996). Reflective practice: The needs of professions and the promise of co-operative education. *The Journal of Cooperative Education*, *29*(2), 103–131.

Varga, A. (Ed.). (2009). *Universities, knowledge transfer and regional development: Geography, entrepreneurship and policy*. Cheltenham, UK: Edward Elgar.

Vernon, C., Kirby, F., Allen, M., & Sargeant, J. (2003). A mixed learning technology approach for continuing medical education. *Medical Education Online, 8*(5).

Victoria University. (2010). *Problem-based learning*. Retrieved February 16, 2010, from http://www.vu.edu.au/ higher-ed-and-tafe/higher-education /health-engineering-and-science/ problem-based-learning

Viravaidya, W., Panyakul, W. R., & Thonglek, S. (2007). Relative significance of admission criteria for the chemical engineering practice school (ChEPS). In Z. J. Pudlowski (Ed.), *Monash engineering education series: 10th UICEE Annual Conference on Engineering Education* (pp. 147-150). Melbourne, Australia: Monash University.

Virolainen, M. (2007). Workplace learning and higher education in Finland: Reflections on current practice. *Education + Training, 49*(4), 290-309.

WACE. (2010). *Advancing work integrated learning globally*. The World Association for Cooperative Education (WACE). Retrieved June 11, 2010, from http://www.waceinc.org/

Wall, J., & Ahmed, V. (2004). *E-learning and pedagogical challenges in construction management: Bridging the gap between academia and industry*. Paper presented at the 20th Annual Conference Association of Researchers in Construction Management ARCOM, (pp. 603 – 613).

Waterkamp, D., & Reinhard, K. (2006). System of industry sponsorship in education. In Bagusat, A., & Hermanns, A. (Eds.), *Management-handbuch Bildungssponsoring* (pp. 111–126). Berlin, Germany: Erich Schmidt Verlag.

Weaver, K. A. (1993). Research and employer oriented learning objectives. *Journal of Cooperative Education, 28*(3), 7–11.

Webster-Wright, A. (2006). *Understanding continuing professional learning.* Unpublished doctoral dissertation, University of Queensland, Brisbane.

Welters, A. T. J., & van de Wetering, H. T. J. (2009). From competence to knowledge based competence. *Proceedings of 37th SEFI Annual Conference.* Rotterdam.

Wenger, E., McDermott, R., & Snyder, W. M. (2002). Seven principles for cultivating communities of practice. *HBSWK*. Retrieved from http://askmecorp.com/pdf/7Principles_CoP.pdf

West, R. (1997). *Learning for life: Review of higher education financing and policy.* Canberra, Australia: Review Committee on Higher Education Financing and Policy, Department of Employment, Education, Training and Youth Affairs.

Williams, A., & Lees, T. (Eds.). (2009). *Knowing your market: Strategy, communication and relationships – employer engagement series two, engaging employers.* Salford, UK: University of Salford.

Williams, A., & Thurairajah, T. (Eds.). (2009). *Working the curriculum: Approval, delivery & assessment–employer engagement series two, work-based learning.* Salford, UK: University of Salford.

Wilson, J. W., & Lyons, E. H. (1961). *Work-study college programs-appraisal and report of the study of cooperative education.* New York, NY: Harper & Brothers.

Wilson, J. W. (1988). Research in cooperative education. *Journal of Cooperative Education, 24*(2-3), 77–89.

Winckler, G. (2009). Universities in the 21st century. *Information Services & Use, 29*(3), 65–72.

Wiredu, G. O. (2005). *Mobile computing in work-integrated learning: Problems of remotely-distributed activities and technology use.* Unpublished doctoral dissertation, University of London, United Kingdom.

Wolstenholme, A. (Ed.). (2009). *Never waste a good crisis–a review of progress since rethinking construction and thoughts for our future.* UK: Constructing Excellence.

Woodworth, R. S., & Marquis, D. G. (1949). *Psychology: A study of mental life.* London, UK: Methuen & Co Ltd.

Wynn, M., Jones, P., Roberts, C., & Little, E. (2008). Innovation in the construction and property management industries - case studies of the knowledge transfer partnership scheme. *Property Management, 26*(1), 66–78. doi:10.1108/02637470810848903

Yin, R. K. (1994). *Case study research: Design and methods* (2nd ed.). Thousand Oaks, CA: Sage.

Young, J. (1997). *Comparative international study of co-operative education employers: The value of employing co-op students.* Cape Town, South Africa: World Association of Cooperative Education.

Zabalza, M. A. (2000). El prácticum y las prácticas en empresas en la formación universitaria. *VIII Congreso Universitario de Innovación Educativa en las Enseñanzas Técnicas.* Santiago de Compostela.

Zegwaard, K., Coll, R. K., & Hodges, D. (2003). Assessment of workplace learning: A framework. *Asia Pacific Journal of Cooperative Education*, *4*(1), 9–18.

Ziguras, C. (2001). Educational technology in transnational higher education in South East Asia: The cultural politics of flexible learning. *Journal of Educational Technology & Society*, *4*(4), 8–18.

Zimmerman, B. J. (2002). Becoming a self-regulated learner: An overview. *Theory into Practice*, *41*(2), 64–70. doi:10.1207/s15430421tip4102_2

About the Contributors

Patrick Keleher (Senior Lecturer, Faculty of Sciences, Engineering and Health, CQUniversity, Australia). Patrick's experience of, and involvement with, Work Integrated Learning/Practice Based Learning commenced 25 years ago when he completed a postgraduate Diploma in Teaching (Physics/Mathematics) at the University of Queensland. Over the 12 years of Patrick's career as a secondary teacher he was a mentor, supervisor and assessor of pre-service science teachers operating within a Work Integrated Learning/Practice Based Learning model. Over the past 13 years, as a university lecturer, Patrick has led the development and delivery of programs at undergraduate (Director: Built Environment Programs) and postgraduate (Director: Maintenance Management, Environmental Management, Process Engineering Programs) level with the central theme of a Work Integrated Learning/Practice Based Learning philosophy. In 2009, Patrick undertook sabbatical at the Practice Based-Professional Learning Centre of Excellence in Learning and Teaching (PBPL-CETL) at The Open University, Milton Keynes, UK to further his WIL/PBL research interests and has served as a Fellow of the PBPL-CETL. Patrick is the convenor of the Work Integrated Learning – Special Interest Group within CQUniversity's Learning & Teaching Education Research Centre (LTERC).

Arun Patil (Lecturer, Faculty of Sciences, Engineering and Health, CQUniversity, Australia) Arun is a lecturer in Engineering at the Mackay campus of CQUniversity, Australia. Prior to this position, Arun was a research scholar in the Faculty of Engineering of Monash University, Australia. His PhD research project focuses on the Development of a Global Accreditation Model for the Quality Assurance of Higher Education, with Engineering as a Case Study. He has over 18 years of teaching and administrative experience in India and Australia in higher and further education. Arun has published widely, and his publications include refereed journals, conference papers, book chapters and academic books. He is a Founder Editor-in-Chief of the "International Journal of Quality Assurance in Engineering and Technology Education". His recently published book, Engineering Education Quality Assurance: A Global Perspective is highly accepted and cited scholarly publication in engineering education community around the globe. Arun has organised over a dozen of international conferences in engineering education in various parts of the globe.

Roberta (Bobby) Harreveld (Former Director, Learning & Teaching Education Research Centre (LTERC), Faculty of Arts, Business, Informatics & Education, CQUniversity, Australia) Roberta (Bobby) is also the Acting Deputy Dean, School of Education at CQUniversity Australia and an Adjunct Fellow with the Centre for Educational Research at the University of Western Sydney in Australia. As Director of LTERC, she is responsible for establishing the strategic direction and operational management plan for education research related activities into learning and teaching at CQUniversity. Bobby's current teaching and research is investigating questions around work oriented curriculum design, technologically mediated pedagogies and epistemologies of teaching and teacher education. This work is located among the education systems of universities, vocational colleges, schools, workplaces and community learning settings. She is supervising a number of research higher degree students in this area. Her interests in work integrated teaching and learning is trans-national, cross-sectoral, multicultural and inter-systemic in orientation. Bobby is actively engaged in successful competitive research grants with colleagues (e.g. ARC Linkage Grant 2007-2010); commissioned research projects; (c) research higher degree supervision; and a sustained output of scholarly work.

* * *

Thomas Akins (Executive Director (newly retired), Division of Professional Practice, Georgia Institute of Technology, USA.) Thomas earned a Bachelor's of Industrial Engineering degree from the Georgia Institute of Technology through the cooperative education program. After working in industry for a brief period of time, Mr. Akins returned to his alma mater to work as Assistant Director of the Co-op Program, eventually becoming Director. Under his leadership, the department expanded staff and services to become an umbrella organization for experiential education, including undergraduate and graduate co-op programs, an extensive internship program, and a strong work abroad program. Over 4,000 students and 1,000 employers are currently participating in these programs at Georgia Tech. Mr. Akins has been active in professional organizations as well, and was awarded the Clement J. Freund Award from the American Society of Engineering Education for his meritorious contributions to the field of cooperative education. In March 2010, he retired from active service at Georgia Tech.

Bruce Calway (Associate Dean International, Associate Dean Learning & Teaching, Faculty of Higher Education, Swinburne University of Technology, Australia) Bruce entered academia in 1991 from a 20 year career in business and ICT design and development. He completed a PhD in Information Systems and a 2nd PhD in Education. Bruce is a Professor at Swinburne University of Technology and he has researched and published extensively in aspects of work-integrated learning, eLearning and learning environment development.

Jean-Marie Chatelet (Assistant Manager - IESP Department, Engineering College (Polytech'Lille), University of Lille, France) Jean-Marie worked in a Research Centre of metallurgy before becoming a certified mechanical engineering professor. He has been an educator for more than 35 years and his experience encompasses working internationally in France, Romania, Spain and Germany and writing about quality management, delivering curriculum on safety, quality, CAD-CAM and robotics. He is involved in leadership of a Masters of Health-Safety-Quality-Environment and in a chartered Engineer training department. In 1982 he was a pioneer in in Computer Aided Teaching and was the author of a laservision disc about robotics edited by the Industry ministry (1988). He is accredited on 3Dstudio (Autodesk Training Centre) and was recognised with an award at the World Submarine Movie Festival (Antibes 1995). He works with the industry and has developed a game JIT-TQ with snecma (Safran group) in 1999. He has been also involved in European projects (Road Map to EU Machinery Directive – 2002, Developing of European Work Based Learning Approaches and Methods - 2006).

Gopinath Chattopadhyay (Director – Engineering Postgraduate Programs, Faculty of Sciences, Engineering and Health, CQUniversity, Australia) Gopi has a Bachelor of Engineering (Mech.), Master of Engineering (Industrial Engineering), MBA, Grad Cert. of Education (Higher Edu.) from Queensland University of Technology (QUT) and PhD in Mechanical Engineering from the University of Queensland (UQ). He has 14 years of experience as engineer and manager in design and development and operations management including maintenance and 17 years of experience in Universities in the area of Engineering Management, Asset Management, Maintenance and Reliability in UQ, QUT and CQUniversity Australia. Gopi has collaborative projects with QUT, Monash, Rail CRC, Qld Rail, Swedish Rail, Lulea University of Technology, Sweden, Saarland University, Germany, Indian Institute of Technology India, Qld Health and many other reputed organisations. Research outputs include more than $2.5 million dollar research grants, 10 research student completions and over 130 refereed international journal and conference publications and book chapters.

Conny Christl (RTD Manager, Innovation Service Network GmbH, Austria) Conny is responsible for national and international RTD projects at ISN- Innovation Service Network and was working within the FP 6 EU Integrated Project APOSDLE. Currently she is writing her PhD in the research field of technology enhanced work-integrated learning at Karl Franzens University Graz (Austria) at the department for information science and information systems. She is especially interested in accompanying innovative learning systems from research to practice. Until 2007 she worked as a project assistant at EURICE (European Research and Project Office) in Saarbrücken (Germany). Conny received her master's degree in information science from Saarland University (Germany), Department for Information Science in 2007.

Kaye Clark (Former Director – Built Environment programs, Faculty of Sciences, Engineering and Health, CQUniversity, Australia) Kaye has amassed broad experience in education and educational administration over thirty years, from a motivational Health and Safety Education teacher and administrator for Red Cross (State Director of Training-Tasmania) to University and TAFE lecturer in Architecture in Tasmania then Queensland. Kaye is a registered Architect with experience in Tasmania, Jakarta (Indonesia), remote Central Australia and Queensland. In her architectural roles as a sole practitioner or in work for other Architects or builders Kaye combined creative, award-winning design skills with practical expertise in documentation including structural and design detail drawing and full contract administration. Previous administrative architectural specialisations (for various employers) included; intensive problem solving, ensuring sustainability, also build-ability while design integrity is maintained; the implementation of standards and quality assurance (QA); job tracking; also the organisation of people and tasks (which is a long-term expertise). Kaye's current role at CQUniversity is Head of Built Environment Programs and lecturer.

David Crowe (Learning & Development Manager, BAM Contractors Ltd, Ireland) David is a degree qualified engineer with 10 years experience in the design, delivery and management of training & development solutions in the manufacturing, automation, pharmaceutical and construction sectors. David is highly motivated and adaptable to changing business needs and has a strong interest in training evaluation.

Urbano Domínguez (Professor in Mechanical Engineering, Industrial Engineering School, University of Valladolid, Spain) Urbano has held other professorial positions and undertaken other leadership positions in Spain at the University of Salamanca, Basque Country University, and Autonomous and Technical Universities of Madrid. Urbano's fields of interest are optimum design of machine elements, quality and metrology, renewable energies and energy analysis. He is also involved in engineering education research, internationalization and the development and implementation of innovative curricula. He has organized several international Conferences on Engineering Education. He has published several books, research papers and communications to Congresses on these topics. He is a member of the European Society for Engineering Education (SEFI) Administrative Council, and he chairs the SEFI Curriculum Development Working Group (CDWG).

Brian Graham (Lecturer, Faculty of Engineering, Waterford Institute of Technology, Ireland) Brian lectures on undergraduate and postgraduate programmes in Construction Management, Quantity Surveying & Civil Engineering in the areas of Information Technology (IT) in construction, project management, research methods and dissertation supervision. He is currently undertaking a PhD on Knowledge Management (KM) within the leading Irish construction organisations, specifically dealing with the integration of Continuing Professional Development with existing Human Resource Management and IT practices. He has published a number of peer-reviewed journal and conference papers in the area of KM in construction. Other teaching/research interests include the development of IT in the Irish construction industry and work-based learning for construction professionals.

Debbie Gulick (Interim Executive Director, Division of Professional Practice and Director (April 2010), Work Abroad Program, Division of Professional Practice, Georgia Institute of Technology, USA) As DoPP Interim Director, Debbie manages the Georgia Tech Co-op, Internship, Graduate Co-op and Work Abroad Programs, which includes developing program strategies, planning budgeting and resource allocation; coordinating division efforts with other Georgia Tech academic and administrative units; overseeing recruitment efforts of students and employers; and monitoring and reporting division participation and results. In her role as Director of the Work Abroad Program, she is responsible for developing and sustaining a large, broad-based program of international internships and cooperative education opportunities to prepare Georgia Tech students to be leaders with top employers, worldwide. She works with undergraduates and graduates of all majors in the Work Abroad Program. Debbie started the Work Abroad Program in 2005, and during that year the program sent 31 students abroad to gain international work experience. In 2009-2010, Georgia Tech Work Abroad students were sent on 212 global work terms in 35 countries.

Steven Hutchinson (Director, Director of the Postgraduate Certificate in Education, Faculty of Education and Language Studies, The Open University, UK) Steve oversees a unique pre-service course in initial teacher education for graduates and since its inception in 1992 it has contributed 5000 new teachers to the profession; many of whom can only train through this route. Key features of the programme are flexible start and finish points; individualized programmes resulting from a web-based Needs Analysis process; and an Assessment Only route. Steve's research uses Activity Theory to investigate the problems, dilemmas and opportunities facing student teachers as they move between the university and schools; as they cross boundaries between and within each of these settings. Prior to his appointment to the Open University in 1997 he taught music in Secondary schools in Birmingham and the West Midlands. Steve directed the Open University's Practice-Based Professional Learning Centre for Excellence in Teaching and Learning from 1st October 2007 to 30th September 2008.

Hong-min Ku (Program Director - Chemical Engineering Practice School (ChEPS), Faculty of Engineering, King Mongkut's University of Technology Thonburi, Thailand) Hong-ming was born in Taiwan, grew up in Thailand, and educated in the US. He holds a B.S degree from the University of Illinois at Urbana-Champaign and M.S. and Ph.D. degrees from Northwestern University, all in chemical engineering. Upon graduation in 1989, Hong-ming went to work for Aspen Technology, Inc., a leading software company in process simulation in Cambridge, Massachusetts, where he held a number of positions in Technical Services, Applications, and Sales Divisions. Hong-ming has been a faculty member in the Department of Chemical Engineering at King Mongkut's University Technology Thonburi in Bangkok, Thailand, since 1997. He currently is the Director of the Chemical Engineering Practice School (ChEPS), an international Master's practice-based curriculum modeled after MIT's David H. Koch Chemical Engineering Practice School. Dr. Ku's research interests are in engineering education, process simulation, and optimization.

Stefanie Lindstaedt (Division Manager Knowledge Services, Knowledge Management Institute (KMI), Know-Center and Graz University of Technology (TUG), Austria) Stefanie is also scientific coordinator of the MIRROR IP on reflective learning at work. She is responsible for the management of many large, multi-firm projects and the scientific strategy of the division. Stefanie is also senior researcher at the Knowledge Management Institute of the Technical University Graz. In this role she is involved within the MATURE IP responsible for the knowledge maturing services to be developed. For more than 15 years she has been leading interdisciplinary, international projects in the fields of knowledge management, technology enhanced learning, semantic systems, and software engineering. She holds a PhD and a M.S. both in Computer Science from the University of Colorado (CU) at Boulder (USA). She is member of the Centre for LifeLong Learning and Design and the Institute of Cognitive Science at CU, has published more than 100 scientific papers, has chaired numerous workshops and served on a multitude of program committees.

Jesús Magdaleno (School of Industrial Engineering, University of Valladolid, Spain) Jesús is currently Associate Professor in Mechanics of Materials at the University of Valladolid, where he was formerly Assistant Teacher. His fields of interest are finite element method (FEM) applied to structural mechanics, elastoplasticity and 2D beam element analysis. He is also involved in engineering education research and innovative curricula. He is an active member of Curriculum Development Working Group (CDWG) of the European Society for Engineering Education (SEFI). He has been Vice- Director for Quality and Innovation between 2000 to 2004 quality and innovation.

Gerald Murphy (PhD candidate, Faculty of Higher Education Lilydale, Swinburne University of Technology, Australia) Gerald was formerly the Certification Program Manager, Executive Officer national PDAustralian Computer Society (ACS) responsible for the Masters level distance education applied learning programs for members; ACSLearn an eLearning portal; Education across the nation. Gerald was Deputy Head of School of Information Technology (Education Programs) at Swinburne University. He was consultant/manager/technical advisor to IMS Computer Services; Olivetti Australia;

Southern Cross Assurance; Turner Manufacturing Company; International Computers; and Independent consultant. He has held the position from 1996-2000 as Chair of Australian Cooperative Education Society. In1997 he was awarded the Donald Maclaren Prize for outstanding contribution to Cooperative Education internationally. He has published extensively on Professional Development; Information Systems Curriculum Development; Quality in Education; and Cooperative Education. He has conducted professional development workshops for: ACS; Australian Society of Accountants 1976-1979; Institute of Chartered Accountants; Australian Dental Association; Australian Optometrist Association; and conducted "in-house" for various organisations.

Walter Nuninger (Head- IESP Department, Engineering College (Polytech'Lille), University of Lille, France) After graduating as a chartered engineer of the ENSEM (mechanics, electricity, automatic) Walter obtained a PhD dealing with robust diagnosis based on redundancy methods. He worked as a researcher engineer for an ALSTOM Transport project (friction, train traction). He has demonstrated project leadership with applied research in an innovative environment dedicated to industrial goals. He has been associate professor at the university since 1999 with a position at the engineering college IAAL dealing with food-industry (mathematics, computer science, automatic, data mining). He has been responsible of the 1st year of the training. In 1998, he became head of the IESP department dealing with LLL and production field. Through the years Walter was involved in several activities dealing with pedagogy (WBL, evaluation, curriculum, industrial project and internship), organization (school, network, conference committee, work groups) and managing (studies, department). His research interests focus on e-learning, pedagogy and international activities.

Aidan O'Connell (Contracts Manager, BAM Contractors Ltd, Ireland) Aidan joined BAM in 1993 and has worked on a wide variety of projects from Public Sector through to high-tech projects to commercial works. He has a proven track record of delivering high quality projects, reflected as the overall winner of the Irish Concrete Society for the Carlow Arts Centre. He has recently completed a number of projects within Dublin City Centre where he gained valuable experience in delivering challenging projects. Aidan was promoted to his current position as Contracts Manager in 2005 and is currently working on the new NUI Galway Engineering Building.

Karin Reinhard (Head of Department of the International Business Program, Faculty of Business, Baden-Württemberg State University of Cooperative Education, Germany) Karin teaches Marketing and Internet Management plus Media Economics. Karin has held the Head of Department position since 2007. Her research interests focus upon cooperative education and social marketing. Before coming to the Baden-Württemberg Cooperative State University she was Executive Head of Division Information and Communications Technology Policy Section at the Bavarian State Chancellery. Among other things she was responsible for the launch of the Bavarian Virtual Market Place, www.baynet.de and the Bavarian Virtual Campus www.vcb.de. Karin published her doctoral thesis on Industry Sponsorship, a comparative study between Germany and the United States. She studied Business Administration at Mannheim University, Germany where she completed a master's degree. In addition she received a postgraduate degree in Educational Science. Until 2008 she was chair of the German chapter of ISBE/SIEC - International Society for Business Education/ Société Internationale d'Education Commerciale, where she still enjoys her membership and Karin is a member of the Academy of International Business.

Shalini Singh (Senior lecturer, Faculty of Management Sciences, Durban University of Technology, Republic of South Africa) Shalini has also lectured at the University of Basle (Switzerland), Berufsakademi in Loerrach (Germany) and Dualhoschule in Ravensburg (Germany). She has co-lectured to exchange students from the University of Beijing. She is also employed as a moderator, examiner and supervisor in a number of universities in South Africa and is a member of the Higher Education Quality Forum for Quality. Shalini is responsible for supervising Master's and Doctoral research in the Department of Operations and Quality Management at her university. Published poster and publications for World Congress on Environmental Health 2004, International Business and Economics Discussion Papers (2009) and presented in Quality related conferences nationally and internationally. She currently chairs a research group which investigates Quality related matters.

Kin Wai Michael Siu (Laboratory Leader of Public Design Laboratory, School of Design, The Hong Kong Polytechnic University, Hong Kong SAR, China) Michael is also a Visiting Professor of the Academy of Arts and Design, Tsinghua University. He is a chartered designer and registered professional engineer. He is fellow and honorary fellow of different international, academic, design and engineering organizations and associations. He was Academic Visitor of the University of Cambridge, Fulbright Scholar at the Massachusetts Institute of Technology, ASIA Fellow of the National University of Singapore, and Visiting Scholar of the UC Berkeley. Michael's major research and design interests include: design culture and theory, public design, design and engineering education. He has provided consultancy services in public design and design and engineering programmes and he has published more than 150 refereed papers. He has won numerous international design and invention awards, and owned more than 50 US and Asian design patents.

Ken Thomas (Head of Department, Faculty of Engineering, Waterford Institute of Technology, Ireland) Ken has been Head of Department of Construction & Civil Engineering at Waterford Institute of Technology since 2002. The Department includes education and research programmes in Construction Management & Engineering, Quantity Surveying, Civil Engineering and Construction Project Management. As a direct result of his PhD studies he co-founded the Construction IT Alliance (CITA) in Ireland in 2001 and he continues to be a Director of this Industry-Academia organization. Ken is passionate about Industry and Academia working closely together, particularly at postgraduate level, for their mutual benefit.

Saranya Thonglek (Doctorate Candidate, School of Chemical Engineering, University of Queensland, Australia) Saranya has been the administrator of the Chemical Engineering Practice School (ChEPS) at King Mongkut's University of Technology Thonburi (KMUTT) since 1997. Saranya received her undergraduate degree in chemical engineering and a Master's degree in biotechnology from the School of Bioresources and Technology, both from KMUTT. She is currently a doctorate candidate in the School of Chemical Engineering at the University of Queensland in Brisbane, Australia. Her research focuses on a comparative study in Work-Integrated Learning (WIL) in chemical engineering between Australia and Thailand.

Patrick Troy (Head of Human Resources and Industrial Relations, BAM Contractors Ltd, Ireland) Patrick is an experienced HR leader with a proven ability to lead the implementation of systems, policies and processes and also to drive cultural and organisational change. He proactively pursues a partnership approach with the operational side of the business to achieve company objectives. Patrick is highly-experienced in industrial relations, negotiations and employment legislation.

John Wall (Head of School of Education, Waterford Institute of Technology, Ireland) John is also the Programme Director for the MSc in Construction Project Management. He is currently continuing that role while on secondment as Head of the School of Education. John has a number of research interests including 'The use of technology in the delivery of lifelong learning in the construction industry', 'Marketing in construction organisations' and 'Project Management in Construction.' Through his research and education activities John has developed a wide range of contacts with industry and higher education providers in Ireland and overseas, including a visiting professorship at Queensland University of Technology in Australia.

Index

Z

CPSIA information can be obtained at www.ICGtesting.com
260174BV00008B/105-142/P